W9-BFA-010

Annual Review of Clinical Biochemistry Volume 1

Annual Review of Clinical Biochemistry Volume 1

Edited by

David M. Goldberg, M.D., Ph.D., F.R.I.C., F.R.C.Path.
Biochemist-in-Chief
Hospital for Sick Children
and
Professor and Chairman
Department of Clinical Biochemistry
University of Toronto
Toronto, Canada

A Wiley Medical Publication

JOHN WILEY & SONS
New York • Chichester • Brisbane • Toronto

ISBN 0-471-04036-3
ISSN 0195-8488

Printed in the United States of America

10 9 8 7 6 5 4 3 2 1

Contributors

Chester A. Alper, M.D.
Professor of Pediatrics
Harvard Medical School
Children's Hospital Medical Center
Scientific Director
Center for Blood Research
Boston, Massachusetts

Peter M.G. Broughton, F.R.C.Path.
Deputy Director
Wolfson Research Laboratories and Department of Clinical Chemistry
Queen Elizabeth Hospital
Birmingham, England

Alvin E. Davis, III, M.D.
Assistant Professor of Pediatrics
Harvard Medical School
Children's Hospital Medical Center
Boston, Massachusetts

Stanley M. Deming, Ph.D.
Associate Professor of Chemistry
University of Houston
Houston, Texas

Gordon G. Forstner, M.D.
Professor of Pediatrics
University of Toronto
Director
Kinsmen Cystic Fibrosis Research Center
Hospital for Sick Children
Toronto, Canada

Arthur R. Henderson, M.B., Ch.B., Ph.D.
Associate Professor of Clinical Biochemistry
University of Western Ontario
Director of Clinical Chemistry
University Hospital
London, Ontario, Canada

Norman B. Javitt, M.D., Ph.D.
Professor of Medicine and Chief of Gastroenterology
The New York Hospital-Cornell Medical Center
New York, New York

Donald J.R. Laurence, Ph.D.
Scientific Staff
Ludwig Institute for Cancer Research
London, England

Barry Lewis, M.D., Ph.D.
Professor of Chemical Pathology and Metabolic Disorders
St. Thomas's Hospital Medical School
London, England

Choh H. Li, Ph.D.
Professor of Biochemistry and Experimental Endocrinology
Director of Hormone Research Laboratory
University of California
Berkeley, California

Arne Lundblad, M.D.
Associate Head
Department of Clinical Chemistry
University Hospital
Lund, Sweden

Beverley E.P. Murphy, M.D., Ph.D.
Professor of Medicine and Professor of Obstetrics and Gynecology
McGill University
Director
Reproductive Physiology Unit
Montreal General Hospital
Montreal, Canada

Alexander M. Neville, M.D., Ph.D.
Professor of Experimental Pathology
University of London
Administrative Director
Ludwig Institute for Cancer Research
London, England

Per-Arne Öckerman, M.D.
Professor and Head
Department of Clinical Chemistry
University Hospital
Lund, Sweden

Harry L. Pardue, Ph.D.
Professor of Chemistry
Purdue University
West Lafayette, Indiana

Charles E. Pippenger, Ph.D.
Assistant Professor of Neuropharmacology
College of Physicians and Surgeons
Columbia University
New York, New York

Janakiraman Ramachandran, Ph.D.
Professor of Biochemistry
University of California
San Francisco, California

George Steiner, M.D.
Associate Professor of Medicine and Physiology
University of Toronto
Director
Diabetes Clinic and Lipid Research Clinic
Toronto General Hospital
Toronto, Canada

Paul L. Wolf, M.D.
Professor of Pathology
University of California
San Diego, California

Bernard Zinman, M.D.
Associate Professor of Medicine
University of Toronto
Toronto, Canada

Preface

Clinical biochemistry is a discipline that has so infiltrated the practice of medicine that its influence is patently at work in most subspecialties, from management of the patient at the bedside to the sophisticated research needed to promote our understanding of human disease. Paradoxically, the "tumor" is in danger of losing its identity to the "host." Many of the major advances that affect the practice of clinical biochemistry are made by investigators who would be regarded, primarily, as clinicians. The laboratory scientist, overwhelmed by demands for his managerial skills, is often only belatedly aware of the advances being made by his clinical colleagues. By the same token, the busy clinician devoted to patient care does not have time to keep pace with developments in the diagnostic laboratory, which must ultimately affect the way in which he investigates his patients; the efficient use of expensive laboratory resources is becoming one of the dominant objectives of our day.

This series seeks to bring together the professional clinical biochemist and the metabolically oriented clinician in an enterprise that will chart the important developments in the subject. It is intended to have wide appeal in the ward as well as in the laboratory. In fact, all the professionals who perform, order, or interpret the tests that make up the diagnostic armamentarium of the chemistry laboratory should profit from the distillation of information contained in its pages.

Because the knowledge needed for a comprehensive understanding of clinical biochemistry is so widely scattered in the scientific and clinical literature, the initial challenge issued to the contributors was to summarize in a critical fashion the important papers in their area over a single calendar year so that, in time, the reader of this series could be told: "Take *Annual Review of Clinical Biochemistry*, and you will have no further need to read any other periodical in the subject." To facilitate this objective, a roster of contributors has been established on a three-year basis to guarantee continuity and stability, qualities that will certainly be needed to meet this objective.

In fact, the contributors have not universally adopted this approach. Some sought selectivity at the expense of comprehensiveness, and in any event the overall size limitations of the volume imposed constraints that were not compatible with the full realization of the original aim. Nevertheless, it is probably true that no current publication approaches this series in the breadth of its scope which, if not complete, covers most of the areas that clinicians and chemists need to know about in the daily practice of their respective roles. As the series progresses, we hope that a more uniform approach will be adopted by the contributors leading to closer fulfillment of the objectives that led to its creation.

Initiating a new venture calls for unstinting effort on the part of many people. I have been fortunate with the cooperation received from the publishers, in particular John DeCarville, Maura Fitzgerald, and Marilyn Zirke. Many colleagues, too numerous to name individually, assisted with the proofreading of various chapters. The challenge of preparing the camera-ready copy was undertaken by Marj Fleming with unequalled flair and devotion. However, without the support and tolerance of my wife and two daughters, this publication would never have seen the light of day.

In a venture of this kind, no one, least of all the editor and contributors, stand to make a "profit" in the commercial sense. Our reward will be the satisfaction of performing a service that we deem helpful to our colleagues. In the long run, this can only be achieved by meeting the requirements of those in the field. All comments, however critical, will be warmly welcomed and, hopefully, will serve to orient our future efforts in a direction where they can be of most "profit" to you, the readers.

D.M. Goldberg

Contents

Chapter 12 Biochemical Aspects of Genetic Disease
Per-Arne Öckerman and Arne Lundblad

Chapter 13 Therapeutic Drug Monitoring
Charles E. Pippenger

1
Quality Control, Laboratory Management, and Reference Values

Peter M.G. Broughton

LABORATORY MANAGEMENT

Although all clinical chemistry laboratories appear to serve the same basic functions, they are in fact complex organizations differing widely between different countries and locally within the same country. They are, or should be, more than analytical factories producing biochemical data, but the amount of direct clinical involvement is variable, depending on a number of local circumstances. The job of the laboratory manager is to see that the laboratory functions efficiently. However, laboratory management is rarely listed as a subject in the scientific literature, perhaps because it is considered more of an art than a science, and is more often practiced (and sometimes preached) than written about. The management of the laboratory is clearly important, and although it is difficult to assess its quality in practice, it undoubtedly varies from good to bad. As a subject, laboratory management surely merits far more serious attention and study.

Leonard (1) has reviewed some aspects of laboratory management, including its overall objectives, organizational structure, and the interrelationships of the personnel involved. Many other topics could be included in the subject, but this review will concentrate on those that have been covered in the literature during the last year.

Costing of Laboratory Tests

In the graphic words of Conn (2), "During the last 30 years the clinical laboratory has evolved from a small crowded room in the hospital basement to an organisation of marvellous complexity." In several countries it is growing at a compound rate of 15% per year and in 1975 (presumably in the United States) $12 billion was spent on clinical laboratory services. Several authorities are now questioning the benefits of this expenditure for biochemical screening and profiling, for example, and the relationship between cost and quality (3). Many of the problems are the result of the rapid introduction of new technologies, changes in methods of medical education, and the rapid expansion of the scientific bases of medicine (2). The handling of laboratory information and the economics of laboratory operations are becoming increasingly complex, and new approaches are needed to deal with escalating laboratory tests and costs (4).

Although the overall cost of a laboratory is relatively easy to calculate, allocation of costs among different tests, or for comparison of different analytical procedures in the same or different laboratories, is much more difficult. This is largely because of the lack of agreement about which costs should be included and how they should be calculated. Some authors include only the cost of reagents and labor directly involved in performing the test (5). These items can be compared only with data for alternative systems calculated in exactly the same way; they do not represent true costs. As a result there is, for example, disagreement about the comparative costs of performing radioimmunoassays with diagnostic kits (6,7) and in carrying out these tests in centralized laboratories using locally prepared reagents.

Several authors have drawn attention to the need for agreement on methods of costing laboratory procedures (2). Krieg et al (8) have made a valuable and detailed analysis of this problem and have based the total cost of a test on the following factors:

* Direct labor costs inside the operational work station (i.e., the cost of labor directly involved in performing the test)
* Labor costs allocated from a higher level work station (not directly involved in performing the test)
* The cost of materials directly identified with the test
* The allocated cost of materials distributed from a higher level work station (e.g., syringes, specimen tubes, etc.)
* Equipment depreciation
* Hospital overheads (building depreciation, administration, etc.)

This list is not necessarily complete, but the costs that should be included depend on the objectives: some require considerable fine detail, whereas others do not (8). Nevertheless, it is clear from the analysis given by these authors that the problem is neither simple nor trivial. After using their cost analysis system for two years, they reviewed its benefits, many of which were unexpected; these benefits included:

* Increased understanding of actual operations by laboratory management
* Better appreciation of material costs and purchasing practices
* Identification of costs associated with development activities, which were unexpectedly high in terms of both labor and materials
* Better insight by all supervisory personnel into the costs of labor and materials associated with specific tests
* Obtaining objective data concerning overhead costs within the laboratory, as well as losses due to wasting time and materials.

Several authors have discussed methods of limiting the excessive and inappropriate use of expensive diagnostic tests, including the use of economic incentives and sanctions, legislation, and better methods for educating medical students in the use of laboratory information (2,9). The overall cost of running a laboratory is determined largely by its workload, and Taylor (10) has made a useful study of the factors influencing the day-to-day variation in the number of tests performed. A potentially controllable factor was the order (requesting) habits of individual clinicians, which seemed to be more significant than less controllable factors such as the number of routine and emergency admissions, occupied hospital beds, or public holidays.

Kassirer and Pauker (9) have discussed the general principles concerning the proper use of diagnostic tests and have pointed out that the cost and risk of a test must be weighed against the cost and benefit of potential therapy. The usefulness of a test is a function of the frequency of false positive and false negative results (i.e., its efficiency). Not infrequently, tests that were initially regarded or claimed to be valuable were later found to be of little or no use. Ranoshoff and Feinstein (11) have identified two reasons for this. First, an appropriately broad spectrum of diseased and nondiseased patients must be chosen for the study: if this is not done, the test may appear to have a falsely high specificity and sensitivity. Second, unless the interpretation of the test and the establishment of the true diagnosis are made independently, bias may falsely elevate the efficiency of the test. Archer (12) has suggested that the assessment of the predictive value of laboratory tests was made easier by the use of odds, rather than probabilities. Perhaps this will encourage wider application of previously published work on the efficiency of laboratory tests: assessment by odds may not affect costs but will certainly increase benefits.

Work Organization

Several detailed aspects of the problem of organizing work have been considered in the literature, including the preservation and storage of urine specimens for toxicological analyses (13) and for screening tests for aminoacidurias (14). Weaver et al (15) have described an evaluation of a pneumatic tube system for transporting blood specimens. The use of a computer to deal with specimen problems, such as insufficient material or hemolysis, has been described by Jomain and Owen (16). Members of the laboratory staff were convinced that the use of a computer increased efficiency by reducing the number of telephone calls and by ensuring that requests for further specimens were sent to those responsible for collecting them.

Selection of laboratory methods and equipment is an important aspect of management that will be considered later. Taylor (17) has used a computer to study the selection of test combinations in multichannel analyzers that would reduce laboratory manpower and increase the efficiency of specimen preparation. With the increasing use of microprocessors, some types of laboratory equipment can be operated by relatively unskilled staff. Minty and Barrett (18) have described a study in which 100 untrained persons used an automated blood gas analyzer. Most of these operators were able to use the analyzer after less than one minute of instruction, and 86% of the results obtained were within the specified acceptable limits of error. In general, the results were better than those obtained in a survey of blood gas laboratories staffed by trained technologists using a variety of instruments.

Although "stat", or urgent, tests are an important part of the work of the hospital laboratory, the literature is almost devoid of signifcant discussion on this subject. What tests should be offered and in what clinical situations? Barnett et al (19) have surveyed the current practice in a number of laboratories and, as may be expected, found wide variations, both in the tests that were offered and the speed with which results were reported. It is hoped that this survey will stimulate others to try to define what is required clinically and what should be provided by the hospital laboratory that serves patients outside the hours of 9 A.M. to 5 P.M.

Safety

The health and safety of the laboratory staff have recently received considerable attention, and in some countries legislation and regular inspection of hospital laboratories has been proposed (20). In the clinical chemistry laboratory, the main risk is the transmission of hepatitis from specimens containing the virus. Leakage of blood from specimen tubes is believed to be a hazard, and a method of testing for leaks has been described (21). This test showed that there was a spontaneous discharge of the contents as many tubes were opened, and that none of the containers presently available seemed to be entirely free of this hazard. Nevertheless, a survey of British hospital laboratories found that only nine cases of hepatitis were recorded over two years (22). This represented an annual rate of 34 cases per 100,000, which was only about a quarter of that found previously, and suggested that control measures in individual laboratories have had considerable success.

Summary and Conclusions

Laboratory management is an ill-defined but important subject deserving far more professional studies leading to scientific publications. Assessment of the cost-benefit of laboratory tests is becoming more important, and objective methods of costing need to be defined and applied to specific tasks. Laboratory managers must be involved in these problems, and they must be more aware of the limitations and benefits of the work of their laboratories. This involves not only implementing the results of published work on efficiency, but also initiating new studies, introducing improved diagnostic procedures, and deleting obsolete and unnecessary tests.

QUALITY CONTROL

The general principles of quality control are now well established (23), but to many laboratory workers the subject has become a routine and rather boring procedure, and the term itself is in danger of losing its meaning. Usually quality control refers to a system in which commercial control samples are analyzed and the results are used to calculate the laboratory staff's accuracy and precision in performing tests. At best, this process measures the quality of performance with control specimens, which is assumed (sometimes incorrectly) to be the same as that with patients' specimens, but it does nothing to control or to improve performance (24). In this review a broader view of the subject will be taken, using the definition of quality control proposed by Büttner et al (25), namely, the study of those errors that are the responsibility of the laboratory and of the procedures used to recognize and minimize them.

Control Materials

It is now becoming apparent that the nature of the control specimens used to assess accuracy and precision are of paramount importance. Ideally these specimens should resemble those of patients as closely as possible, but the commercial materials, which are almost universally used, are prepared from animal sources, and artifacts may be introduced during their preparation and lyophilization.

The nature of the control specimens is particularly important with enzyme assays. Burtis et al (26) have described the preparation and characterization of six lyophilized control materials based on purified aspartate aminotransferase (AST) obtained from human erythrocytes. Some commercial control sera have been shown to behave differently in two alkaline phosphatase methods, because the buffer was unable to maintain the optimum pH of 10.0 (27); modification of the methods removed the discrepancy. The optimum pH of commercial preparations may vary widely (28) so an error of 0.05 pH units in the buffered substrate introduced an error of about 5% in the response between the control sera used for calibration and patients' specimens. Consequently a check should always be made to ensure that the pH optimum of the enzyme in control sera (whether used for calibration or for control of precision) is close to that in human serum. Some kits for enzyme assay give different results when tested with commercial control sera (29). Liquid human serum was found to be satisfactory for the external quality control of T_3, T_4, and digoxin by radioimmunoassay, but not for T_4 by one kit method; ox serum was unsatisfactory for certain methods of T_4 assay (30).

Serum specimen from patients are frequently lipemic, and the use of clear, nonturbid control sera may give the analyst a false sense of security. Lipemia caused interference in some tests used in two commercial analytical systems (31). With one commercial enzyme analyzer, precision was found to be worse when turbid or opalescent reconstituted control sera were used (32). Vigorous centrifugation improved the precision, but unless specifically instructed, it is unlikely that most users would do this with turbid patients' specimens. A water-soluble cholesterol derivative has been described (33), which when added to serum before lyophilization, gives a clear, reconstituted product. However, to specify that control specimens must be clear does not seem to be an adequate solution to interference by turbidity.

For some assays, artificial mixtures are used as control specimens. Dwenger and Trautschold (34) have found that serum pools, an artificial serum, and a buffer-protein mixture give variable precision and recovery values for radioimmunological methods for digoxin, depending on the protein concentration. Whole blood, obtained from cattle that had been fed lead salts, is suitable as a control for some methods for the assay of erythrocyte protoporphyrin, but not others (35). The ascorbic acid added to some control sera during manufacture can result in partial destruction of vitamin B_{12} if high concentrations of ascorbic acid are present. Studies of the long-term stability of glucose in lyophilized sera showed two distinct and opposite trends in many materials (37). Increase in the value averaging 4.7 mg/dl/year, was observed with glucose oxidase methods and a neocuproine procedure, whereas a decrease in the value averaging 3.0 mg/dl/year, was found with ferricyanide, hexokinase, and orthotoluidine methods. It was suggested that these effects were due to a gradual diminution of free and bound glucose and/or to a shift of glucose from the protein-bound state to the free state.

At present a wide variety of materials is used for calibrating methods for the determination of total protein and albumin. Farrance et al (38) have compared a number of preparations using seven assay procedures, and they have concluded that before any preparation can be used as a calibration standard, its suitability should be checked by methods that measure a different functional aspect of the protein molecule. Any differences obtained by all or any of these methods reflect the suitability of the material for calibration purposes. This was particularly important with dye-binding or immunochemical methods, and the measurement of absorbance at 278-280 nm did not, by itself, appear to be a valid measure of protein content if the procedure was to be used for calibrating immunochemical methods.

Commercial preparations are now becoming available for the control of blood pH and gas measurements. In one assessment (39), the mean values found differed from those stated by the manufacturer, although a tonometered buffer system was satisfactory for the control of accuracy and precision of pH, PCO_2 and PO_2. Steiner et al (40) have described the preparation and storage of a hemolyzed blood product that can be used for quality control of pH, PCO_2 and PO_2 measurements, and that give consistent and reproducible results.

Although the accuracy of the stated values of control sera is sometimes questioned (and the method of deriving them is often questionable), the reproducibility of the values for all vials in a batch is usually taken for granted. Jansen et al (41) have given three examples where wide intervial variation was found during interlaboratory trials. In one case, serum potassium results showed a between-laboratory coefficient of variation (CV) of 9.6%, which was about four times greater than that normally found. Creatinine and possibly glucose results also showed unusual variability, and it was concluded that the vials were not homogeneous, possibly due to intermittent contamination. Another product showed unusually wide variations in calcium, creatinine, urate, cholesterol, and glucose results because of the mixing of bottles from two different batches based on the same bulk serum. A third product gave an atypical urea result, which was thought to be due to bacterial infection during manufacture. Contamination can also be a problem with specimens used for trace element control, as, for example, with zinc eluted from the rubber stopper of the vial (42). These variations in results help to emphasize that the consistency of results for one analyte between different vials does not guarantee that the product will be acceptable for other analyses.

Assessment of Analytical Methods

It is now generally agreed that new analytical methods should be evaluated before they are introduced into routine use (43), and there are several published recommendations on how this should be done. Nielsen and Ash (44) have outlined a protocol that includes a useful checklist of the points to be considered, starting with an examination of the clinical need for the test. In a series of papers, Westgard et al (45-47) have considered the various stages of the evaluation in more depth. Details of the design of replication and recovery experiments were discussed, including tests for interference and specificity, and a procedure was described for comparing two methods. They also have described the calculations and statistical methods used to analyze the experimental data and have discussed the scope and limitations of various statistical procedures, with worked examples of the calculations, a glossary, and a set of related problems for self-study (together with the answers). A variety of statistical methods has been used to compare the results of two methods. Cornbleet and Shea (48) have studied the use of the product moment and rank correlation coefficients for this purpose and have concluded that all types of correlation coefficient were markedly influenced by the range of the results and were of no value in analyzing method comparison data.

With the rapid proliferation of commercial analytical systems and diagnostic kits, there is considerable interest in methods for their evaluation. Protocols for the evaluation of diagnostic kits have been published (29,49,50), which include an examination of the contents of the kit, labeling, and the manufacturer's claims for its performance, together with tests for precision, accuracy, linearity, specificity, and stability. Kim and Logan (50) have proposed a "diagnostic index" that can be calculated from accuracy and precision data, and they have applied this index to the results obtained in an assessment of 19 kits measuring serum creatine kinase activity. Evaluations of urine dipstick procedures (51-54) showed wide variations among different products; precision was often poor, mainly because of the variable results obtained by different technologists. In an assessment of three kits for the determination of urea concentration (55), one was found to give inaccurate results with quality control materials and to provide insufficient information about the method of blank correction for interfering compounds. In a comparison of six commercial thyrotropin radioimmunoassay kits (56), considerable differences were found in the slope of the calibration curves, so sensitivity was sometimes inadequate in the clinically important range, resulting in large intra-assay variations and inconsistent recoveries.

Similar procedures are used for the evaluation of multichannel analyzers. Caragher and Grannis (57) have described the use of sets of specimens with quantitative linear relationships that can be used to evaluate instrument calibration and to identify a variety of analytical biases. A more limited protocol for the rapid assessment of within-batch performance of multichannel analyzers has been applied to 15 instruments (58).

In practice, the evaluation of analytical methods presents several major difficulties, which have yet to be resolved. A full evaluation is lengthy and time consuming, and important points may be missed if shortcuts are taken. When the evaluation is completed, there are problems in publishing the results, since most journals are reluctant to accept evaluation reports unless they are of more than transient interest. With commercial analytical systems and diagnostic kits, the manufacturer may make modifications before the assessment is complete and the results are available. Consequently, the prospective user or purchaser needs to review all the available information and then make a limited range of tests before reaching a decision.

Internal Quality Control

Werner et al (59) have presented some interesting ideas on imprecision, which they divided into two components: additive errors and multiplicative errors. In additive errors, the standard deviation (SD) is approximately constant at different concentrations and in multiplicative errors, the SD decreases as concentration increases. The former may be due, for example, to fluctuations in signal detection, and the latter typically arise from pipetting errors, but the two may occur together. The contribution that each makes to the overall error (imprecision) was calculated from measurements of the SD at different concentrations. It was found that with most current methods the overall errors were mixed, although predominantly additive or predominantly multiplicative errors could occur as well. The authors have suggested that this technique can provide a useful insight into the sources of errors and should be added to routine quality control procedures. Errors with "blind" quality control samples (which were introduced by the supervisor without the knowledge of the analyst) were larger than those with "open" samples, which were inserted into the analytical run by the analyst, and this was largely due to the multiplicative and not the additive error component.

Boerma et al (60) have described the use of bench controls,

duplicate specimens, and other conventional techniques for monitoring the accuracy and precision of serum cholesterol assays. A modified cumulative sum (cusum) technique has been applied to radioimmunoassay quality control data (61) and was found to be about 50% more efficient than conventional charting techniques in detecting changes in the mean and variance of control specimens. The method described depends on the use of a V shaped mask to determine when the slope of the cusum chart changes significantly. Manual calculation of radioimmunoassay results can be an important source of imprecision, and methods of automating this calculation have been discussed (62).

External Quality Control

External quality control usually starts when an enthusiast circulates samples of the same specimen to a number of laboratories for analysis and then compares the results: almost invariably these results show poor agreement. If the results are analyzed in more detail, useful information about the causes of the disagreement can often be obtained. This has led to the development of a wide variety of external quality control schemes, which now form an important part of routine quality control procedures and are becoming powerful tools in the investigation of analytical error.

The Expert Panel on Nomenclature and Principles of Quality Control of the International Federation of Clinical Chemistry (IFCC) has published (63) a comprehensive review of the principles of external quality control, in which the objectives, advantages, and limitations of external quality control, the materials used, the design of surveys, and the procedures for evaluating the results, are discussed, together with a checklist of the remedial action to be considered in various circumstances. An initial objective of external quality control is often to measure the "state of the art" for a test. Recent interlaboratory surveys have shown this to be unsatisfactory for the analysis of estriol and total estrogen in pregnancy urine (64), nickel in urine (65), erythrocyte protoporphyrin (66), lead and cadmium in blood and urine (67), and several drugs in serum (68,69). In most cases, the "true" value for the specimen is not known, and the target value is then usually taken as the mean result of all participants after removal of outlying values. Differences from this consensus value are used as a measure of inaccuracy of the individual laboratory or analytical method. However, this consensus value may not be identical with the true value. In an important paper, Gilbert (70) has compared the consensus values obtained by participants in the College of American Pathologists' survey program with definitive values found by isotope dilution mass spectrometry methods. The results suggested close agreement for serum calcium, chloride, iron, lithium, magnesium, potassium, and sodium values, particularly when the concentration of analyte was close to the critical clinical decision level. The degree of the inaccuracy was related to the concentration of the analyte for calcium, iron and magnesium, and, less markedly, for lithium, potassium, and sodium. These findings suggest that consensus values can be effectively used for assigning target values, for comparing the results of interlaboratory surveys, and for calibrating and evaluating analytical methods, although they may be less useful at extremes of concentration. Provided specimens are stable, results seem to be repeatable over a period of months and possibly years. These specimens have been termed "Survey Validated Reference Materials" (SVRMs) and can be used to assign values to other serum quality control specimens and to investigate the accuracy of analytical methods (71).

The results of many external quality control schemes are now providing valuable information about the reliability of commonly used analytical methods. With radioimmunoassays of hormones and other substances, the need for standardization of methodology and the use of matched reagents has been stressed by several authors (77-74). Many of the basic concepts of quality control of radioimmunoassays, including standardization and assay design, have been reviewed by Hall et al (75), who suggested the use of "precision profiles" in which SD is plotted against concentration, to illustrate variations within assays, between assays, and between laboratories. Regression analysis (76) has shown that a major part of between-laboratory variation of cholesterol and triglyceride results is due to systematic error, arising from incorrect calibration, the use of inappropriate blanks, nonlinearity, and nonspecificity. Other authors (77,78) have described how a group of 12 lipid research laboratories achieved a high degree of accuracy and precision by rigid standardization and control of the entire procedure, using both internal and external quality control methods. Another study (79) described in detail how two laboratories obtained comparable results for plasma cholesterol and glucose by a thorough standardization of the analytical methods used, including the use of common control materials, exchange of patient samples, and a comparison of all details of laboratory procedures. These papers provide a good illustration of the detailed work that is necessary to obtain good interlaboratory agreement.

Paule and Mandel (80) have made a statistical analysis of the results of surveys for plasma calcium, potassium, and urea and have found considerable differences in precision between methods. The dispersion of results between laboratories was wider than the long-term within-laboratory precision, which suggested that the principal sources of error were within laboratories and not between laboratories. Aronsson et al (81) have made a detailed analysis of the factors that influence between- and within-laboratory variation. Three commercial enzyme reagents from one manufacturer were found to give larger variations in results than similar reagents from another company. They calculated the ratio of the between-laboratory to the within-laboratory variation for different analytical methods and suggested that high values of this ratio indicated the need for investigation of calibration procedures. Grannis and Massion (82) have used linearly related serum samples in surveys of five enzyme assays and have claimed that these provided participants with an objective evaluation

and comparison of the linearity, precision, and calibration of their methods. Differences between laboratories seemed to be primarily due to proportional biases, which in some cases were clearly identified as due to instruments or reagents rather than to the methods used.

For some methods, primary calibration standards give more consistent results than commercial serum standards (81). Hainsworth and Simpson (83) have surveyed the accuracy and precision of weighing techniques in 101 laboratories. Ninety-four percent of weighings surveyed were subject to errors of not more than 1.7 mg at the 2 gm level, and not more than 1.0 mg at the 100 mg level. However, 9% of the laboratories consistently obtained results that showed larger errors, which seemed to be partly due to the lack of recent servicing of their analytical balances. Some causes of error are well known, such as hemolysis (84) and evaporation of specimens (85), but external quality control can provide information about more subtle errors. Skendzel et al (86) have used a questionnaire to obtain data on the factors that may influence the results of cholesterol determinations, including reagents and equipment, blanking and calibration procedures, and sample preparation. This information was useful in helping to identify some of the factors responsible for small differences in test results.

Burtis et al (26) have found that the most consistent results for AST in serum were obtained by laboratories that supplemented optimized reaction solutions with pyridoxal phosphate. For serum creatinine (87), manual and automated systems gave comparable results, but centrifugal analyzers showed a consistently high bias. The use of Lloyd's reagent resulted in a very slight reduction in the creatinine concentration of lyophilized sera, although this does not, of course, mean that the reagent will have no effect on results with patients' specimens. AutuAnalyzer methods uniformly gave the highest results for serum iron (88), whereas the lowest values were reported by participants using methods that did not rely on protein removal, but it was not possible to determine which was correct. It was concluded that protein-based standards should be used when a method is found to be affected by protein. The use of hydroxylamine as a reducing agent seemed to give lower results than other reducing agents. A survey of the results of protein-bound iodine and thyroxine by column chromatography (89) has demonstrated wide interlaboratory variance among specimens with values in the euthyroid range, which, the authors suggested, seemed to relegate these tests to the estimation of a nonhormonal iodine. A survey of total serum protein analysis (90) has revealed a definite negative bias by the users of one commercial analytical system. In another survey (91) results for 13 serum proteins showed considerable between-laboratory variations (CV 8 to 29%), it was concluded that a common calibration standard was needed, together with better standardization of the titer and specificity of commercial antisera. The results of a urine chemistry survey for 11 constituents, involving more than 450 laboratories has been described by Glen (92). Performance was found to be deficient for amylase, 17-oxosteroids, and total protein, and severe bias in glucose results was noted when neocuproine and ferricyanide methods were used.

Before calculating consensus values, it is usual to remove outlying values; that is, those results that lie outside two to three SD from the initial mean. The frequency of these outlying values can be a useful quality control statistic. Gilbert and Sheehan (93) have evaluated the characteristics of outlying values and of the laboratories that reported them. They found that some outlying values were caused by the use of different units, errors in decimal transcription, or reversal of the results of two specimens. Others were caused by differences between the number of decimal places used for the target values and for the reported result. Laboratories that reported outlying values more frequently were termed "noncore" laboratories. The causes of these errors, and their frequency in different laboratories and for different tests, clearly merit further investigation. It also provides a good example of the type of problem that is revealed by, and can be investigated with, external quality control techniques.

The daily mean of all patients' results obtained by a laboratory is frequently used as an internal quality control procedure. Changes in the mean value indicate either a shift in analytical accuracy or an alteration in the biological mean value of the specimens analyzed. Grannis and Lott (94) have described a pilot study to identify the benefits and limitations of making interlaboratory comparisons of the results obtained routinely on clinical specimens. Each laboratory provided results from specimens from healthy persons, patients, and control specimens. The control specimen results were found to be indicators of analytical bias that affected the clinical assays, but they did not always correctly show the kind or magnitude of the bias. Using pattern recognition techniques it was shown that a laboratory's clinical assays had characteristic distributions that were related to the populations served, as well as to the analytical precision. White (95) used urea, creatinine, and electrolyte results from patients in two hospitals as a method of comparing laboratory performance. Reference values were derived from the patients' data obtained by the two laboratories, and it was found that changes in the mode or apparent reference range correlated well with changes in accuracy and precision, as indicated by an external quality control scheme. This type of study may prove useful in confirming that errors indicated by the analysis of control specimens also occur with patients' specimens.

Summary and Conclusions

Important advances have recently been made in two aspects of quality control. Most quality control techniques assume that errors that are detected and measured by analysis of control specimens indicate similar errors in the results of patients' specimens. This assumption has rarely been tested, but it is now clear that artifacts and errors arising during manufacture can produce misleading results for control specimens. These specimens are usually derived from animal sources, and the enzymes in these may behave quite differently from those in normal or pathological human serum. It is also worth noting that drug interferences (which are discussed below) are a potentially large source of error with some patients' specimens but will not be detected by any of the control specimens that are currently available.

External quality control has been shown to be a powerful technique that can be used for a variety of purposes. This makes it important that its objectives are clearly understood, and the survey designed to give the maximum information (63). The finding that consensus values of several inorganic analytes are close to the definitive values (70) has added a new dimension to the capabilities of large interlaboratory survey programs. It is now possible to measure accuracy in objective terms and to evaluate analytical methods on a collaborative basis. Important information about the reliability of individual analytical methods is emerging from the results of external quality control procedures, but the quality of performance of individual laboratories will not improve until this information is made more widely known and used.

REFERENCE VALUES

The term *normal range* has a superficial simplicity that, in the light of current knowledge, is becoming increasingly misleading. To the clinician, normality means freedom from disease, whereas to the statistician it usually implies that results have a Gaussian distribution. In practice, the "normal" population is made up of many subpopulations with differences in at least some of their biochemical characteristics. Furthermore, accurate measurement of these "normal" values reveals distributions that cannot be adequately described in terms of the standard deviation. An understanding of the many different types of variation that occur in healthy persons is not only important in its own right but is essential for the proper interpretation of the changes that occur in disease.

General Principles

In an important paper, the IFCC Expert Panel on Theory of Reference Values (96) has proposed a series of definitions that should permit unambiguous description and discussion on the subject:

* *A reference individual* is a person selected by using defined criteria.
* *A reference population* consists of all possible reference individuals. It may consist of only one member, that is, a person may serve as a reference for himself or another individual.
* *A reference sample group* is an adequate number of reference individuals selected to represent the reference population.
* *A reference value* is the value obtained by observation or measurement of a particular type of quantity in a reference individual.
* *A reference distribution* is the distribution of reference values, which is not necessarily Gaussian.
* *A reference limit* is derived from the reference distribution and

is used for descriptive purposes. It may be defined so that a stated fraction of the reference values is less than, or equal to, the limit with a stated probability.

* *A reference interval* is the interval between, and including, two reference limits.
* *An observed value* is a value of a particular type of quantity, obtained by observation or measurement, to be compared with reference values, reference distributions, reference limits, or reference intervals.

The interrelationship between these terms is shown in Figure 1.

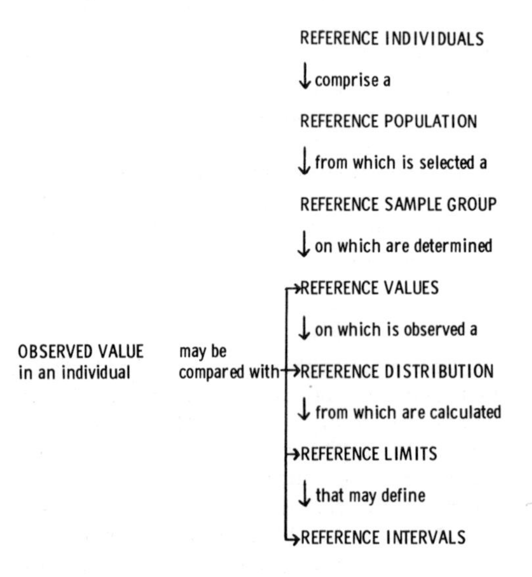

Figure 1. Concept of reference values: relationship of recommended terms. (Reproduced with permission from Grasbeck ref. 96.)

Reference values are only meaningful when the method used to obtain them is adequately described, and this may require details of the following:

1. Characterization of the reference population with respect to age, sex, genetic and socioeconomic factors, and so on
2. The criteria used for including or excluding individuals from the reference sample group
3. The physiological and environmental conditions under which the reference population was studied, including, for example:
   * Time and date of specimen collection
   * Intake of food and drugs (including alcohol)
   * Posture (including time spent in this posture)
   * Smoking
   * Degree of obesity
   * Pregnancy, stage of menstrual cycle
4. The specimen collection procedure used, including preparation of the individual and handling of the specimen

5. The accuracy and precision of the analytical methods used

Examples can readily be found to illustrate the importance of each of these (and other) factors on different types of reference values (97,98). Apart from analytical variation, the factors that can influence biological variation can be classified as

* Intraindividual: postural, dietary, hormonal, diurnal, and seasonal

* Interindividual: age, sex, race, economic, environmental, genetic, physical type, stress, medication, and life-style

Some of these are reviewed below.

Specimen Collection

Many of the factors that can influence a result, such as diet, posture, exercise, timing, and technique of specimen collection, can be standardized. Fasting may have pronounced effects on serum uric acid, lipids (99), and electrolyte concentrations (100). Posture at the time of blood collection affects the plasma concentrations of many protein-bound substances. Thus, plasma levels of total protein, albumin, calcium, alkaline phosphatase, bilirubin, and cholesterol all increase when the subject is standing (101). Consequently, when specimens from blood donors are used to obtain reference values, it is important to specify the length of time the donors are supine before the specimens are collected (102). Although arterial-venous differences are usually well known or predictable (103), it may be particularly important to indicate the point of blood withdrawal in patients with stabilized or centralized circulation (104).

Age and Sex

McPherson et al (105) have made a detailed study of the effects of age, sex, and other factors on the plasma concentrations of 19 commonly determined constituents in a group of 1,000 healthy blood donors. The effects of age and sex were shown to be the rule and to interact in many instances. Plasma concentrations of creatinine, urea, glucose, cholesterol, potassium, and globulin tended to increase with age, whereas total protein, albumin, calcium, inorganic phosphate, and iron levels tended to fall progressively. Some examples are shown in Figures 2 and 3. Age alone had an effect on the blood glucose level, and only sex differences were noted for the plasma concentration of protein-bound iodine. A simple combination of the effects of age and sex was shown for the plasma creatinine concentration, but in all other instances the sex differences varied at different ages. Sex differences, dependent on age, have also been reported for alkaline phosphatase levels (106). Plasma aldosterone (107) and blood alanine levels fall with age, whereas blood glycerol, nonesterified fatty acids, and ketone body concentrations increase (108).

Erythrocyte δ-amino levulinic acid dehydratase activity in males is about 10% lower than in females (109), but xanthurenic acid excretion shows no sex differences, although it increases with age (110).

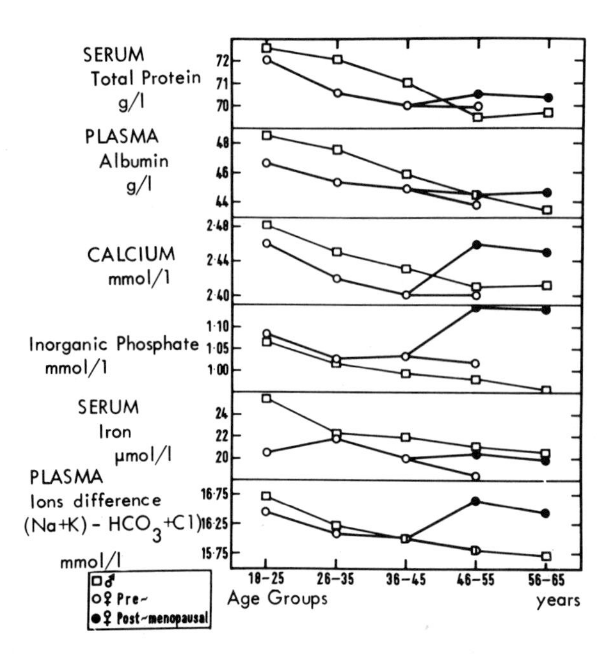

Figure 2. Plots of mean levels of blood constituents that show a tendency to fall significantly with increasing age. (Reproduced with permission from McPherson, et al, ref. 105.)

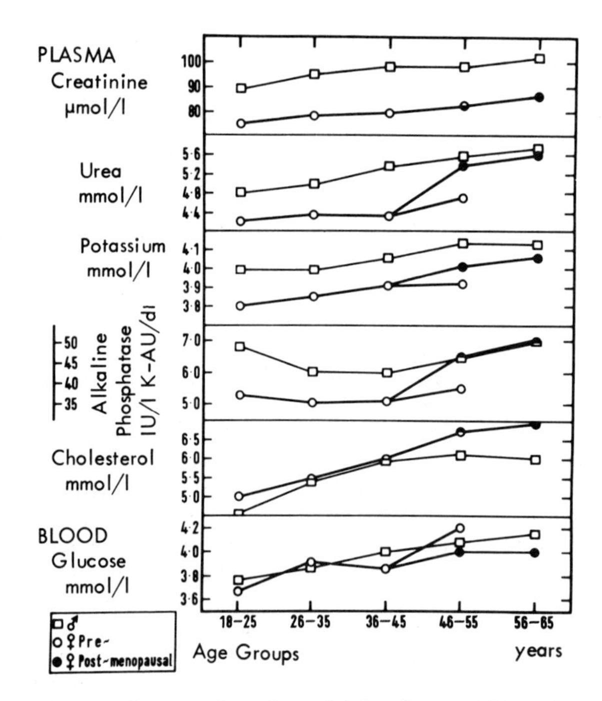

Figure 3. Plots of mean levels of blood constituents that show a tendency to rise significantly with increasing age (Reproduced with permission from McPherson et al, ref. 105.)

Cherian and Hill (111) measured the concentrations of 14 serum constituents in 1,062 children and adolescents aged 4 to 20 years. A significant relationship to age and sex was found for inorganic phosphorus and creatinine and to age only for total protein. For all other constituents, they found no significant relationship to age and sex, although other authors have noted that both serum uric acid and uric acid excretion increase throughout childhood (112). In a further study (113), it was found that serum alkaline phosphatase increased markedly from 5 to 14 years in boys and from 5 to 12 years in girls. The peak at puberty was more pronounced in boys, and values then decreased toward adult levels. AST levels showed a gradual decrease in both sexes from the age of 5 but, in contrast, the serum creatine kinase concentration was relatively constant between the ages of 5 and 12 years.

Urinary excretion of 3-methoxy-4-hydroxy mandelic acid in children is highest in infancy and then decreases with age until 10 to 15 years (114). Serum triiodothyronine (T_3) and thyroxine (T_4) levels increase after birth, and throughout infancy and childhood T_4-binding globulin levels are higher than those in middle-aged persons (115). There is a progressive increase in serum T_4 concentration with age in men and a smaller increase in women, which is concealed by the raised values resulting from the use of oral contraceptives (116). In general, changes in thyroid hormone concentrations with age are relatively minor, although they may correlate with each other (117). According to Evered et al (116), some of the effects previously reported as due to age may be related to the high frequency of undetected thyroid disease in the subjects studied and to the failure to exclude those taking drugs. A method has been described for assessing serum T_3 concentration that is independent of the subject's age and variations in serum binding proteins (118). This method uses a map on which serum T_3 and T_3-binding capacity values are plotted. For each age group, patients with thyrotoxicosis were effectively separated from euthyroid patients, including those who were pregnant or receiving estrogens.

Apart from the influence of oral contraceptives (see below) significant changes in blood chemistry occur in postmenopausal women (105), many of which are due to hormonal changes. In all instances when the menopause has a significant influence on blood chemistry there is an increase in concentration of the component concerned. The range of distributions remains unchanged with increasing years, suggesting that the findings are due to physiological or environmental factors, rather than to the early development of some pathological process. Significant changes in calcium metabolism occur after menopause (119). The negative calcium balance in postmenopausal women can be reversed with estrogens, probably because of an enhancement of intestinal absorption and improved renal conservation of calcium. Plasma calcitonin concentration is much lower in women than in men, but during pregnancy or the administration of estrogens, the levels may be higher than in men (120). Deficiency of this hormone may therefore have a role in postmenopausal osteoporosis. Women show two different types of urinary steroid metabolic profiles (121). Identification of the steroid pattern during adult life may provide an indication of a late fetal, or possibly neonatal, event, but it remains to be determined whether the type of steroid pattern affects a woman's susceptibility to specific diseases.

Pregnancy

α-Fetoprotein (AFP) measurements in serum and amniotic fluid are now widely used as screening tests for neurotubal defects in the fetus. Webb et al (122) have reported amniotic fluid AFP levels at 21 to 24 weeks gestation in pregnancies with a normal outcome, which supplement data on the more commonly used tests made at 15 to 20 weeks.

Measurements of urinary estrogen excretion and of plasma estriol are used to monitor fetal well-being in the third trimester of pregnancy, but so far there is little agreement on which test is preferable. Minute-to-minute fluctuations over 90 minutes and day-to-day variations in plasma estriol levels suggested that fluctuations of up to ± 45% are within the "normal range" (123). In pregnancies with a normal outcome, the estrogen to creatinine ratio in an early morning urine specimen showed smaller day-to-day variations than either the ratio in a 24-hour specimen or the 24-hour estrogen excretion (124). A significant fall in the ratio in an early morning specimen could therefore be detected more easily. Data from patients who delivered healthy babies provided reference ranges, which, after logarithmic transformation, increased linearly with the period of gestation. A fall in the logarithm of the estrogen to creatinine ratio exceeding 40% of this range was found to be unusual in pregnancies with a normal outcome, and suggested the likelihood of impaired fetoplacental function. Maternal plasma and urine concentrations of β-microglobulin and γ-glutamyl transpeptidase are constant throughout pregnancy and can therefore be used to assess renal tubular function (125). Maternal plasma and amniotic fluid concentrations of prolactin in normal pregnancy have also been recorded (126).

Body Weight

It is becoming apparent that body weight is an important attribute in defining reference ranges of several blood constituents, after age and sex dependencies have been taken into account. Munan et al (127) have studied a total of 2,368 subjects, aged 10 to 96, of both sexes in an attempt to examine the relationships between body weight and biochemical features. Three patterns were identified:

1. Positive weight dependencies were found with serum urate, glucose, lactate dehydrogenase (LDH), and cholesterol concentrations; most subjects with high concentrations of these constituents were in the higher weight groups and vice versa. Similar associations with body weight were found for plasma creatinine, total protein, AST, and blood hemoglobin concentrations, but only in men. Foster et al (108) also noted that serum insulin and blood alanine concentrations were related to body weight.

2. Inverse relationships with body weight were found for plasma concentrations of inorganic phosphate in both sexes, and for

serum calcium in women only; that is, subjects with low body weight tended to have higher concentrations of these components.

3. There was no association, in either sex, between body weight and plasma concentrations of sodium, potassium, chloride, urea, bilirubin, or alkaline phosphatase.

Genetic Factors

Increasing attention is being given to the influence of genetic factors on reference ranges. McPherson et al (105) showed that persons with blood group O or B had higher serum alkaline phosphatase levels than those with group A or AB. Blood hemoglobin concentrations in blacks have been reported to be on average about 0.5 gm/dl lower than those in whites (128). In white populations, the frequency distribution of blood glucose levels is unimodal. However, in a study of a population inhabiting the small island of Nauru in the central Pacific, it was found that distributions were unimodal between the ages of 10 and 19, but groups of both sexes over 20 years of age showed bimodal frequency distributions (129). This was probably a reflection of the high incidence of diabetes in the population.

Drugs

A wide variety of drugs can influence the results of chemical analyses, either by analytical interference (which is method dependent) or by a real effect on the constituent measured. Therapeutic concentrations of ascorbic acid interfere with several tests, depending on the method used (130). Van Steirteghem et al (131) gave 3 gm of ascorbic acid daily to a number of healthy adults and found both analytical and biological interferences; uric acid seemed to exhibit both effects. Panek et al (132) have studied the effects of 84 drugs on a variety of tests, and have found that the most commonly affected procedure was the phosphotungstate method for uric acid, but tests for total protein, albumin, LDH, and alkaline phosphatase were also affected by some drugs. Usually, interference is studied *in vitro* by adding the drug to the specimen, but the importance of *in vivo* testing was illustrated by the finding that metabolites of spironolactone gave spurious digoxin results with two diagnostic kits (34).

Several studies have been made of the effect of alcohol on biochemical results (133-135). Increases in gamma glutamyl transpeptidase, triglycerides, uric acid, and AST levels were frequently found, although other liver function tests did not seem to be affected. Test results showed progressively higher values with increasing alcohol consumption, and the effects seemed to be more pronounced with beer than with wine (136). These markers were extremely sensitive, and elevated levels were frequently found in persons whose alcohol intake would be regarded as normal, and who were in no sense alcohol dependent.

Oral contraceptives are known to have pronounced effects on several tests, such as those to determine thyroid hormone concentrations (116). They produce significant decreases in plasma phosphate and albumin concentrations and increases in protein-bound iodine and globulin fractions (105,137).

Salway (138) has discussed possible solutions to the problem of drug interference and has stressed the need for quantitative data. Friedman et al (138) have described a trial in which a computer was used to generate reports of all potential interactions between the drugs administered to a patient and his laboratory tests. Physicians found that the system had both educational and clinical value, and a review of patients' charts suggested that automatic reporting of possible drug interference resulted in an alteration of the patients' therapy in a significant number of cases.

Biological Rhythms

The nature of biological rhythms and their relevance to clinical biochemistry have recently beeen discussed in an authoritative review by Daly (140), illustrated by many examples of the different types of biological variation that can occur with time. Knowledge of circadian changes (which occur every 24 hours) is particularly important in deciding the best time to collect a blood specimen. For example, plasma cortisol levels show wide circadian variations (141), which are unaltered in obesity (142). Serum iron concentration also varies considerably during the day, but no consistent pattern could be identified and the changes were not sex related (143). Serum magnesium levels show different types of circadian rhythm in various groups of subjects (144). A significant circadian rhythm was detected in young men and in elderly men and women, but not in elderly insane persons. The 24-hour mean values were higher, and the rhythm amplitude larger, in elderly persons than in young ones. The location of the peak time (the acrophase) differed throughout the 24 hours in different types of subjects.

Important changes in hormone levels occur during the menstrual cycle. In a recent study (145), it was found that the follicular phase of the cycle is characterized by apparently random fluctuations and the excretion of free corticosteroids. However, after the midcycle estrogen peak, there is a significant fall in corticosteroid excretion, which then rises to a peak 8 to 10 days after ovulation and is synchronous with the second estrogen peak occurring during the luteal phase.

A detailed study has been made of the circadian and circannual (i.e., about 1 year) changes in plasma hormone concentrations in a group of healthy young men (146). Annual changes were found in the 24-hour mean levels of plasma FSH (with an annual peak time in February), LH (March), thyroxine (September), cortisol (February), renin activity (April), testosterone (October), urinary excretions of 17-hydroxy corticosteroids (March), aldosterone (March), andpotassium (May), as well as in sexual activity (September). Plasma prolactin concentrations did not show an annual variation. In addition to these findings, annual changes were noted in the circadian acrophase for plasma thyroxine, cortisol, testosterone levels, and renin activity, and the urinary excretion of aldosterone and potassium. Plasma levels of testosterone and prolactin have been reported to be lower in infertile men than in fertile ones (147).

Miscellaneous Factors

Reference intervals have been reported for a wide range of laboratory tests (148), and for components of cerebrospinal fluid (CSF) (149), serum IgE (150), and albumin (151). Other studies have been made of plasma prostaglandin (152) and vitamin A (153) levels in premature and full-term infants. The vitamin B_6 status of a group of female adolescents was assessed using erythrocyte alanine aminotransferase (ALT) as an index of deficiency (154), and between 13 and 31% of the subjects showed some deficiency, depending on the criteria used.

Trace metals and lipids provide good examples of some of the difficulties of defining reference values. Both are currently of major interest, and in neither case is it possible to define levels or ranges that are associated with health and with sickness. Most values lie in the gray area in between, and their precise location seems to be a measure of the risk to the person.

Determination of reference values for blood levels of toxic substances present in the environment is made particularly difficult by the analytical problems of measuring traces of these substances, especially contamination (155), and the multiplicity of factors that can influence concentrations. For example, serum zinc levels vary with sex, age, length of fasting, time of day, and geographic area (156). Tobacco smokers have higher concentrations of blood cadmium (157), serum copper, and ceruloplasmin (158). Bogden et al (159) have compared the plasma concentrations of a number of trace metals in maternal and cord blood and have concluded that the levels of some nutrient metals may be important in determining the probability of low birth weight. There is also considerable interest in the effects of lead in domestic water supplies, and one recent study (160) has shown that mothers and their children living in houses with lead water pipes have higher blood lead levels than those living in houses with copper pipes. In such instances, the problem is not to define what is "normal" but to delineate the lowest levels that are associated with pathological changes in the various reference groups.

Plasma lipid levels are affected by age (161) and may be influenced by diet and exercise (163). Regional differences have also been reported (164). Frerichs et al (165) have found that newborn white infants have higher average levels of cholesterol and β-lipoproteins than black infants. Stress at delivery, birth weight, socioeconomic status, and season all had effects, but the magnitude and direction of the relationships between lipids were similar to those of preschool and school-aged children in the same community, suggesting that basic biochemical relationships are established at birth. Other authors (166) have noted that children from parents with high cholesterol or triglyceride levels seemed more likely to have high levels than other children, thus simplifying the problem of identifying children who might develop atherosclerosis in later life.

Arntzenius et al (167) have made a survey of the risk factors for coronary heart disease in women aged 40 to 41 in Leiden, Netherlands. They found that the mean serum high-density lipoprotein (HDL) cholesterol level was significantly higher in women than in men. There was no association between HDL cholesterol

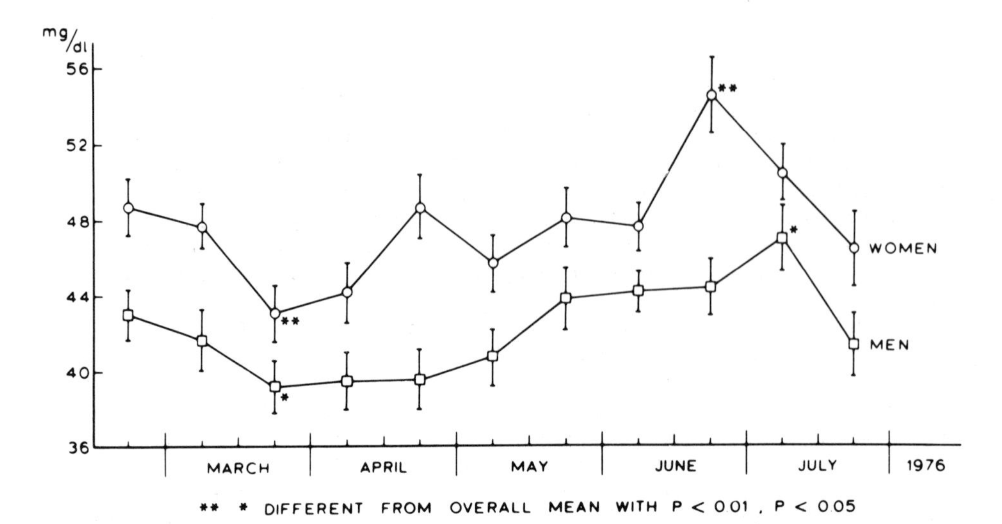

Figure 4. Mean HDL cholesterol levels in 40-year-old citizens of Leiden, Netherlands, in biweekly periods. (Reproduced with permission from McPherson et al, ref 105.)

and blood pressure, obesity, or electrocardiograph changes, but cigarette smoking and the use of oral contraceptives were strongly associated with reduced serum HDL concentrations. Other authors (168,169) have also found significantly higher HDL cholesterol levels in women, with little variation with age but with no significant correlation with smoking habits. It was also noted (170) that HDL cholesterol levels seemed to be lower in spring than in midsummer, possibly because of a rise in the temperature of the environment (see Fig. 4).

In most of these surveys of HDL cholesterol, differences among the various groups were small, and without data on the relative accuracy of the methods used, it is difficult to compare some of the results reported for different populations. Moreover, the coefficient of variation (CV) of HDL cholesterol analysis has been reported to be 6.17% (168), whereas Goldberg (171) has asserted that a laboratory that could not achieve a CV of less than 5% at a concentration of 30-40 mg/dl should not offer this test. However, it is doubtful whether even this level of precision is adequate for a reliable assessment of the small differences between groups, and this lack of precision, together with the relatively large biological variations between persons, produces problems in using this test for assessing risks in the individual person on a routine basis.

Methods of Obtaining and Expressing Reference Values

Three different types of population have been used to obtain reference data: apparently healthy persons, patients who visit their family doctors or a hospital clinic, and hospital inpatients (98). In each case, the criteria used for excluding any individual need to be carefully specified (96,97).

Currently, most laboratories use data obtained from apparently healthy persons. Since the distribution of these values may not be Gaussian, reference intervals may be calculated after logarithmic transformation (105) or may be expressed as percentiles (106,111). It is, however, insufficient merely to describe these persons as "healthy" because none of the definitions of this state is entirely satisfactory (96). Moreover, reference values obtained from healthy subjects under carefully controlled conditions may not be suitable for making clinical decisions on a patient in a hospital where specimens may be collected under different conditions.

To overcome this, several authors have proposed methods of deriving reference intervals from data obtained from hospitalized patients. Leonard and Westlake (172) used data obtained in an admission profile study. The case notes of all patients with results outside 5 and 95% limits were examined and, when evidence was obtained that the illness affected the results, these were excluded. Harwood and Cole (173) used a similar procedure and expressed the results as percentile rankings. Horvath (174) used data from patients with a short hospital stay and a short laboratory record. A statistical program was used repeatedly to exclude results with excessive variation, thereby progressively narrowing the range. In this way the most consistent data were included in the calculation of reference intervals. The statistical properties of the data sets, as well as comparison with "normal ranges" for healthy subjects and a review of patient records, suggested that serviceable reference ranges were obtained in this way. This method is claimed to be suitable for monitoring reference ranges of a laboratory at frequent intervals. Other authors (175) have described a technique for calculating the maximum number of persons in a patient population who will give normal values and, by difference, the minimum possible number of those who will give values above and below "normal". This technique permits the estimation of the probability that an observed result is within a particular subpopulation, as well as the selection of decision limits for classifying a patient. It is also said to allow for estimating the degree of change required for a subject with an initially normal value to start giving results that are above or below normal values. If properly interpreted, a normal value may help to differentiate between diagnoses of conditions that give normal results with differing frequencies (176).

In a person in a steady state, the observed values usually vary with time, but in most cases a single value is taken for comparison with appropriate reference values. The consequences of this depend on the relative magnitude of the analytical, between- and within-person variation. When the within-person variation is large because of analytical error, improved analytical technique or repeated sampling is necessary (177). Williams et al (178) tested the assumption that a reference group of persons, characterized by age and sex, gives a narrower range of variation than does a larger mixed population. If this were true, the demographic set of values would provide a more sensitive reference than the customary "normal" range for interpretation of values within the person. They defined a ratio, R, of within-person to between-person SDs that, for the 16 serum constituents measured, was similar for defined age/sex groups. When the results were adjusted to remove the average analytical variation, R values were less than 0.80 for all constituents except creatine kinase, which indicates that these were all strong discriminators of individuality. Other authors (179) found that the ratio, R, was less than 0.10 for alkaline phosphatase, cholesterol, dopamine-β-hydroxylase, and LDH and thus showed a high individuality. They found that the ratio was greater than 1.0 for serum sodium, carbon dioxide, and creatine kinase. More than 30% of the total variance for sodium, carbon dioxide, calcium, total protein, and chloride was due to analytical factors. They also presented limits for the expected difference between measurements of the same analyte in the same person on different days. These results suggested the need for individual rather than population-based reference ranges, even if the latter are obtained from persons of similar age and the same sex.

Summary and Conclusions

In discussing the diagnosis of hyperprolactinemia, Jeffcoate (180) provided a good example of the type of problem that laboratory managers often face. Many clinicians find the significance of raised prolactin levels difficult to understand, and interpretation is made more complex because of the many endogenous factors that can alter prolactin secretion. The problems of selecting a suitable reference interval are made more difficult by the between- and within-laboratory errors of this assay.

Reference values are affected by many different factors, and there must be few measurements for which a single, universally applicable "normal range" can be used. Nevertheless, it must be recognized that statistical differences between reference values in different populations are not necessarily of clinical importance. Realizing the complexity of the problem, laboratory managers now need advice on whether they should *(a)* use well-authenticated published reference intervals or *(b)* measure their own by one of the (different) published procedures. At the same time, it must be remembered that the overzealous use of different reference intervals may merely confuse the clinician, instead of helping him to interpret results obtained on his patients (98).

In normal hospital practice, a compromise will usually be made by quoting reference intervals adjusted for age and sex only when these are clearly of statistical and clinical significance (181). However, with increasing knowledge, the use of age- and sex-related reference intervals will become more common, and this should help to improve the standard of interpretation of laboratory results which, in general, has failed to match improvements in laboratory efficiency and precision over recent years (105).

REFERENCES

(Note: the more important references are indicated by asterisks.)

1. Leonard PJ: In *Scientific Foundations of Clinical Biochemistry*, vol. 1. (eds): DL Williams, RF Nunn, V Marks. Heinemann, London, 1978, p 48.

2. Conn RB: *N Engl J Med* 298: 422, 1978.

3. Plueckhahn VD: *Med J Aust* 1:113, 1978.

4. Rabkin MT: *N Engl J Med* 298:454, 1978.

5. Rollason JG: *Med Lab Sciences* 35:187, 1978.

6. Barnard DJ, Bingle JP, Garratt CJ: *Brit Med J* 1:1463, 1978.

7. Chapman C, Hayter C: *Brit Med J* 2:830, 1978.

8. *Krieg AF, et al: *Amer J Clin Pathol* 69:525, 1978.

9. Kassirer JP, Pauker SG: *N Engl J Med* 299:947, 1978.

10. Taylor HW: *Clin Biochem* 11:179, 1978.

11. Ransohoff DF, Feinstein AR: *N Engl J Med* 299:926, 1978.

12. Archer PG: *Amer J Clin Pathol* 69:32, 1978.

13. Rockerbie RA, Campbell DJ: *Clin Biochem* 11:77, 1978.

14. Vidler J, Wilcken B: *Clin Chim Acta* 82:173, 1978.

15. Weaver DK, et al: *Amer J Clin Pathol* 70:400, 1978.

16. Jomain PA, Owen JA: *Ann Clin Biochem* 15:276, 1978.

17. Taylor HW: *Clin Biochem* 11:87, 1978.

18. Minty BD, Barrett AM: *Int J Anaesth* 50:1031, 1978.

19. Barnett RN, McIver DD, Gorton W.: *Amer J Clin Pathol* 69:520, 1978.

20. Editorial: *Brit Med J* 1:871, 1978.

21. Lewis SM, Wardle JM: *J Clin Pathol* 31:888, 1978.

22. Grist NR: *J Clin Pathol* 31:415, 1978.

23. Broughtcn PMG: In *Scientific Foundations of Clinical Biochemistry* vol 1, (eds): DL Williams, RF Nunn, V Marks. Heinemann, London, 1978, p 459.

24. Annino JS: *N Engl J Med* 299:1130, 1978.

25. *Buttner J, et al: *Clin Chim Acta* 63:F25, 1975.

26. Burtis CA, et al: *Clin Chem* 24:916, 1978.

27. Longlands MG, Wiener K: *Ann Clin Biochem* 15:164, 1978.

28. Fowler RT, Walmsley TA: *Clin Chim Acta* 87:159, 1978.

29. Kim EK, Logan JE: *Clin Biochem* 11:238, 1978.

30. Ratcliffe WA, Logue FC, Ratcliffe JG: *Ann Clin Biochem* 15:203, 1978.

31. Hindriks FR, Goren A: *Clin Chem* 24:2062, 1978.

32. Fraser CG, Fudge AN, Penberthy LA: *Ann Clin Biochem* 15:121, 1978.

33. Proksch GJ, Bonderman DP: *Clin Chem* 24:1924, 1978.

34. Dwenger A, Trautschold I: *J Clin Chem Clin Biochem* 16:587, 1978.

35. Penton Z, Bissell G: *Clin Chem* 24:504, 1978.

36. Block E, Akerkar A, Moholick M: *Clin Chem* 24:176, 1978.

37. Lawson NS, Haven GT, Morre TD: *Amer J Clin Pathol* 70:523, 1978.

38. Farrance I, et al: *Ann Clin Biochem* 15:31, 1978.

39. Evans JR: *Ann Clin Biochem* 15:168, 1978.

40. Steiner MC, et al: *Clin Chem* 24:793, 1978.

41. Jansen AP, et al: *Clin Chim Acta* 84:255, 1978.

42. Urquhart N: *Clin Chem* 24:1652, 1978.

43. Broughton PMG: *Progress in Clin Pathol* 7:1, 1978.

44. Nielsen LG, Ash KO: *Amer J Med Technol* 44:30, 1978.

45. Westgard JO, et al: *Amer J Med Technol* 44:290, 1978.

46. Westgard JO, et al: *Amer J Med Technol* 44:420, 1978.

47. Westgard JO, et al: *Amer J Med Technol* 44:552, 1978.

48. Cornbleet PJ, Shea MC: *Clin Chem* 24:857, 1978.

49. Lloyd PH: *Ann Clin Biochem* 15:136, 1978.

50. Kim EK, Logan JE: *Clin Biochem* 11:244, 1978.

51. James GP, Bee DE, Fuller JB: *Amer J Clin Pathol* 70:368, 1978.

52. James GP, Bee DE, Fuller JB *Clin Chem* 24:1934, 1978.

53. Simpson E, Thompson D: *Clin Chem* 24:389, 1978.

54. Simpson E, Thompson D: *Ann Clin Biochem* 15:241, 1978.

55. Fraser CG, Fudge AN: *Ann Clin Biochem* 15:265, 1978.

56. Leicht E, Biro E: *J Clin Chem Clin Biochem* 16:551, 1978.

57. Caragher TE, Grannis GF: *Clin Chem* 24:403, 1978.

58. McLelland AS, Fleck A, Burns RF: *Ann Clin Biochem* 15:12, 1978.

59. *Werner M, Brooks SH, Knott LB: *Clin Chem* 24:1895, 1978.

60. Boerma GJM, Stylblo K, Leijnse B: *Clin Chim Acta* 83:211, 1978.

61. Kemp KW, et al: *J Endocrinol* 76:203, 1978.

62. Challand GS: *Ann Clin Biochem* 15:123, 1978.

63. *Buttner J, et al: *Clin Chim Acta* 83:189F, 1978.

64. Huis In't Veld LG, et al: *J Clin Chem Clin Biochem* 16:119, 1978.

65. Adams DB, et al: *Clin Chem* 24:862, 1978.

66. Jackson KW: *Clin Chem* 24:2135, 1978.

67. Paulev P-E, Solgaard P, Tjell JC: *Clin Chem* 24:1797, 1978.

68. Buhl SN, Kowalski P, Vanderlinde RE: *Clin Chem* 24:442, 1978.

69. McCormick W, et al: *N Engl J Med* 299:1118, 1978.

70. *Gilbert RK: *Amer J Clin Pathol* 70:450, 1978.

71. Grannis GF: *Amer J Clin Pathol* 70:580, 1978.

72. Haven GT, Hansell JR, Haven MC: *Amer J Clin Pathol* 70:532, 1978.

73. Hall PE: *Hormone Res* 9:440, 1978.

74. Rohle G, Breuer H: *Hormone Res* 9:450, 1978.

75. Hall PE, et al: In: *Radioimmunoassay and Related Procedures in Medicine, 1977*, vol 2. (Vienna) International Atomic Energy Agency, 1978.

76. Munster DJ, Lever M, Walmsley TA: *Clin Biochem* 11:194, 1978.

77. Ahmed S, et al: *Clin Chem* 24:330, 1978.

78. Lippel K, et al: *Clin Chem* 24:1477, 1978.

79. Boerma GJM, et al: *Clin Chem* 24:1126, 1978.

80. Paule RC, Mandel J: *Amer J Clin Pathol* 70:471, 1978.

81. Aronsson T, et al: *Scand J Clin Lab Invest* 38:53, 1978.

82. Grannis GF, Massion CG: *Amer J Clin Pathol* 70:487, 1978.

83. Hainsworth IR, Simpson E: *Clin Chim Acta* 84:233, 1978.

84. Frank JJ, et al: *Clin Chem* 24:1966, 1978.

85. Nielsen LG, et al: *Amer J Med Technol* 44:112, 1978.

86. Skendzel LP, et al: *Amer J Clin Pathol* 70:584, 1978.

87. Chinn EK, et al: *Amer J Clin Pathol* 70:503, 1978.

88. Itano M: *Amer J Clin Pathol* 70:516, 1978.

89. Thiessen MM, et al: *Amer J Clin Pathol* 70:511, 1978.

90. Burkhardt RT, Batsakis JG: *Amer J Clin Pathol* 70:508, 1978.

91. Landaas S, Skrede S, Eldjarn L: *Scand J Clin Lab Invest* 38:295, 1978.

92. Glenn GC: *Amer J Clin Pathol* 70:513, 1978.

93. Gilbert RK, Sheehan L: *Amer J Clin Pathol* 70:481, 1978.

94. Grannis GF, Lott JA: *Amer J Clin Pathol* 70:567, 1978.

95. White JD: *Clin Chim Acta* 84:353, 1978.

96. *Grasbeck R, et al: *Clin Chim Acta* 87:459F, 1978.

97. Siest G, Petitclerc C: *Ann Biol. Clin* 36:217, 1978.

98. Wilding P, Bailey A: In *Scientific Foundations of Clinical Biochemistry*, vol 1. DL Williams, RF Nunn, V Marks (eds): Heinemann, London. 1978, p 451.

99. Gumaa KA, et al: *Brit J Nutr* 40:573, 1978.

100. Mustafa KY, et al: *Brit J Nutr* 40:583, 1978.

101. Dixon M, Paterson CR: *Clin Chem* 24:824, 1978.

102. Picard RA: *Clin Chem* 24:1083, 1978.

103. Morena E, de la, et al: *Biochem Med* 20:15, 1978.

104. Rommel K, Koch C-D, Spilker D: *J Clin Chem Clin Biochem* 16:373, 1978.

105. McPherson K, et al: *Clin Chim Acta* 84:373, 1978.

106. Gardner MD, Scott R: *J Clin Pathol* 31:1202, 1978.

107. Lijnen P, et al: *Clin Chim Acta* 84:305, 1978.

108. Foster KJ, et al: *Clin Chem* 24:1568, 1978.

109. Davis JR, Avram MJ: *Clin Chem* 24:726, 1978.

110. Dodu JM, et al: *Clin Chim Acta* 90:13, 1978.

111. Cherian AG, Hill JG: *Amer J Clin Pathol* 69:24, 1978.

112. Stapleton FB, et al: *J Pediatr* 92:911, 1978.

113. Cherian AG, Hill JG: *Amer J Clin Pathol* 70:783, 1978.

114. Haymond RE, Knight JA, Bills AC: *Clin Chem* 24:1853, 1978.

115. Stubbe P, et al: *Horm Metab Res* 10:58, 1978.

116. Evered DC, et al: *Clin Chim Acta* 83:223, 1978.

117. Kagedal B, Sandström A, Tibbling G: *Clin Chem* 24:1744, 1978.

118. Mardell R, Gerson M: *Clin Chem* 24:1792, 1978.

119. Heaney RP, Recker RR, Saville PD: *J Lab Clin Med* 92:953, 1978.

120. Hillyard CJ, Stevenson JC, MacIntyre I: *Lancet* 1:961, 1978.

121. Pfaffenberger CD, Horning EC: *Anal Biochem* 88:689, 1978.

122. Webb T, et al: *Lancet* 2:578, 1978.

123. Kirkish LS, et al: *Clin Chem* 24:1830, 1978.

124. Philips SD, et al: *Clin Chim Acta* 89:71, 1978.

125. Kelly A.M., McNay MB, McEwan HP: *Brit J Obst Gyn* 85:190, 1978.

126. Yuen BH, et al: *Brit J Obst Gyn* 85:293, 1978.

127. Munan L, et al: *Clin Chem* 24:772, 1978.

128. Dallman PR, et al: *Amer J Clin Nutr* 31:377, 1978.

129. Zimmet P, Whitehouse S: *Diabetes* 27:793, 1978.

130. Siest G, et al: *J Clin Chem Clin Biochem* 16:103, 1978.

131. Van Steirteghem AC, Robertson EA, Young DS: *Clin Chem* 24:54, 1978.

132. Panek E, Young DS, Bente J: *Amer J Med Technol* 44:217, 1978.

133. Whitehead TP, Clarke CA, Whitfield AGW: *Lancet* 1:978, 1978.

134. Whitfield JB, et al: *Ann Clin Biochem* 15:297, 1978.

135. Whitfield JB, et al: *Ann Clin Biochem* 15:304, 1978.

136. Martin-Boyce , et al: *Lancet* 2:529, 1978.

137. Kamyab S, Baghdiantz A, Motamedi H: *J Ster Biochem* 9:811, 1978.

138. Salway JG: *Ann Clin Biochem* 15:44, 1978.

139. Friedman RB, Young, DS, Beatty ES: *Clin Pharm Therap* 24:16, 1978.

140. Daly JR: In *Scientific Foundations of Clinical Biochemistry*, vol 1. Williams DL, Nunn RF, Marks V (eds). Heinemann, London p440, 1978.

141. Knoll E, Wisser H, Rebel FC: *J Clin Chem Clin Biochem* 16:567, 1978.

142. Copinschi G, et al: *Clin Endocrinol* 9:15, 1978.

143. Long R, Delaney KK, Siegel L: *Clin Chem* 24:842, 1978.

144. Touitou Y, et al: *Clin Chim Acta* 87:35, 1978.

145. Walker MS, McGilp I: *Ann Clin Biochem* 15:201, 1978.

146. Reinberg A, et al: *Acta Endocrinol* 88:417, 1978.

147. Pierrepoint CG, et al: *J Endocrinol* 76:171, 1978.

148. Scully RE, McNeeley BU, Galdabini JJ: *N Engl J Med* 298:34, 1978.

149. Breebaart K, Becker H, Jongebloed FA: *J Clin Chem Clin Biochem* 16:561, 1978.

150. Azavedo J de, et al: *Irish J Med Science* 147:262, 1978.

151. Adewoye HO, Fawibe JF: *Brit J Nutr* 40:439, 1978.

152. Mitchell MD, et al: *Prostaglandins* 16:319, 1978.

153. Brandt RB, et al: *J Pediatr* 92:101, 1978.

154. Kirksey A, et al: *Amer J Clin Nutr* 31:946, 1978.

155. Reimold EW, Besch DJ: *Clin Chem* 24:675, 1978.

156. Bjorksten F, et al: *Acta Med Scand* 204:67, 1978.

157. Ward RJ, Fisher M, Tellez-Yudilevich M: *Ann Clin Biochem* 15:197, 1978.

158. Davidoff GN, et al: *Amer J Clin Pathol* 70:790, 1978.

159. Bogden JD, et al: *J Lab Clin Med* 92:455, 1978.

160. Elwood PC, Thomas H, Sheltawy M: *Lancet* 1:1363, 1978.

161. Kritchevsky D: *Postgrad Med* 63:133, 1978.

162. Askevold R, et al: *Acta Paed Scand* 67:157, 1978.

163. Yates BJ, Wingo WD, Lopez CA: *J Amer Diet Assoc* 72:398, 1978.

164. Ricci G, et al: *Atherosclerosis* 31:125, 1978.

165. Frerichs RR: *Pediat Res* 12:858, 1978.

166. Morrison JA, et al: *Paediat* 62:468, 1978.

167. Arntzenius AC, et al: *Lancet* 1:1221, 1978.

168. Bradby GVH, Valente AJ, Walton KW: *Lancet* 2:1271, 1978.

169. Schaefer EJ, et al: *Lancet* 2:391, 1978.

170. Van Gent CM, Van Der Voort H, Hessel LW: *Clin Chim Acta* 88:155, 1978.

171. Goldberg JM: *Clin Chem* 24:2061, 1978.

172. Leonard JV, Westlake AJ: *Clin Chim Acta* 82:271, 1978.

173. Harwood SJ, Cole GW: *JAMA* 240:270, 1978.

174. Horvath BM: *Amer J Clin Pathol* 69:398, 1978.

175. Grannis GF, Lott JA: *Clin Chem* 24:640, 1978.

176. Gorry GA, Pauker SG, Schwartz WB: *N Engl J Med* 298:486, 1978.

177. Morgan DB: *Ann Clin Biochem* 15:49, 1978.

178. *Williams GZ, Widdowson GM, Penton J: *Clin Chem* 24:313, 1978.

179. *Van Steirteghem AC, Robertson EA, Young DS: *Clin Chem* 24:212, 1978.

180. Jeffcoate SL: *Lancet* 2:1245, 1978.

181. Bold AM, Wilding P: *Clinical Chemistry Companion* (Oxford) Blackwell, 1978.

2
Instrumentation and Computers

Harry L. Pardue
Stanley N. Deming

Because virtually all clinical chemical analyses involve some type of instrumentation, and because many instruments now involve some type of computer application, a totally comprehensive chapter reviewing the use of instrumentation and computers would include almost all papers in the clinical chemistry literature. Thus, it was necessary to establish some guidelines and criteria for the selection of papers to be included in this review. The criteria we used were that the paper should *(a)* represent a major new development in clinical chemical instrumentation or computer applications, *(b)* represent a new application or a substantial and significant modification of existing technology, *(c)* provide significant new information about the limitations and capabilities of existing systems, *(d)* represent the adaptation of promising technology from another area to clinical chemistry, and *(e)* suggest future trends or potentially useful new directions for the advancement of clinical chemistry.

MULTILAYER FILM SYSTEM

One of the novel developments reported during 1978 was a multilayer film element for chemical analyses applied to clinical problems (1,2). In this system, reagents in matrices of hydrophilic polymers are coated in one or more layers on top of a transparent base, and the layered film element is coated with a white, isotropically porous polymer with an 80% void volume. When a drop of fluid is applied to the porous layer, the fluid spreads, filling a void volume equivalent to the drop volume. Low-molecular-weight compounds, including water in the sample fluid, diffuse from the spreading layer into the reagent layer or layers, where appropriate chemical reactions take place to produce changes in absorbance that are related to analyte concentration or activity. Absorbance changes are monitored by reflection densitometry from the white background of the spreading layer. Two companion papers (1,2) describe operational and performance characteristics of the system for a variety of clinical determinations including glucose, urea, α-amylase, bilirubin, and triglycerides.

Data reported in these papers indicate that quantitative reliability of results can be as good as those for other, more conventional systems, but they do not demonstrate any improvement in the quantitative reliability of data for any of the analytes examined. The discussions suggest some potential operational advantages for the system. Perhaps the most obvious of these potential advantages is the low sensitivity of the final result to the sample volume used. It is suggested that the fluid sample applied to the spreading layer spreads uniformly filling the void volume as it spreads. This suggestion is supported by data that show that for sample volumes between four and 11 μl, the amount of sample per unit area ($\mu l/cm^2$) changes at a rate of only about 1% per μl of sample. In other more conventional sampling methods, there is a one-to-one correlation between analyte result and sample volume.

The fact that only small molecules diffuse into and through the reaction layers may represent an advantage in some situations and a disadvantage in others. One potentially advantageous feature is the natural separation process that takes place between small analyte molecules and large molecules that could interfere with a determination. A potentially disadvantageous feature from the manufacturer's point of view is that it appears to dictate that reactions involving large analyte molecules such as enzymes must take place at or near the interface between the spreading layer and the first reagent layer to produce small molecules that can diffuse through other reagent layers if subsequent reaction steps are needed. It appears that the packaging of reagents into compact, ready-to-use "slides" may be advantageous under some circumstances.

One feature of the system that should be neither overemphasized nor ignored is the fact that the reflection density, D_R, which is measured directly, does not change linearly with concentration. The effect, which is analogous to stray light in conventional absorbance measurements, results in a continuously decreasing sensitivity with increasing concentration. While it is possible to use a mathematical transformation to compute a function, D_T, which changes linearly with concentration, two points appear important to us. First, if the D_R measurement uncertainty is fixed throughout, then it will become progressively more difficult to differentiate between similar concentrations as the analyte concentration increases. Second, three parameters used in the linearization process are determined empirically from calibration data. Because each sample will involve a different reagent slide or observation cell, these parameters could differ from one sample or standard to another. While data presented in the referenced papers strongly indicate that these are not serious problems, we believe the first merits clarification with an error analysis similar to that applied for conventional absorbance measurements, and that the second merits continuous attention.

As a final note, it appears highly improbable that this technology can be implemented independently by persons in their own laboratories, as have some other technologies. In general, we believe this new approach to chemical analyses merits the attention of analytical and clinical chemists alike.

FLOW INJECTION ANALYSIS

Most readers will be familiar with continuous-flow analytical systems that use air-segmented flow streams. Betteridge (3) has reviewed a relatively recent development of an unsegmented continuous-flow analytical system that has been called flow injection analysis (FIA). In this system, the sample is introduced as a "slug" into the flowing reagent stream and mixing is primarily by radial diffusion. As the sample and reagent mix by diffusion, chemical reactions take place, and changes in physical properties such as light transmission, electrical potential, conductance, are monitored at an appropriate point downstream from the mixing point. Measurements usually are made before the mixing process reaches steady state, and equations that describe changes in concentration profiles as the sample slug moves through the flow stream have been developed and are summarized in the review. These equations show well-defined relationships between concentration and variables that are easily measured.

It is stated that the absence of air segmentation leads to higher sample throughput, that the reproducibility is good (1 to 2%), that there is no sample carry-over, and there is no need to introduce and to remove air bubbles. In the review, several applications are tabulated, including glucose in serum; and 48 references are included. Commercial instrumentation is discussed briefly. The review and references included in it are highly recommended reading for those interested in potential new approaches to chemical instrumentation.

RIA WITH MAGNETIZABLE PARTICLE SEPARATION

Although there were many applications of radioimmunoassay (RIA) methods reported during 1978, one unique approach used magnetizable particles to separate bound and unbound antigens in a flow system (4). The magnetizable material consisted of Fe_3O_4 embedded in a cellulose matrix. The material was ground to a size of 1 to 2 micrometers, and antibody for the species of interest was coupled to the particles by methods described previously. In a typical analysis, a labelled hormone such as $^{125}I\text{-}T_3$ is incubated with the antibody attached to the magnetizable particles and the sample (serum) containing binding protein. The amount of $^{125}I\text{-}T_3$ bound to the magnetizable particles is inversely related to the number of binding sites in the sample. The solution containing the $^{125}I\text{-}T_3$ equilibrated between sample binding proteins and the magnetizable particles is then passed over magnets where the magnetizable particles and the $^{125}I\text{-}T_3$ attached to them are held up, while $^{125}I\text{-}T_3$ bound to sample proteins passes through to a waste container. After the magnetizable particles are washed thoroughly (two steps), they are released by removing the magnetic field and are flushed into a chamber where the radioactivity is counted.

The method was applied for triiodothyronine (T_3), thyroxine (T_4), and thyroglobulin (T_g), and results by the new method were compared with results by more conventional manual separation methods. Information related to these comparisons were presented only in terms of correlation coefficients ($r \simeq 0.97$), and a broader interpretation is not possible. The authors state that the "results are regarded as satisfactory and the values for the commercial quality control sera lie within the range quoted by the manufacturers." This clever approach to the separation step in RIA would appear to merit continued development and utilization.

KINETIC METHODS

Gustafson et al (5,6) have discussed the fact that albumin reacts more rapidly with bromcresol green than do some other proteins and have suggested that albumin levels should be determined from what they call the "immediate reaction". In the proposed method, the progress curve is plotted for an extended time period and extrapolated back to zero time, and albumin levels are computed from the zero-time absorbances obtained in this fashion.

It is judged that this is one of the many reactions for which kinetic data can be used to differentiate among species that are involved in similar reactions with different rates. For example, Landis and Pardue have studied the kinetic behavior of the reactions of unconjugated and conjugated bilirubins with p-diazobenzenesulfonic acid (7) and have shown that the kinetic differences are such that the levels of the two species can be determined in the presence of one another (8) with appropriate mathematical procedures.

Figure 1(*a*) shows typical response curves for equimolar amounts of the two bilirubin species at pH 7.5, and it is clear that unconjugated bilirubin reacts more rapidly than does conjugated bilirubin at this pH. Figure 1(*b*) shows the response curve for a mixture of the two bilirubins with a change in the scale of the time axis to emphasize the slow and rapid phases of the reaction. A linear least squares method applied to absorbance at 250 nm plotted against time during a 687-msec period was used to resolve mixture response curves into concentrations of the individual components. Results for 56 sera processed by the new kinetic method and by the more common two step equilibrium method were compared giving regression equations of $y = (1.01 \pm 0.05)x + (6.6 \pm 8)$ for total bilirubin (TB), $y = (1.04 \pm 0.03)x - (2.7 \pm 3)$ for unconjugated bilirubin (UCB), and $y = (0.91 \pm 0.04)x + (3.8 \pm 4)$ for conjugated bilirubin (CB) where uncertainties are quoted at $\pm$ 1 SD. Standard errors of estimate were 28, 14, and 20 µmol/l respectively. Slopes and intercepts for TB and UCB and the intercept for CB are not different from unity and zero at the 95% confidence level; however, the slope for CB is different from unity. This difference as well as the scatter reflected in the standard errors of estimate could arise from either or both methods, and additional work is needed to resolve the question. Nevertheless, the data show that two component analyses based on kinetic differences are feasible.

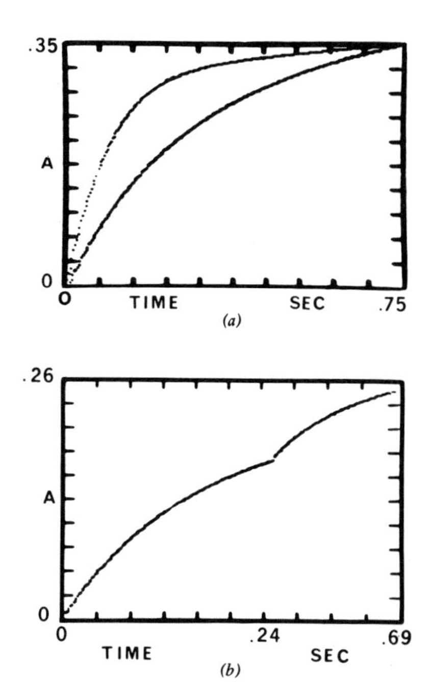

Figure 1. Response curves at 530 nm for UCB and CB reactions with p-diazobenzenesulfonic acid. (*a*) UCB (upper curve) and CB (lower curve) on same time scale. (*b*) Mixture of UCB and CB with modified time scale. UCB: unconjugated bilirubin, CB: conjugated bilirubin. (Reproduced with permission from Landis and Pardue, ref. 8.)

Another feature of this work that is unusual in clinical practice is the use of the stopped-flow mixing methods that permits the rapid reaction rates illustrated in Figure 1 to be monitored. This type of mixing system could be useful for the albumin/bromcresol green reaction discussed above.

One criticism of kinetic methods applied to the determination of substrates of enzymes and other metabolites is that these methods are more dependent on experimental variables than are the more common equilibrium methods. Mieling and Pardue (9) have described a kinetic method that is insensitive to variables such as temperature, pH, enzyme activity. The concept of the method is illustrated in Figure 2, where the open diamonds represent measured values of absorbance and the solid curve represents a fit of these data to a first order kinetic model. Simply stated, the method consists of using absorbance plotted against time during the early part of the reaction in a regression program to predict the total change in absorbance, $\Delta\hat{A}$, that would occur if the reaction were monitored to completion. The calculated absorbance change is proportional to analyte concentration and is relatively independent of variables that affect rate constants. The method was evaluated with a model reaction used for the determination of thiocyanate.

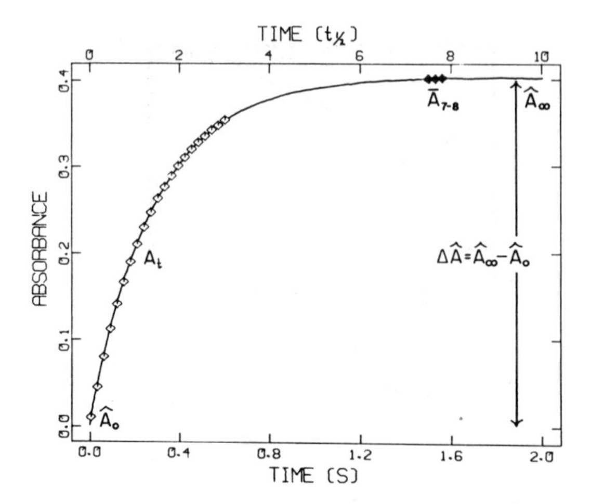

Figure 2. Illustration of the multiple-regression method with experimental data. (Reproduced with permission from Meiling and Pardue, ref. 9)

This point is illustrated for data collected at 24 and 26°C for which the regression equation is $\Delta\hat{A}_{24} = 1.002\Delta\hat{A}_{26} - 0.005$, with standard error of estimate of 0.001. Whereas the temperature coefficient estimated from the slope is about 0.1%/°C for this method, it would be about 8%/°C for a conventional kinetic method. The similarity between results obtained by the kinetic method and those obtained by an equilibrium method was further confirmed by regressing 278 pairs of results obtained by the kinetic method and by an equilibrium method for a wide variety of conditions, including three different temperatures, different pH values, different ionic strengths, and so on. The resulting equation was $y = 1.006x - 0.002$, demonstrating excellent agreement between kinetic (y) and equilibrium (x) values of ΔA.

Although these data were obtained for an analyte (SCN^-) that is of little interest in clinical chemistry, it should be possible to extrapolate the results to any other reaction system that can be adjusted to first-order kinetics. The method could remove or minimize the dependency of enzymatic methods for substrates on enzyme activity and variables that influence it.

Makin et al (10) have described an instrument system for glucose that uses an oxygen-sensing amperometric electrode in conjunction with operational amplifier circuitry that computes the first derivative of the response curve. The circuitry is shown in Figure 3, where IC1 and IC2 are operational amplifiers. Amplifier IC1 samples the current and provides gain control via the range selector switch and gain control potentiometer, while IC2, in conjunction with the input and feedback capacitors and resistors, computes the first derivative of the amplified response curve. The computed slope is proportional to glucose concentration in the range from 0-20 mmol/l. Results for sera and plasma correlated reasonably well with results obtained with a continuous-flow analyzer.

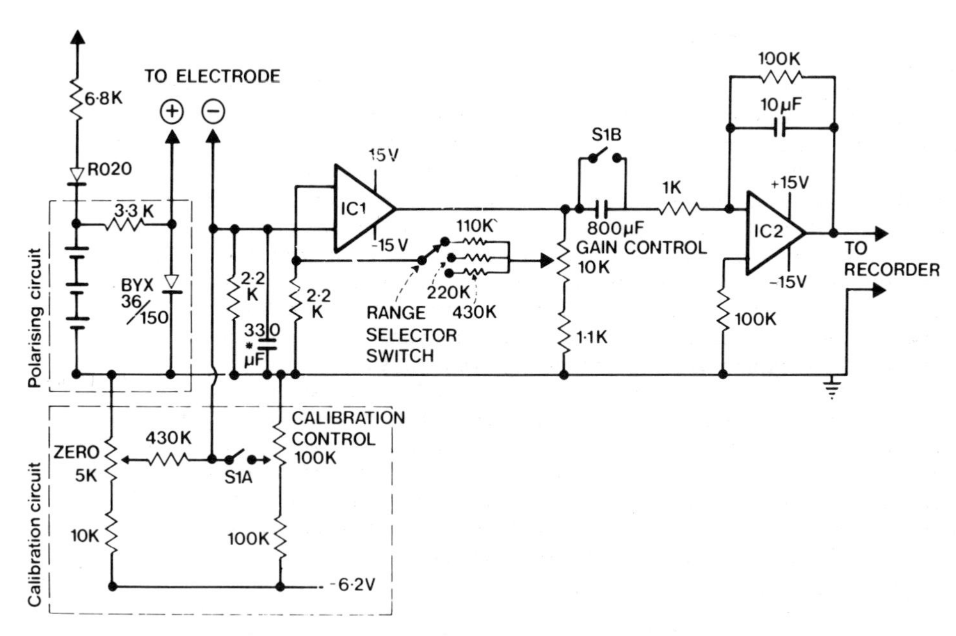

Figure 3. Differentiator circuit for glucose determination (Reproduced with permission from Makin et al, ref. 10)

ELECTROCHEMICAL METHODS

A novel development during this review period is that of a new electrochemical method for glucose based on the catalysis by nickel of the electrochemical oxidation of glucose (11). The nickel catalyst is plated onto a lead dioxide electrode, and the electrode is protected from protein adsorption by a filter secured with a rubber O-ring. When placed in a glucose solution, the electrode requires about 10 seconds to equilibrate to a stable electrolysis current, which is linear in glucose concentration in the range from 10^{-7} to 10^{-3} mol/l. Comparison studies showed good correlation with control serum values and with results obtained in a local clinical laboratory. This electrode permits the *direct* measurement of glucose without any preliminary or accompanying chemical reactions and appears to offer significant advantages over many other commonly used physicochemical methods. It appears to us that it could offer significant advantages of simplicity and speed as the measurement step in feedback systems.

Fenn et al (12) have described a liquid chromatography detector based on thin-layer electrochemistry with application to the determination of catecholamines in blood plasma. Detection limits for epinephrine and norepinephrine standards were 0.2 - 0.3 pg, or 2 - 3 ppt, with 100 μl injections.

Adams and Carr (13) have described a coulometric pH-stat system for use with immobilized enzymes, and have reported results involving urea determinations with urease. A glass electrode was used to detect pH changes, and the pH was maintained within narrow limits by electrochemical generations of H^+ or OH^-. Data presented for several commercial reference sera exhibited reasonable agreement among the pH-stat results for urea by this method and results by other methods.

Garrenstrom et al (14) have described a unique approach to using ion selective electrodes in complex matrices. A computer was used to prepare a solution, the composition of which matched the sample in terms of species that influence electrode response. Results reported for the simultaneous determination of sodium and potassium showed errors of less than 1% for both species.

Caserta et al (15) have described a computer-controlled system for bipolar pulse conductivity measurements and have demonstrated its applicability to chemical rate measurements with stopped-flow mixing. While no clinical applications were reported, the system may have potential for such applications.

Himpler et al (16) have described a new sensor for CO_2, in which a gas-permeable membrane separates the sample from a thin film of water between a pair of platinum electrodes used to measure the conductance. Measured conductance is proportional to the square root of sample CO_2 concentration. The response time is about 1 minute, and the system was applied to CO_2 determinations in the range of importance in blood gas analysis.

Imaging Detectors and Trace Elements

Talmi et al (17) have discussed the use of optoelectronic imaging detectors for fluorescence spectrometry, and Hoffman and Pardue (18) have discussed an autoranging amplifier system for a vidicon-based rapid scanning spectrometer. While neither of these papers includes specific clinical applications, both include references to clinical applications, and it is expected that these detectors will be used extensively in clinical instrumentation in the future.

Most methods for trace element determinations used routinely in clinical laboratories use instrumentation designed for one or two elements at a time. Situations may arise in which it is desirable to determine multiple elements on a single sample, and this can be done most efficiently with instrumentation designed to determine several elements simultaneously. Felkel and Pardue (19,20) have reported on a system that uses an echelle spectrometer to disperse optical spectra in two dimensions and an imaging detector to monitor many spectral lines in the spectrum simultaneously. The system was applied for simultaneous determinations of several elements by both atomic absorption and atomic emission spectrometry.

The concept of the system is illustrated in Figure 4, which shows the emission spectra for mercury recorded with two different detectors, an image dissector and a silicon vidicon. The peaks on each plot represent emission line intensities at each of several wavelengths ranging from 250 nm near the front of each figure to about 580 nm at the back. In a multi-element experiment, spectral lines of the different elements that occur at other wavelengths in the two-dimensional pattern would be monitored and intensities would be related to concentration. Some typical data for simultaneous determination of eight transitional metals are included in Figure 5, and similar data for alkali and alkaline

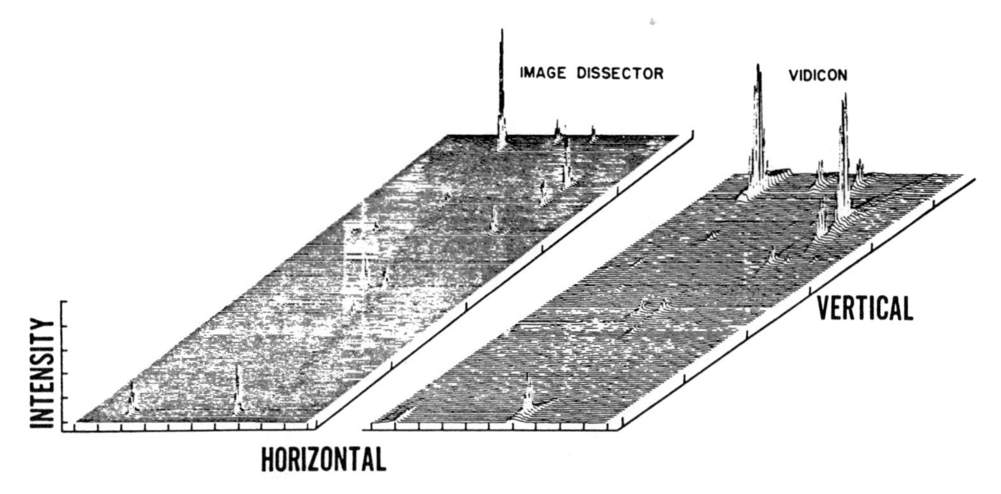

Figure 4. Three-dimensional representations of the spectrum from a mercury pen-lamp, recorded with an image dissector (left) and a silicon target vidicon. (Reproduced with permission from Felkel and Pardue, ref. 19)

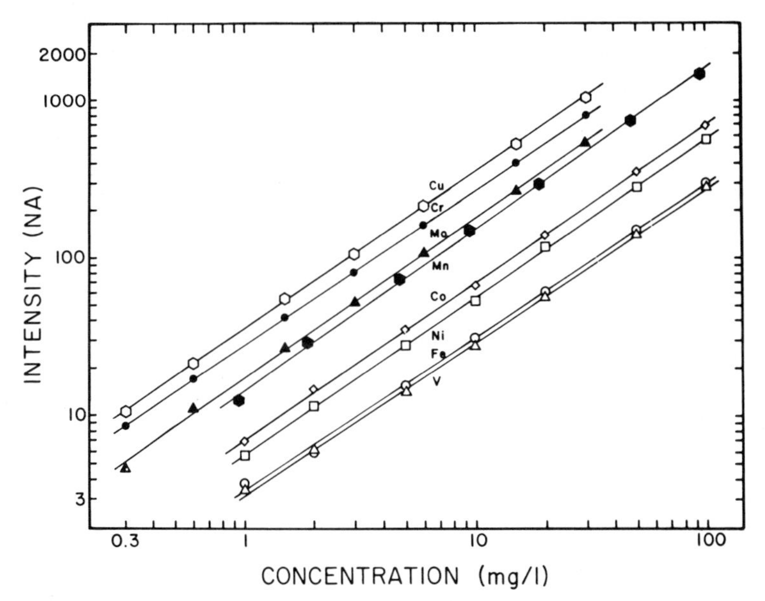

Figure 5. Linearity plots for the simultaneous determination of copper, chromium, molybdenum, manganese, cobalt, nickel, iron, and vanadium by atomic emission spectrometry. Cu: 3247.54 Å, Cr: 4254.35 Å, Mo: 3798.25 Å, Mn: 4030.76 Å, Co: 3453.51 Å, Ni: 3619.32 Å, Fe: 3737.13 Å, V: 4379.24 Å. (Reproduced with permission from Felkel and Pardue, ref. 20)

earth metals are included in the paper (20). These data illustrate good linearity over about two to three orders of magnitude for each element, and numerical data show detection limits ranging from a low of 0.004 mg/l (4 ppb) for copper to a high of 0.04 mg/l (40 ppb) for potassium. These detection limits are comparable to those reported with the same DC plasma excitation source used with conventional optics and detectors. The advantage of the present system is that analytical and background wavelengths are selected electronically under computer control, and choices of elements to be determined and lines to be monitored are made via computer instructions rather than mechanical changes in the system as are required with conventional instruments.

CHROMATOGRAPHIC DETECTORS

Traditionally, the most commonly used clinical chemical methods have been based on relatively simple separation methods such as precipitation and extraction and have depended heavily on the selectivity of chemical reactions to provide the needed specificity. In the interest of speed and simplicity, many procedures now eliminate the deproteinization step and depend on enzymes such as cholesterol oxidase, glucose oxidase, or hexokinase to provide selectivity. Such procedures are effective but seldom permit the determination of more than one analyte at a time.

In recent years there has been a growing interest in the use of more powerful separation methods followed by the use of detectors with varying degrees of selectivity. There probably is no combination in which the combined selectivities of a separation process and a highly selective detector are used more effectively than in combined gas chromatography/mass spectrometry (GC/MS) systems. In a series of three papers, Gates et al (21-23) have discussed the principles and applications of GC/MS systems for the quantitative profiling of organic metabolites in body fluids. The first paper in the series discusses the role of a computer in the collection and statistical evaluation of data; the second paper discusses a complete procedure for the mechanized separation, identification, and quantitation of up to 150 components in each sample; and the third paper presents data for 134 acids in the urine of adult and juvenile subjects. These papers demonstrate remarkable capabilities for this combination of separation/detection systems. Several other papers have described applications of mass spectrometry to clinical problems (24-28).

There have also been some variations on the conventional approaches to mass spectrometry. One of these involves negative ion mass spectrometry (29). Whereas most MS instruments monitor positive ions, it is possible to monitor negative ions as well. Mårde and Ryhage (29) have described the application of negative-ion chemical-ionization mass spectrometry to some amphetamine-like compounds and have compared the negative ion spectra with positive ion spectra. It is clear from the comparisons that this newer approach offers complimentary information that can extend the capabilities of this already powerful tool.

The GC/MS combination does not represent the only combination that can use the selectivity of both the separation and detect-

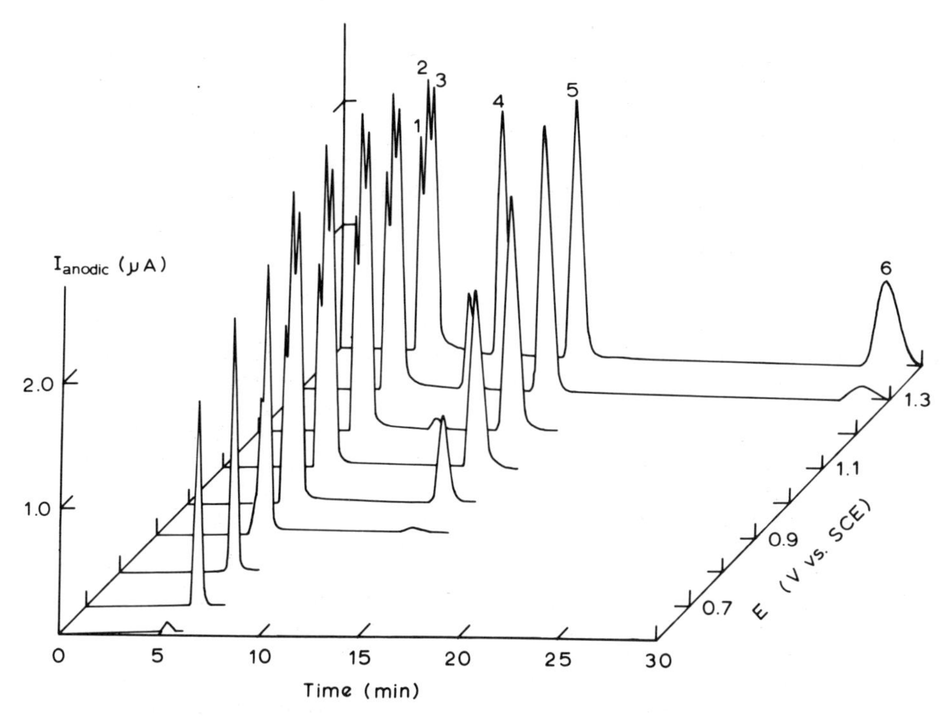

Figure 6. Chromato-voltammogram of methylxanthines at glassy carbon detector. 1: 1.77 μg 7-methylxanthine, 2: 1.61 μg 3-methylxanthine, 3: 1.59 μg 1-methylxanthine, 4: 3.67 μg theobromine, 5: 4.09 μg theophylline, 6: 7.27 μg caffeine. Detection by *dc* amperometry; column A; eluent 1. (Reproduced with permission from Lewis and Johnson, ref. 30)

ion steps. Lewis and Johnson (30) have described an amperometric system for the detection of methylxanthines in serum after separation by so-called reversed-phase liquid chromatography. They have discussed several characteristics of the system, and Figure 6 shows how different polarizing potentials can be used to improve the selectivity of different incompletely separated peaks. As an example, all the components are detected at 1.4 V; however at 1.15 V theobromine and caffeine are not detected. If a faster throughput for theophylline is desired, the electrolysis potential can be set at 1.15 V and theophylline in subsequent samples can be determined without interference from caffeine or theobromine from a previous sample. The reader can visualize other selectivities from the plot.

In recent years several papers have been published on the use of imaging detectors as multiwavelength monitors for liquid chromatography, but apparently no papers on the subject were published in the clinical literature during 1978.

In addition to these reports that emphasize the combined capabilities of the separation process and selective detectors, there have been numerous applications of both gas chromatography (31-40) and liquid chromatography (41-52) that have used only the selectivity of the separation process. Applications cover a wide range including drugs, drug metabolites, normal metabolites, and isoenzymes. It is our view that these separation methods, which were essential in solving some complex problems, will prove useful for more common determinations when there is reason to sacrifice speed and simplicity for improved selectivity and reliability.

IMMOBILIZED ENZYMES

During the last decade there have been numerous applications of immobilized enzymes for substrate determinations. One of the most recent developments in this area has been the immobilization of enzymes on the inner walls of nylon tubes, with the resultant ease of use with continuous flow systems. In two papers, Sundaram et al have described applications of immobilized urease for urea (53) and immobilized uricase for uric acid (54). The tube reactors were used with a commercially available continuous-flow analyzer using common reaction sequences. The urease system was also used for an indirect determination of citrulline in serum. Results obtained by this method correlated well with other procedures in common use.

Freeman and Seitz (55) have described a system in which an enzyme (peroxidase) is immobilized on the end of a fiber optic element, and the chemiluminescence light emitted when hydrogen peroxide and luminol react is transmitted to a detector. The system requires approximately 4 seconds to reach steady state and should be applicable as a monitor for reactions that produce hydrogen peroxide. Unfortunately, response appears to be closer to second order than to first order in H_2O_2 concentrations.

SERUM PROTEIN

Holzer et al (56) have described a novel approach to the determination of serum protein. The method is based on the observation of a linear relationship between protein concentration and solution density. Solution density is determined from its influence on the resonant frequency of a U-shaped glass tube to which the solution is added. Reported data show a good correlation between results by the test method and a biuret method. The principal advantage of the method appears to be the simplicity of a direct physical measurement; the principal disadvantage appears to be its total lack of specificity.

TEMPERATURE CONTROL

One of the critical problems in many clinical analyses is temperature control, and one of the critical problems in temperature control is calibration of temperature sensors. A recent paper (57) describes the design of a device that uses the melting point of purified gallium as the reference for calibrating thermometers. The device appears to offer a practical solution to a difficult calibration problem in clinical laboratories, and its role in enzymology is discussed in the Chapter by Henderson in this volume.

LASER NEPHELOMETRY

There is growing emphasis on the use of laser nephelometry to monitor antibody-antigen reactions. Commercial nephelometers for determinations of IgG, IgA, IgM, and C'3 have been evaluated in two papers (61,62), and in a third paper (60) the use of a commercial instrument to study conditions for the determination of IgG, IgA, and IgM has been described. All papers show good correlation between the laser nephelometric results and radial immunodiffusion methods, with slopes of regression plots in the range of 0.92 to 1.05 and intercepts in the range of 10 to 30 mg/dl for IgG, 7 to 22 mg/dl for IgA and 3 to 4 mg/dl for IgM and a slope of 1.04 and intercept of 0.9 mg/dl for C'3. Insufficient data were given to permit an evaluation of these estimates of the slopes and intercepts. The last paper mentioned (60) gives information related to the effect of reaction conditions on the kinetic behavior of the reactions.

PIPETTING DEVICE

Kenney et al (61) have described an electronically controlled pipetting device that they judge to be superior to syringe driven systems for samples containing proteins and dissolved gases. The device consists of an inverted (to-deliver) pipet connected at the bottom to a T connector which permits gravity flow to fill the device from a reservoir through one solenoid-controlled valve and to drain the device partially through a second valve. The volume delivered is controlled by a pair of photodetectors located at equal distances below the delivery tip and the calibration mark. Quoted precision is 0.2% or better for volumes between 2 and 50 ml.

OSMOTIC PRESSURE

It is suggested that colloid osmotic pressure is useful in assessing risks of pulmonary edema, and a new device for its measurement has been described (62). This device uses a membrane, selectively impermeable to molecules of molecular mass greater than 30,000, between the sample chamber and a reference chamber filled with saline solution, and a pressure transducer to determine the osmotic pressure. The sensor chamber accommodates samples between 50 and 300 μl, and equilibrium is established within 2 minutes. Measurements on a control sample of human serum albumin gave an average value of 25.9 gm/cm^2 and a standard deviation of 0.4 gm/cm^2. Comparative measurements on 114 paired samples with the new instrument and a reference method had a mean difference of 4.28×10^{-2} gm/cm^2, and a t test showed no significant difference. The paper includes details of the design of the membrane sensor and electronic measurement circuitry and a discussion of parameters that influence osmotic pressure measurements.

MISCELLANEOUS

Some other topics that merit attention include potential applications of magnetic resonance spectroscopy (63), neutron activation analysis (64), calorimetry (65,66), and fluorescence polarization (67), a technique that may complement conventional fluorescence methods.

There have been many evaluations of commercial instrument systems, including a comparison of five coagulation instruments (68), flame photometers (69,70), a six channel analyzer (71), a micro CO_2 analyzer (72), and a discussion of the use of linearly related specimens in the evaluation of multichannel analyzers (73). Also, there have been a number of evaluations of chemical methodologies on existing instruments. Some of these are ammonia on a discrete sample analyzer (74); EMIT reagents on automatic kinetic analyzers (75-77); a kinetic method for triglycerides (78); a dye-binding method for albumin (79); and glucose procedures (80).

Some straightforward but potentially useful developments are the modification of an enzyme analyzer to handle the so-called lag phase in enzyme reactions (81); a disk electrode that is sensitive to ammonia (82,83); and a simple photometer circuit (84).

COMPUTER APPLICATIONS

Distributed Laboratory Computing

Clinical laboratory computing is an important application of computers to clinical chemistry. Three approaches have often been used: (*1*) developing a laboratory computing package using the internal resources of the laboratory and hospital; (*2*) purchasing a turnkey system from a vendor; or (*3*) working with a vendor to integrate a vendor-supplied package into the laboratory or hospital system The first approach often results in the expenditure of significantly more manpower and financial resources than were first anticipated. The second approach is achieved at a fixed cost, but the resulting system is often too inflexible for adaptations to growing and changing laboratories.

Wycoff and Wagner (85) have reported on the successful integration of a vendor-supplied laboratory computer system (Laboratory Computing, Inc.) into a large hospital information system: in 1976-1977 more than 2.5 million results were reported to the 800 physicians serving 900 inpatients and 1200 outpatients each day at the University of Iowa Hospitals and Clinics complex. Preliminary planning of the computerized system, design finalization, software development, hardware description, intersystem communications, system operation, error resolution, physician access, postimplementation evaluation, and financial impact have been clearly discussed.

The authors estimated that the project consumed approximately 2.5 pathologist years and eight analyst/programmer years over the first five calendar years since the beginning of the project. The vendor-supplied hardware and software required approximately $310,000 in 1972. A significant portion of the capital investment was returned through recovery of laboratory charges that were previously lost using manual billing procedures.

The authors' overall evaluation of the system was highly favorable. They concluded: "Based upon attainment of initial objectives, the computerized laboratory system . . . must be considered a success. Beyond the advantages documented above, we have experienced an intangible benefit in the form of enhanced quality of patient care. This is the natural result of an improvement in the overall efficiency of clinical laboratory services. Finally, in a time of ever-increasing hospital costs, it is refreshing to install a system that genuinely contributes to cost containment."

Glucose Controlled Insulin Infusion

Although it has become commonplace to use computers to control analytical instrumentation and to process analytical data, relatively few papers have appeared in the clinical chemistry literature describing systems in which the patient is a part of a closed-loop feedback system. One such system involves a glucose analyzer used in the control of insulin infusion in diabetic subjects (86). The concept of the system is illustrated by the block diagram in Figure 7. Diluted blood from the patient is pumped into an analyzer and analytical data are processed by a computer to control the rate of insulin infusion. Blood glucose values are available within 2 minutes after the blood leaves the patient. The principal emphasis in the paper is on the glucose analyzer. It uses an electrochemical cell to monitor the hydrogen peroxide produced in the glucose oxidase catalyzed oxidation of glucose by oxygen. The electrochemical cell is isolated from the reaction mixture by a membrane in a manner analagous to that used in the Clark oxygen electrode.

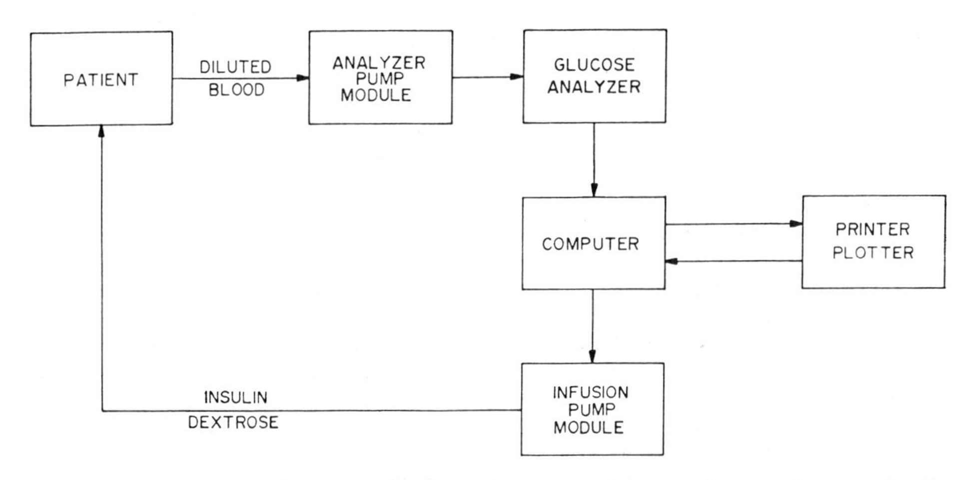

Figure 7. Block diagram of Biostator operation and computer control. (Reproduced with permission from Fogt et al, ref. 86)

Standard addition experiments (so-called recoveries) demonstrated very good linearity for the analyzer system, and results obtained in a comparison of the analyzer data with the proposed national glucose reference method that uses the hexokinase/glucose-6-phosphate dehydrogenase coupled system gave a slope of 1.00 and an intercept of 0.049 gm/l, with a standard error of estimate of 0.079 gm/l, for 61 patient samples containing glucose in the range from about 0.2 to 5.0 gm/l. Whole blood samples to which known amounts of glucose were added after complete glycolysis were used to evaluate the system over a range from 0.25 to 9.00 gm/l and the guidelines of the Food and Drug Administration's (FDA) proposed Product Class Standard for Glucose were used in the evaluation. The bias between the proposed and reference method was within the guidelines (5%, or 50 mg/l) throughout the range, and the imprecision was borderline below 1 gm/l, well within the guidelines between 2 and 7 gm/l, and outside the guidelines above 8 gm/l. Slope and intercept values were 1.01 and -0.0195 gm/l, respectively. The standard error of estimate and imprecision of the slope and intercept were not given for these data. The method was shown to be free of interference from bilirubin, creatinine, uric acid, dextran, and sodium salicylate. In a related application, Noy et al (87) used a double lumen catheter to sample blood from a vein and a continuous-flow analyzer to determine glucose, lactate, alanine, and 3-hydroxybutyrate.

Optimization Studies

Although automated analyzers and computers have been used extensively for routine analyses in clinical chemistry, there has been relatively little emphasis on the use of computer-controlled systems to characterize chemical reactions and/or to optimize analytical procedures. Olansky and Deming (88) have described work in which a continuous-flow analyzer has been used with an on-line computer and with simplex optimization methods to optimize conditions for kinetic creatinine determinations with alkaline picrate. While the new information obtained for the creatinine-picrate reaction is significant, the most important feature of this paper is that the system described offers the means of developing analytical procedures that will be well understood and optimized for multiple parameters.

Figure 8 is a schematic diagram of the modified flow system used in this work. Synthetic samples or control sera were mixed with computer-controlled amounts of sodium hydroxide and picric acid in a mixing coil before the flow cell where absorbance was monitored. In this study the flow was stopped when the sample and reagents were in the flow cell so the kinetic data could be collected. The amounts of NaOH and picrate added to each sample were adjusted via computer control of the speeds of motors in the peristaltic pumps for the two reagents.

The study consisted of three major phases: (*a*) simplex optimization of hydroxide and picrate concentrations in the reagent stream, to provide optimum sensitivity to creatinine; (*b*) a mapping study in the region of the simplex optimum, to determine the effects of picrate and hydroxide concentrations on creatinine sensitivity and on albumin, glucose, and acetone sensitivities; and (*c*) a

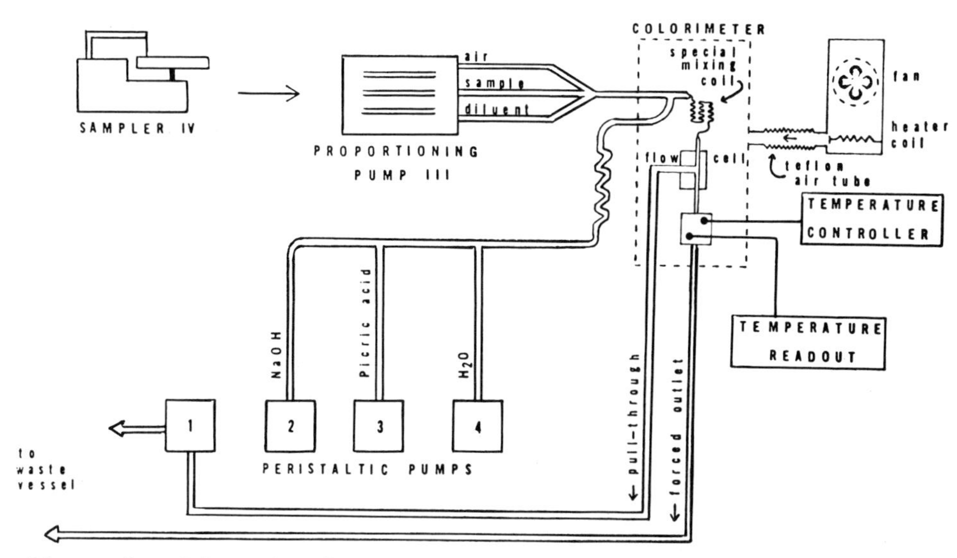

Figure 8. Schematic diagram of a modified AA-II system. (Reproduced with permission from Olansky and Deming, ref. 88)

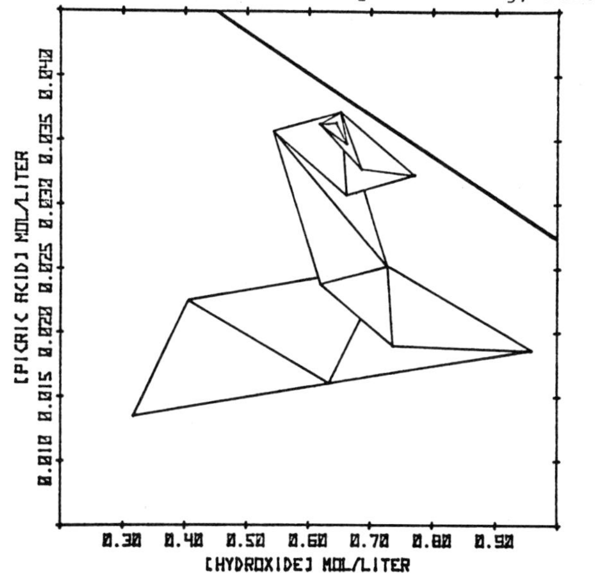

Figure 9. Simplex progress. Diagonal line at upper right indicates total reagent flow boundary. (Reproduced with permission from Olansky and Deming, ref. 88)

verification study in which a single reagent containing the determined optimum amounts of hydroxide and picrate was used. The response, R, that was optimized was the sensitivity of the rate of absorbance change with respect to creatinine concentration ($R = \partial\ (dA/dt)/\partial$ [creatinine]).

Figure 9 shows the progress of the simplex optimization of the response, R, in picrate-hydroxide space from a starting point near the lower left-hand corner of the figure to the experimental optimum near the top center of the figure. The procedure involves a variable step size that permits rapid approach to the region of the optimum followed by good resolution of the optimum region. The paper includes a contour map of the sensitivity as functions of picrate and hydroxide concentrations.

Optimization that considers combined effects of two or more variables is superior to the one-variable-at-a-time methods, which are in more common use. It is clear that computer controlled instrumentation is an efficient way to implement the approach.

Sequential Testing

Cembrowski et al (89) have extended their work on computer-controlled instrument systems for sequential chemical testing to include an application to liver function assessment. The results of an initial SMA 12/60 screen are used to automatically decide which additional tests should be run to further define the nature of the liver disease. These additional tests are then immediately carried out by the computer-controlled instrument system and are presented to the clinician in the initial report. By using a computer to anticipate and have ready the results of tests that the clinican would probably have ordered, it is possible to avoid most of the usual time delays caused by the "Ping-Pong" relationship between the clinican and the laboratory.

GC-MS-Computer System

The combined use of a gas chromatograph, a mass spectrometer, and a computer system provides one of the most powerful analytical chemical tools available today, primarily because of its dual capabilities of separation and identification. Although a number of papers have appeared in which the use of the gas chromatograph-mass spectrometer-computer system to identify compounds in a mixture is described, the paper by Gates et al (90) demonstrates an example of the state-of-the-art capabilities of computerized pattern searches for GC-MS-computer systems (see also references 21-23).

As the authors pointed out, with urine samples there is a relative lack of qualitative variation from sample to sample. Thus, two features are of interest to the clinical chemist: *(1)* "quantitative differences in one profile compared to a profile from another source or the same source at a different time" and *(2)* "the appearance of one or more highly unusual constituents in a particular sample". A massive search of library-file computer-stored spectra is not necessary for the identification of the components in the mixture; a smaller, "local" library of mass spectra is sufficient. The low resolution GC-low resolution MS-computer system was designed to provide automated qualitative and quantitative analysis of 100 or more components in a complex organic mixture.

Instrument Controllers

Instrument controllers can be divided into four general categories: *(1)* mechanical devices consisting largely of cam actuated switches; *(2)* hardwired logic systems that are designed using combinatorial and sequential logic methods, component minimization techniques,

and other digital design methods; *(3)* computer-based software systems where sequencing and decision making are programmed into memory; and *(4)* "state-driven" systems, which are implemented in hardware but very much resemble software flow charts. While instrument designers are moving away from mechanical and hardwired logic systems toward computer (especially microcomputer) systems for instrument control, state-driven systems today provide a simple means of designing control systems without the time-consuming complexities of hardwired logic design or microcomputer programming. Crouch et al (91) provide an excellent introduction to state-driven systems and their applications in analytical and clinical chemical instrument control.

Active Filters

O'Haver (92) has shown how a microprocessor-based, linear-response time, low-pass filter can be used to provide high noise-filtering capability with a very short effective response time; the theoretical response time is one half that of a simple resistor-capacitor filter with the same noise filtering capability. The application of this type of filter to instruments giving "rectangular" output signals (e.g., atomic absorption, autoanalyzer) are suggested.

Titration Systems

Three papers demonstrate the increased capabilities available with the introduction of microcomputer technology into automatic titrators (93-95). Wu and Malmstadt (93) have described a versatile microcomputer-controlled titrator for quantitative determinations. It is capable of handling UV-visible spectrophotometric, potentiometric, and amperometric end-point detection and acid-base, redox, complexímetric, and precipitation reactions. The instrument is constructed from standard laboratory modules which are controlled by a microcomputer programmed in Altair BASIC language.

Leggett (94) has described a microcomputer controlled automatic potentiometric titration system for fundamental solution equilibria studies. The system can produce data that lead to very precise and accurate equilibrium constants; for example, the pK_a of ascorbic acid was determined to be 4.00 in 0.100 M KCl at 25.0°C.

Busch, Freyer, and Szameit (95) have reported on a highly accurate microprocessor-controlled differential titrator for acids or bases that uses two identical glass electrodes to compensate for zero-point drift. Data are presented which demonstrate excellent repeatability and agreement with theory.

REGRESSION DATA USED IN METHOD COMPARISONS

A time-honored procedure for investigators who are developing new instrumentation or methodologies or are evaluating existing instrumentation or methodologies, is to compare the results of the new approach with results obtained by a "reference" approach using least-squares regression data. One fact that has been impressed upon us in preparing this review is that relatively few investigators include adequate information to permit a valid evaluation of the regression data. For example, what is usually included is the slope, intercept, and correlation coefficient ($y = bx + a$; $r = 0.???$). What this type of presentation does not account for are the facts that *(1)* the numerical values of a and b generated by the method of least squares are only estimates, and a user needs some measure of the imprecision of each parameter to interpret it intelligently; and *(2)* the correlation coefficient is only an indirect and insensitive indicator of the quality of a fit. Data that are more helpful in evaluating regression results are the standard deviation of the slope, the standard deviation of the intercept, and the standard error of estimate. Because these data are so consistently absent from papers published in the clinical literature, it is worthwhile to use specific examples to illustrate their importance.

Finley et al (96) have evaluated a new multiple-channel chemistry analyzer, the Beckman Astra-8, that is controlled by a microprocessor. While the authors present extensive data to support their conclusion that the instrument performed well and that the several performance characteristics studied and ease of operation recommend its use, our discussion focuses on the comparison data they presented. These authors present sufficient data to permit an unambiguous interpretation of regression results by the informed reader; our emphasis is on some ambiguities that could have arisen had the authors not been diligent in reporting appropriate statistical data.

One example involves the data for aqueous chloride standards. The regression equation, without uncertainties, is $y = 1.0009x + 0.003$ with $r = 0.9998$, which suggests a remarkably small intercept. However when the $\pm$ 2 SD uncertainties are included, the regression equation becomes $y = (1.000 \pm 0.006)x + (0.0 \pm 0.6)$, and the uncertainty in the intercept is observed to be about $100 \times 0.6 \div 5$, or 12% of the smallest concentration examined, a fact that would be completely obscured without the standard deviation of the intercept.

Another example involves the slopes for the glucose data, which are given as 1.0090 for aqueous standards and 1.00090 for the patient comparison study. Given no other information about the slopes (as is usually the case in comparison studies) one might easily conclude that the slope for the patient data is better than that for the aqueous standards. However, when the standard deviations for the slopes are considered, it is observed that the $\pm$ 2 SD value for sera is 0.012, which is 0.012/0.0009, or about 13 times larger than the apparent deviation of the slope from unity; thus the uncertainty for the standards (0.004) is about three times better than that (0.012) for the sera. The correlation coefficients of 0.9999 and 0.994, respectively, for these data reflect the fact that the first is a better fit; however, the standard errors of estimate ($S_{y/x}$) of 15 and 58 represent much more direct measures of the fact that the scatter about the regression line is about four times larger for the patient samples than for the aqueous standards.

An example from one of the present authors' laboratories (8) quotes the slope of total bilirubin in sera by a new kinetic method compared to a conventional method as 1.01. This appears to suggest excellent agreement until it is noted that the $\pm$ 2 SD range is ± 0.1. It is our view that regression slopes and intercepts should not be accepted for publication without standard deviations and standard errors of estimate any more than averages of repeat runs should be accepted without standard deviations or coefficients of variation. We strongly commend Finley et al (96) for their detailed data presentations and encourage others to follow their example.

CONCLUSIONS

This review has identified several highly original projects that will almost certainly lead to major advances in instrumental approaches to clinical chemistry. Probably the most surprising aspect of the literature survey is the relatively small number of publications that emphasize new instrumental approaches and computer applications. It is our belief that there are technologies available to chemists today that permit us to implement analytical strategies that would have been prohibitively difficult and time consuming in the recent past. For example, the two dimensional data in Figure 6 illustrate how the combined characteristics of the separation and detection steps can be used to obtain degrees of selectivity that could not be obtained with either step alone. This experiment, and many others discussed in this review, would be difficult to implement manually. Properly utilized small computers and microprocessors can handle the control, data acquisition, and data processing operations in these very complex strategies with greater ease and speed than an analyst is able to handle some of the simplest strategies of the recent past. We believe instrumental developments and computer applications in the future need to be guided less by strategies of the past that were often dictated by limitations in instrumental technologies available to us, and more by the increasingly powerful capabilities of newer technologies that are being developed. The literature survey conducted in the process of preparing this review convinces us that there are excellent opportunities for scientists with innovative ideas to apply these new technologies to problems in clinical chemistry.

REFERENCES

1. Curme HG, et al: *Clin Chem* 24:1335, 1978.

2. Spayd RW, et al: *Clin Chem* 24:1343, 1978.

3. Betteridge D: *Anal Chem* 50:832A, 1978.

4. Nye L, et al: *Clin Chim Acta* 87:307, 1978.

5. Gustafsson JEC: *Clin Chem* 24:369, 1978.

6. Gustafsson JEC, Uzqueda HR: *Clin Chim Acta* 90:241, 1978.

7. Landis JB, Pardue HL: *Clin Chem* 24:1690, 1978.

8. Landis JB, Pardue HL: *Clin Chem* 24:1700, 1978.

9. Mieling GE, Pardue HL: *Anal Chem* 50:1611, 1978.

10. Makin, HL, Warren PJ, Edridge JD: *Clin Chim Acta* 84:137, 1978.

11. Schick KG, Magearu VG, Huber CO: *Clin Chem* 24:448, 1978.

12. Fenn RJ, Siggia S, Curran DJ: *Anal Chem* 50:1067, 1978.

13. Adams RE, Carr PW: *Anal Chem* 50:944, 1978.

14. Gaarenstroom PD, et al: *Anal Chem* 50:811, 1978.

15. Caserta KS, et al: *Anal Chem* 50:1534, 1978.

16. Himpler HA, Brand SF, Brand MJD: *Anal Chem* 50:1623, 1978.

17. Talmi Y: *Anal Chem* 50:936A, 1978.

18. Hoffman RM, Pardue HL: *Anal Chem* 50:1458, 1978.

19. Felkel HL Jr, Pardue HL: *Clin Chem* 24:602, 1978.

20. Felkel HL Jr, Pardue HL: *Anal Chem* 50:602, 1978.

21. Gates SC, Sweeley CC: *Clin Chem* 24:1663, 1978.

22. Gates SC, Dendramis N, Sweeley CC: *Clin Chem* 24:1674, 1978.

23. Gates SC, et al: *Clin Chem* 24:1680, 1978.

24. Weinham RJ, et al: *Clin Chem* 24:45, 1978.

25. Muskiet FAJ, et al: *Clin Chem* 24:122, 1978.

26. Muskiet FAJ, et al: *Clin Chem* 24:1899, 1978.

27. Muskiet FAJ, et al: *Clin Chem* 24:2001, 1978.

28. Ehrhardt J, Schwartz J: *Clin Chim Acta* 88:71, 1978.

29. Mårde Y, Ryhage R: *Clin Chem* 24:1720, 1978.

30. Lewis EC, Johnson DC: *Clin Chem* 24:1711, 1978.

31. Nuret P, et al: *Clin Chim Acta* 82:9, 1978.

32. Malkus H, Jatlow PI, Castro A: *Clin Chim Acta* 82:113, 1978.

33. Hengen N, Hengen M: *Clin Chem* 24:50, 1978.

34. Vandemark FL, Adams RF, Schmidt GJ: *Clin Chem* 24:87, 1978.

35. Trocha P, D'Amato NA: *Clin Chem* 24:193, 1978.

36. Nielsen LG, Ash KO, Thor E: *Clin Chem* 24:348, 1978.

37. Smith SL, Novotny M, Weber EL: *Clin Chem* 24:545, 1978.

38. Sasa S, et al: *Clin Chem* 24:1491, 1978.

39. Blair D, Rumack BH, Peterson RG: *Clin Chem* 24:1543, 1978.

40. Sampson EJ: *Clin Chem* 24:1805, 1978.

41. Weatherburn MW, Trotman RBB, Jackson SH *Clin Biochem* 11:159, 1978.

42. Bohlen P, et al: *Clin Chem* 24:256, 1978.

43. Tria M, Ku L, Abrahams I: *Clin Chem* 24:168, 1978.

44. Powers JL, Sadee W: *Clin Chem* 24:299, 1978.

45. Anzano MA, Naewbanij JO, Lamb AJ: *Clin Chem* 24:321, 1978.

46. Inada Y, et al: *Clin Chem* 24:351, 1978.

47. Nilsson-Ehle I, Nilsson-Ehle P: *Clin Chem* 24:365, 1978.

48. Engbaek F, Magnussen I: *Clin Chem* 24:376, 1978.

49. Kiser EJ, Johnson GF, Witte DL: *Clin Chem* 24:536, 1978.

50. De Ruyter MGM, De Leenheer AP: *Clin Chem* 24:1920, 1978.

51. Anhalt JP, Brown SD: *Clin Chem* 24:1940, 1978.

52. Proelss HF, Lohmann HJ, Miles DG: *Clin Chem* 24:1948, 1978.

53. Sundaram PV, et al: *Clin Chem* 24:234, 1978.

54. Sundaram PV, et al: *Clin Chem* 24:1813, 1978.

55. Freeman TM, Seitz WR: *Anal Chem* 50:1242, 1978.

56. Holzer VH, et al: *J Clin Chem Clin Biochem* 16:391, 1978.

57. Sostman HE, Manley KA: *Clin Chem* 24:1331, 1978.

58. Daigneault R, Lemieux D: *Clin Biochem* 11:28, 1978.

59. Conrad VA, Schürmann J, Kreutz FH: *J Clin Chem Clin Biochem* 16:299, 1978.

60. Baruth VB, et al: *J Clin Chem Clin Biochem* 16:397, 1978.

61. Kenney AP, Horne RG, Fallow TM: *Clin Chim Acta* 82:285, 1978.

62. Bisera J, et al: *Clin Chem* 24:1586, 1978.

63. Yang MT, et al: *Clin Biochem* 11:90, 1978.

64. Versieck J, et al: *Clin Chem* 24:303, 1978.

65. D'Ascenzo G, et al: *Clin Chem* 24:119, 1978.

66. Rehak NN, Young DS: *Clin Chem* 24:1414, 1978.

67. Maeda H: *Clin Chem* 24:2139, 1978.

68. Beckals HR, Leavelle DE, Didisheim P: *Am J Clin Path* 70:71, 1978.

69. Jacklyn CL, Kriz CD, MacAulay MA: *Clin Biochem* 11:10, 1978.

70. MacAulay MA, Mathers JJ, Jacklyn CL: *Clin Biochem* 11:23, 1978.

71. Hainsworth IR: *J Clin Chem Clin Biochem* 16:213, 1978.

72. Lam CWK, Tan IK: *Clin Chem* 24:143, 1978.

73. Caragher TE, Grannis GF: *Clin Chem* 24:403, 1978.

74. Hindriks FR, Groen A: *J Clin Chem Clin Biochem* 16:289, 1978.

75. Miller DT, Krieg AF, Demers LD: *Clin Biochem* 11:50, 1978.

76. Malkus H, et al: *Clin Biochem* 11:139, 1978.

77. Vinet B, Zizian L, Gauthier B: *Clin Biochem* 11:57, 1978.

78. Lehnus G, Smith L: *Clin Chem* 24:27, 1978.

79. Pinnell AE, Northam B *Clin Chem* 24:80, 1978.

80. Chua KS, Tan IK: *Clin Chem* 24:150, 1978.

81. Brooks TJ, Sammons HG: *Clin Chim Acta* 88:569, 1979.

82. Schindler VJG, Schindler RG, Aziz O: *J Clin Chem Clin Biochem* 16:441, 1978.

83. Schindler VJG, Schindler RG, Aziz O: *J Clin Chem Clin Biochem* 16:447, 1978.

84. Carroll WE, Jackson RD, Wilcox AA: *Clin Chem* 24:92, 1978.

85. Wycoff DA, Wagner JR: *Am J Clin Path* 70:390, 1978.

86. Fogt EJ, et al: *Clin Chem* 24:1366, 1978.

87. Noy GA, Buckle ALJ, Alberti KGMM: *Clin Chim Acta* 89:135, 1978.

88. Olansky AS, Deming SN: *Clin Chem* 24:2115, 1978.

89. Cembrowski GS, et al: *Clin Chem* 24:555, 1978.

90. Gates SC, et al: *Anal Chem* 50:433, 1978.

91. Crouch SR, et al: *Anal Chem* 50:291A, 1978.

92. O'Haver TC: *Anal Chem* 50:676, 1978.

93. Wu AHB, Malmstadt HV: *Anal Chem* 50:2090, 1978.

94. Leggett DJ: *Anal Chem* 50:718, 1978.

95. Busch N, Freyer P, Szameit H: *Anal Chem* 50:2166, 1978.

96. Finley PR, et al: *Clin Chem* 24:2125, 1978.

3
Kidney Function and Renal Disease

Paul L. Wolf

INTRODUCTION

Any account of the biochemistry of renal disorders is necessarily complex. The kidney does not function in isolation. It is the site of action of many hormonal and homeostatic influences. In turn, its effects upon other organ systems and the internal milieu of the body are widespread. The attempt has been made in this review to make a judicious selection of the 1978 literature in order to highlight various aspects of these complex interrelationships. Comprehensiveness has been sacrificed in favor of selectivity, but the attempt has been made to bring the reader face to face with the outstanding problems in renal physiology and pathology as expressed in biochemical terms, especially at the clinical level.

The Table of Contents represents the structure on which this review was built. The subdivisions are however frequently arbitrary, since many issues such as calcium, phosphorous, and electrolyte metabolism cannot be contained within the artificial boundaries which have been constructed.

ENZYMATIC ABNORMALITIES

Creatine Kinase

The etiology of the presence of creatine kinase I BB (CK BB) in the serum of patients with chronic renal failure is not clear. Previously the presence of CK BB was ascribed to its release from abnormal kidney or possibly from peripheral nerves due to uremic peripheral neuropathy. A number of papers have attempted to clarify this problem.

Creatine kinase isoenzyme I (BB) is generally not detectable in normal serum, and its occurrence in serum has been documented in only a few disease states. In particular, increased activity of this isoenzyme has been reported in association with chronic renal failure (CRF), hemodialysis, and renal transplantation. It has been demonstrated that the apparent creatine kinase (CK) observed in the serum of patients with CRF or kidney transplants and in those undergoing hemodialysis is an artifact observed as a result of measuring CK isoenzymes by fluorescence (1). These observations resemble the artifact in the fluorometric determination of lactate dehydrogenase (LD) isoenzymes in the sera of patients with endstage renal failure. The artifact binds to albumin, is not a protein, and occurs in some normal sera at very low concentrations. Support for this concept was provided by another group of investigators (2). Both papers are described in the Chapter by Henderson. There have been attempts to characterize the fluorescent compounds that cause this possible artifact (3). However, other authors believe that there is a true elevation of CK BB in the serum of patients with CRF (4).

The controversy over the cause of the elevated CK BB in the serum in CRF has not been resolved. the CK BB in the serum may be derived from kidney tissue or from peripheral nerves as a result of uremic neuropathy, a distinct clinical entity resembling the peripheral neuropathy of thiamine deficiency. Uremic neuropathy and thiamine deficiency neuropathy are associated with a decreased erythrocyte transketolase. Uremic serum contains an inhibitor of RBC transketolase, which is removed by dialysis.

Serum Alkaline Phosphatase (ALP)

The cause of the increase in serum ALP in CRF has not been thoroughly resolved. A recent article attempted to clarify this problem. A group of 25 patients with terminal CRF, treated regularly by hemodialysis, was examined. Activity of the bone ALP isoenzyme in serum was significantly elevated in 12 patients with signs of bone disease, either isolated or combined with liver damage, even when total ALP activity was within normal limits. The intensity and incidence of raised bone isoenzyme activity increased with the duration of dialysis therapy. Elevated activity of liver ALP isoenzyme correlated with the other laboratory and clinical signs of liver involvement in 16 patients. Activity of the intestinal isoenzyme was elevated in over half the patients; this elevated activity was unrelated to liver or bone damage and showed an inverse correlation to the total serum calcium level. The assay of total serum ALP activity had no diagnostic value (5).

It is likely that bone is the major source of increased ALP in CRF, because of secondary hyperparathyroidism. However, hepatic disease may occur in these patients for a variety of reasons, especially viral hepatitis caused by exposure to blood products in the severely anemic uremic patient. Finally, uremia may cause prominent uremic erosions and ulcers in the gastrointestinal tract leading to seepage of intestinal ALP into the blood.

Diamine Oxidase

The enzyme diamine oxidase is present in many tissues and plays a role in the metabolism of certain amines, some of which may be toxic. In renal failure, plasma diamine oxidase activity was found to be increased in chronically uremic patients and before and after dialysis therapy in patients undergoing maintenance hemodialysis. Diamine oxidase activity was decreased in the urine of chronically uremic patients, as compared with normal subjects. In chronically uremic rats, diamine oxidase activity was observed to be increased in plasma and to be reduced in the urine, as compared with sham-operated, pair-fed control rats. In the uremic rats diamine oxidase activity was also decreased in kidney and was unchanged in liver and muscle. Total amine levels were elevated in plasma and were reduced in the urine of patients and rats with CRF. Although the clinical significance of abnormal diamine oxidase activity in renal failure is not clear, it is possible that this enzyme may have a pathophysiological role in uremia (6).

Aspartate Transaminase (AST) and Lactate Dehydrogenase (LD)

The triad of elevated serum AST and LD levels, positive blood and urine cultures, and acute renal failure was noted in a patient with severe pyelonephritis. Bilateral medullary necrosis was found on biopsy specimens and at postmortem examination. These findings may help to establish a prompt antemortem diagnosis (7). Both AST and LD are normally present in renal tubules and may be released into the blood with renal tubular necrosis.

Poly C-Avid Ribonuclease

The normal level of serum or plasma poly C-avid ribonuclease activity is 1047 ± 247 U/ml. Serum levels increase proportionately with elevations in serum creatinine, reaching levels of 9,500 to 35,000 U/ml in patients undergoing dialysis. The levels can be returned to normal by successful renal transplantation but not by dialysis. Purified human urinary ribonuclease, a glycoprotein enzyme similar to the serum ribonuclease, was capable of *(1)* inhibiting the incorporation of ^{3}H-thymidine into mitogen-stimulated lymphocytes; *(2)* inhibiting the proliferation and growth of bone marrow red cell colonies; and *(3)* adversely affecting the growth and viability of precursor fat cells (8).

BIOCHEMICAL HEMATOLOGICAL ABNORMALITIES

Serum Ferritin

Many studies have been published about the effect of chronic renal failure on erythropoiesis and myeloid leukogenesis. Serum ferritin is a useful parameter in the investigation of anemia. An excellent study of serum ferritin in CRF has been reported. Eighty-seven patients with endstage renal failure on long-term hemodialysis, 25 patients not on dialysis, and 37 patients with renal transplants were studied. Serum ferritin was measured by radial immunodiffusion and radioimmunoassay. The correlation between the two methods was excellent ($P < 0.001$). In 25 patients on long-term dialysis, a good correlation was found between serum ferritin levels and stainable iron ($P < 0.001$). All patients with adequate iron stores had serum ferritin levels above 60 ng/ml, whereas only 1 out of 10 with decreased or absent iron stores had a higher level (118 ng/ml). According to these criteria the iron stores were decreased in 59% of the patients on long-term hemodialysis, decreased or adequate in 14%, and adequate or increased in 27%. There was no correlation between serum ferritin levels and serum iron and total iron-binding capacity. The distribution pattern of the serum ferritin levels was log normal and did not significantly differ in the three groups studied, although the patients with renal transplants had nearly normal hemoglobin and creatinine levels. Elevated serum ferritin levels in patients on hemodialysis (21%) could be only partly caused by repeated transfusions or chronic infections (9).

Folic Acid

The anemia of CRF is normocytic and usually results from decreased renal synthesis of erythropoietin (EP) or from decreased production of an erythropoietin activator. However, occasionally a macrocytic anemia may occur and is secondary to decreased folic acid levels, which may be due to repeated dialysis or poor nutrition.

A recent investigation examined folic acid levels of uremic patients. Twenty-five uremic patients on long-term hemodialysis were followed over a period of nine months; hemoglobin levels, mean corpuscular volumes, and folate concentrations in serum and erythrocytes were measured. The daily dietary intake of folic acid was estimated to be 80 to 90 μg, and no folic acid supplements were given. None of the patients developed macrocytic anemia. At the end of the period, all patients had a normal erythrocyte folate content. Serum folate was normal in 17 and below normal in 8 patients. These eight patients were in a negative folate balance at the time of investigation. In seven patients dialyzed with a RP VI dialyzer, the maximum loss of folic acid was 75 μg/dialysis, and in six patients on long-term intermittent peritoneal dialysis, the maximum loss was 13 μg/dialysis. Thus, there is no need to give oral folic acid supplements to uremic patients on long-term dialysis provided their daily dietary intake of folic acid is adequate.

2,3-Diphosphoglycerate (2,3 DPG) and ATP

Erythrocyte 2,3 DPG and ATP may fluctuate with serum phosphorus levels. The serum phosphorus level is usually elevated in CRF because of poor excretion and is usually increased during active bone growth in children. An elevated serum phosphorus level may be associated with an increase in RBC 2,3 DPG and ATP and a decreased serum phosphorus level with a decrease in RBC 2,3 DPG and ATP. In a recent study, 2,3 DPG and ATP in children with CRF were considered. Erythrocyte 2,3 DPG and ATP levels were determined in 43 children with CRF on conservative treatment, and in 12 children on regular hemodialysis immediately before and after a hemodialysis session. The results were compared with nonanemic and anemic control subjects. Erythrocyte 2,3 DPG levels in renal failure were similar to those in anemic and nonanemic controls at normal blood pH but rose during dialysis as a result of alkalosis. In contrast, ATP levels were already high at a normal blood pH in renal failure. 2,3 DPG levels correlated with packed cell volume (PCV) in children with renal failure but persisted at lower concentrations compared with controls. The levels of both organic phosphates in the erythrocytes showed a significant correlation with blood pH. The lack of increase in 2,3 DPG, in combination with elevated ATP levels, suggests uremia-induced inhibition of 2,3 DPG synthesis (11).

Erythropoiesis and Erythropoietin (EP)

Several important investigations have been published about the effect of CRF on erythropoiesis. In CRF the serum contains a material(s) that is inhibitory to erythropoiesis *in vitro*. The chemical nature and mechanism of action of this material are unclear at the present time. To study this problem further, normal serum and CRF serum samples were extracted with two organic solvents, chloroform and petroleum ether. The serum samples were studied before and after extraction in a tissue culture system in which dog erythroblasts were stimulated by EP to synthesize heme. Before extraction, cells suspended in CRF serum synthesized less heme than cells cultured in normal serum. Extraction of CRF serum with chloroform, but not petroleum ether, resulted in an improvement in its ability to support heme synthesis. Studies of the material extracted by chloroform from CRF serum showed that it inhibited erythropoiesis. The authors proposed that the erythropoietic inhibitory substance in uremic serum is soluble in chloroform but not in petroleum ether, suggesting that it is a polar lipid (12).

The same authors studied the effect of CRF serum on erythroid colony growth *in vitro*. Considerable evidence suggests that insufficient EP production and the presence of a toxic factor inhibiting erythropoiesis are two major factors responsible for the production of anemia in patients with CRF. The toxic factor can be detected in a number of tissue culture systems. To evaluate its mechanism of action in a proliferation-dependent system, the formation of erythroid colonies was studied in plasma clots containing normal serum and CRF serum, using normal mouse marrow cells as the target organ. Fewer colonies were found in cultures containing uremic serum. This effect became greater as the concentration of serum was increased. No differences were found in the size or morphological characteristics of colonies formed. Addition of urea and creatinine to normal serum did not affect its ability to support colony growth. Uremic sera had no effect on white cell colony growth in the plasma clot system. It was concluded that materials inhibitory to erythroid proliferation are present in CRF serum (13).

Another investigation analyzed EP levels in patients with CRF. The anemia of chronic renal disease has been attributed primarily to a decrease in EP production. Results of this study showed that measureable levels of plasma EP can be demonstrated in a majority of patients with endstage renal failure who are undergoing long-term hemodialysis. These levels were not related to the type of renal disease, nor were they greatly affected by androgen therapy or by nephrectomy. Although elevations in EP levels in such patients were somewhat lower than in nonuremic patients with comparable anemia, the presence of measureable EP levels suggests that impaired end-organ response may play a role in the anemia of CRF (14).

A rabbit model was used in another investigation to determine the role of EP in CRF. EP deficiency may not be a major contributing factor in the early stages of the anemia of renal insufficiency. Serum EP titers are lower in advanced renal failure when compared with those of nonuremic anemic patients suffering from equivalent anemia. With increasing renal insufficiency, a relative deficiency in EP becomes increasingly more important as a pathogenic factor in reduced erythropoiesis. Kidneys without excretory function may still be erythropoietically effective, since a further increase in the anemia occurs after bilateral nephrectomy. However, a basal erythropoiesis is still maintained by extrarenal EP production, which is also enhanced by hypoxia. EP deficiency is compensated after successful renal transplantation. A decreased response of the bone marrow to EP may be another factor contributing to the hypoproliferative state of erythropoiesis in uremia. As demonstrated in a chronic uremic rabbit model, there may be a blockade of further differentiation of the erythroid precursors. The relationship between this blockade in differentiation and the inhibition of heme synthesis is not clear (15).

Granulopoiesis

The results of an excellent investigation pertaining to the presence of an inhibitor of granulopoiesis in the plasma of patients with renal failure have been published (16). Plasma from 31 patients with moderately severe to severe renal failure inhibited granulopoietic colony formation in human marrow *in vitro*. The inhibitory activity could not be removed by *in vitro* dialysis but was present in an ultrafiltered fraction of molecular weight less than 25,000 daltons. It inhibited the production of colony-stimulating activity (CSA) by leukocytes in the culture system but had little or no effect on preformed CSA or on the granulopoietic colony-forming cell itself. The level of plasma inhibitory activity correlated with the degree of azotemia but not with the neutrophil or monocyte counts. Despite the potency with which granulopoiesis was inhibited *in vitro* , none of the patients was severely neutropenic, and only four had mild neutropenia (the neutrophil count was $1.5\text{-}2.1 \times 10^{9}/1$).

Previous studies have suggested a relationship between bone marrow mast cells and osteomalacia in CRF. An interesting study addressed itself to this problem. Marrow mast cells were counted in iliac bone from patients with CRF treated by renal transplantation. Mast cell numbers were initially increased but returned to the normal range in many patients after renal transplantation. Improvement in osteitis fibrosa and osteomalacia after transplantation was not clearly related to this diminution in the number of mast cells. The use of prednisone in renal transplant patients may have some effect in reducing the numbers of mast cells. There is no fully acceptable explanation for the increase in marrow mast cells that occurs in CRF (17).

CALCIUM AND PHOSPHORUS ABNORMALITIES

Mechanisms

The decreased excretion of phosphorus and the consequent increased serum phosphorus concentration are well-known biochemical abnormalities in CRF. The increased serum phosphorus concentration is usually accompanied by a decreased serum calcium level. A few of the explanations offered for the decrease in serum calcium concentration are decrease in serum albumin (an important transport protein for calcium) and decreased hydroxylation of vitamin D as a result of the kidney disease. Several recently published studies offer new insight into mechanisms for the increase in serum phosphorus and the decrease in serum calcium levels.

Micropuncture studies were carried out in rats to determine what changes in tubular transport of phosphate occur in CRF and secondary hyperparathyroidism. Rats underwent subtotal nephrectomy (NX) and were fed a low-calcium, high-phosphorus diet for three to four weeks. Other groups consisted of normal control animals, normal rats infused with sodium phosphate to raise filtered load of phosphate, subtotal NX rats parathyroidectomized (PTX) on the day of experiment, and parathyroidectomized rats infused with sodium phosphate. It was found that the amount of phosphate filtered per nephron is markedly increased in subtotal NX rats due to high single-nephron filtration rates, that proximal tubular fluid plasma phosphate ratios are less than 1.0, and that fractional reabsorption of phosphate is decreased in the proximal tubules. More phosphate was present in the final urine than in distal convoluted tubules. Acute PTX in subtotal NX rats resulted in a striking increase in proximal phosphate reabsorption, and urinary phosphate became approximately equal to that remaining in distal tubules. Phosphate loading in normal rats reduced fractional reabsorption in the proximal tubule, but urinary phosphate was not greater than that at the end of distal tubules. Acute parathyroidectomized phosphate-loaded animals had no significant effect on proximal tubular phosphate reabsorption. These observations suggest that phosphate homeostasis in CRF is achieved by inhibition of proximal phosphate reabsorption, counteracting a greatly enhanced intrinsic capacity for reabsorption. In addition, the large amount of urinary phosphate is consistent either with secretion by the collecting ducts or with a disproportionately high contribution by deep nephrons. The changes in phosphate transport are mediated by parathyroid hormone (PTH) and are completely abolished by acute removal of the hormone (18).

Intestinal malabsorption of calcium and the development of osteomalacia in conservatively treated renal failure is explained by a quantitative deficiency of 1,25-dihydroxycholecalciferol (DHCCF), which also contributes to the development of hypocalcemia. An excess of 25-hydroxycholecalciferol can substitute for this deficiency. The presence and healing of azotaemic osteomalacia is unrelated to the prevailing plasma Ca times P product. The data suggest that vitamin D acts directly on bone mineralization, but the claim that this apparent effect is normally due to 25-

hydroxycholecalciferol is considered unproven. Most of the phenomena of azotaemic osteodystrophy are encountered in simple vitamin D deficiency; as in that condition, deficiency of DHCCF may be of primary significance in causing secondary hyperparathyroidism in renal failure (19).

Hypomagnesemia

The relationship between hypomagnesemia and hypocalcemia in CRF has been analyzed. A decrease in the serum magnesium concentration is unusual in CRF since renal failure usually leads to decreased excretion. Severe hypocalcemia secondary to magnesium depletion has been described in numerous patients with gastrointestinal disorders. The development of profound hypomagnesemia in chronic renal disease is a rare finding. Three patients with advanced renal failure and magnesium depletion were studied. Severe hypocalcemia also was present in these patients. Despite hyperplasia of the parathyroid glands, the levels of immunoreactive PTH in blood were inappropriately low for the degree of renal insufficiency. After administration of magnesium, there was a significant increase in the levels of circulating immunoreactive PTH in serum with a concomitant improvement in the hypocalcemia (20) (Figure 1).

1,25-Dihydroxycholecalciferol (DHCCF)

The therapeutic utilization of the potent vitamin D metabolite DHCCF in patients with CRF to correct hypocalcemia and to improve bone disease has been extensively studied by various groups. Excessive utilization of this vitamin may result in significant hypercalcemia, which would be deleterious. A controlled study of the effects of the DHCCF and vitamin D3 was performed in 18 nondialyzed patients with CRF. Patients with a creatinine clearance below 35 ml/min and mild renal osteodystrophy were selected. After six months' observation of the spontaneous course, the patients were randomly allocated to a six months' oral treatment with either DHCCF or vitamin D3 in initial daily doses of 1 μg and 4000 IU respectively, combined with 0.5 gm calcium. DHCCF quickly corrected hypocalcemia, reduced serum ALP and serum-immunoreactive PTH levels and more than doubled the urinary excretion rate of calcium. Vitamin D3 had similar but less pronounced effects. Seven of eight patients on DHCCF developed hypercalcemia, which necessitated a reduction in dosage. None of the patients on vitamin D3 treatment developed hypercalcemia. The percentage fall in creatinine clearance was greater during treatment than before treatment in all patients on DHCCF ($P < 0.01$) and in seven of nine patients on vitamin D3 treatment (although the group change here was not significant). Deterioration of renal function is a major limitation of the clinical use of DHCCF and vitamin D3 in nondialyzed patients with CRF. In fact, the decreased formation of DHCCF seen in CRF might protect renal function at the expense of abnormalities in mineral metabolism (21).

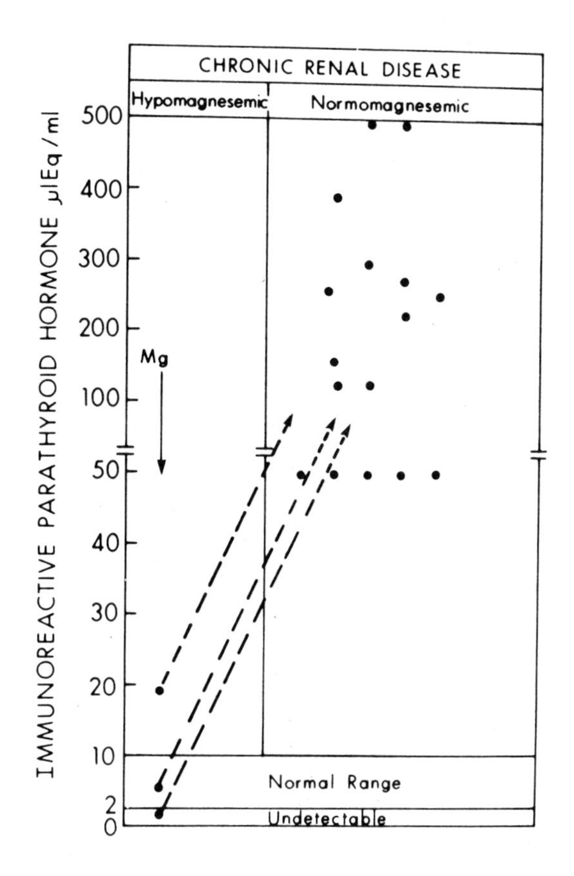

Figure 1. Immunoreactive parathyroid hormone (PTH) concentrations in three uremic hypomagnesemic patients (left panel). In one patient, PTH was slightly elevated, in the second normal, and in the third, undetectable. After Mg administration PTH increased in all three patients to the expected levels seen in patients with uremia (broken lines). For comparison, the levels of PTH in 16 uremic normomagnesemic patients are plotted in the same figure (right panel). From ref. 20 with permission of authors and Annals of Internal Medicine.

DHCCF was studied in a double-blind controlled fashion in patients on long-term dialysis. Serum calcium levels were unchanged in 16 patients on vitamin D3 (400-1200 U/day). In 15 patients on DHCCF (0.5-1.5 µg/day), the serum calcium concentration increased from 9.05 ± 0.15 to 10.25 ± 0.20 mg/dl ($P <$

0.001), returning to 9.37 ± 0.16 mg/dl ($P < 0.001$) in the post experimental period. Patients on vitamin D3 showed no reversible decrease in immunoreactive PTH levels, but patients on DHCCF did, from a control of 1077 ± 258 to 595 ± 213 µl equivalents/ml ($P <$ 0.01), and returned to 1165 ± 271 µl equivalents/ml ($P < 0.005$). Nine of 12 patients on vitamin D3 who underwent serial iliac crest biopsy showed histologic deterioration, and 6 of 7 who received DHCCF were improved or unchanged ($P < 0.025$). Bone mineral and calcium levels decreased in patients on vitamin D3 ($P < 0.05$) but not in those on DHCCF. Hypercalcemia occurred in 5 of 15 patients. It was concluded that DHCCF has a calcemic effect in patients on long-term dialysis, decreases levels of immunoreactive PTH, and is associated with histological improvement in bone disease. Thus, DHCCF is a valuable adjunct in the management of renal osteodystrophy but requires monitoring of the serum calcium levels to avoid hypercalcemia (22).

Hypocalcemia and Phosphate Retention

Hyperplasia of the parathyroid glands and elevated blood levels of PTH are almost always present in patients with CRF. Evidence of this secondary hyperparathyroidism is found even with mild renal insufficiency. Therefore the mechanism(s) underlying this abnormality must be operative in the early stages of the renal disease. All investigators agree that the stimulus for the hyperplasia of parathyroid glands in these patients is hypocalcemia. A great deal of effort has been put into the investigation of the mechanism(s) of the hypocalcemia of renal failure. At present, two postulates have been advanced: phosphate retention and skeletal resistance to the calcemic action of PTH. The proponents of these two notions have provided evidence supporting their concepts. The vigor with which this has been done perpetuated the impression that these two theories are mutually exclusive and, as such, a controversy might have been falsely created (23).

The available data are consistent with the formulation that phosphate retention and skeletal resistance to the calcemic action of PTH are both important in the genesis of uremic secondary hyperparathyroidism, and that these two pathogenic processes are probably linked with each other. In patients with early renal failure, the main effect of phosphate retention is probably not a direct one on serum calcium but a suppression of the production of DHCCF, which would cause skeletal resistance to PTH. The latter would result in hypocalcemia and secondary hyperparathyroidism. In patients with advanced renal failure, the situation is more complicated. In addition to its effect on DHCCF, the severe hyperphosphatemia would also directly lower serum calcium levels. Also uremia *per se* can worsen the skeletal resistance and, as such, aggravate the hypocalcemia. Hyperparathyroidism will, therefore, be more profound.

The formulation described above may have therapeutic implications. It is obvious that secondary hyperparathyroidism could be prevented if both phosphate retention and the skeletal resistance to PTH were corrected. This could be achieved by restriction of the dietary phosphate intake in proportion to the glomerular filtration rate (GFR). However, such a therapeutic approach is difficult to achieve and would not be feasible in an ambulatory patient with early renal insufficiency. If phosphate restriction does not match the decrease in GFR, prevention of secondary hyperparathyroidism may not be achieved. On the other hand, if phosphate restriction is greater than needed, phosphate depletion may ensue. An alternative method would be to supply the patient with DHCCF to disrupt the sequence of events that will lead to secondary hyperparathyroidism.

The authors do not believe that a controversy exists about the pathogenesis of secondary hyperparathyroidism in renal failure. It is their opinion that both phosphate retention and skeletal resistance to the calcemic action of PTH are important, and that these two factors are probably related.

Hypercalcemia Causing Renal Failure

It is unusual for the patient with CRF to manifest hypercalcemia. If hypercalcemia is present in a patient with CRF, the elevated serum calcium concentration may be related to the patient taking excessive calcium-containing drugs, such as is seen in the Burnet syndrome of renal failure due to excessive milk and alkali.

Other causes of hypercalcemia associated with renal failure are tertiary hyperparathyroidism becoming manifest during dialysis or long-term therapy, excessive use of calcium in the dialysis fluid, and various diseases associated with hypercalcemia resulting in renal failure from nephrocalcinosis. A number of papers have been published about this subject.

The influences of hypercalcemia on renal function was studied retrospectively in 13 patients suffering from primary hyperparathyroidism, sarcoidosis, vitamin D intoxication, malignant lymphoma, or chronic lymphatic leukemia. Different treatments, depending on the primary disease, often induced a rapid fall in the serum calcium concentration. The serum creatinine concentration always fell simultaneously. The serum phosphate concentration fell in all but two patients. Changes in the serum calcium and the serum creatinine levels correlated significantly ($P < 0.001$) as did changes in serum calcium and serum phosphate concentrations ($P < 0.05$). Serum calcium/serum creatinine and serum calcium/serum phosphate ratios were significantly higher in patients with primary hyperparathyroidism than in patients with hypercalcemia of nonhyperparathyroid origin ($P < 0.01$, for both). This suggests a different effect of calcium on the GFR in hyperparathyroid and nonhyperparathyroid patients, the latter group being more sensitive to the influence of hypercalcemia (24).

Between 1972 and 1976, 15 patients with CRF of different etiology and varying severity were observed. These patients experienced a total of 23 hypercalcemic periods during treatment with calcium-containing drugs. Twelve instances of hypercalcemia occurred during conservative treatment (serum creatinine: 177-1061 μmole/l, equivalent to 20-120 mg/l), and 11 occurred during chronic hemodialysis (serum creatinine: 707-1061 μmole/l, equivalent to 80-120 mg/l). In 15 cases hypercalcemia was caused by a hexacalcium-hexasodium-heptacitratehydrate complex (Acetolyt), in six cases by the combined use of this drug with calcium ion-exchange resins on a calcium-polystyrolsulfonate base, and in two cases by the use of calcium tablets and calcium polystyrolsulfonate, respectively. The daily doses of these drugs were in the usual therapeutic range in most cases. Deterioration of renal function was observed in two cases and coma in two additional cases. In five patients gastric ulcers were demonstrated. Three patients died. In no patient was there evidence of florid hyperparathyroidism. The results of these observations indicate that treatment with calcium-containing drugs in patients with renal failure should be carried out only with strict control of serum calcium concentrations (25).

LIPID METABOLISM ABNORMALITIES

Triglyceride

When patients with CRF manifest elevated serum cholesterol and triglyceride concentrations, the clinician and biochemist usually associate this combination with the nephrotic syndrome. However, nonnephrotic renal failure may also be associated with an increase in the plasma triglyceride level.

A few important papers have related to this topic. Plasma triglyceride turnover has been studied in 12 patients with chronic nonnephrotic renal failure (creatinine clearance: 3.5-11 ml/min) and in 8 control subjects. The patients showed significantly increased fasting triglyceride concentrations and absolute plasma triglyceride turnover rates, while the fractional turnover rates were significantly decreased. The finding of an elevated triglyceride turnover rate with a near-normal fractional removal rate in the patients with the highest creatinine clearance suggests the involvement of an increased triglyceride synthesis rate in these patients. However, most patients showed an impaired removal mechanism as the major cause of the hypertriglyceridemia (26).

Mechanisms of Hyperlipoproteinemia

Cardiovascular disease is one of the most frequent complications in patients undergoing hemodialysis and after renal transplantation. Studies have shown the frequency of hyperlipoproteinemia as a risk factor for coronary heart disease in both groups and in patients with severe renal insufficiency. In patients with uremia and in patients undergoing hemodialysis, type IV hyperlipoproteinemia is the pattern observed, whereas the type II pattern is more common after renal transplantation. Studies on the cause of these secondary lipid disorders have provided different explanations for the elevated lipoprotein concentrations, with some suggesting increased synthesis and others disturbed catabolism.

To investigate the pathogenesis of hypertriglyceridemia

in patients with renal disease, plasma lipoprotein composition as well as hepatic triglyceride lipase and lipoprotein lipase in postheparin plasma were measured. Three groups with renal disease were studied: conservatively treated chronic uremia, patients undergoing maintenance hemodialysis, and renal-allograft recipients. A selective decrease in the hepatic triglyceride lipase activity with normal lipoprotein lipase levels was found in conservatively treated uremic patients and in patients undergoing hemodialysis. Elevated concentrations of very low-density and increased triglycerides in low-density lipoproteins occurred in these patients. In contrast, after renal transplantation, hepatic triglyceride lipase and lipoprotein lipase levels were normal in patients who had type II hyperlipoproteinemia as a common lipoprotein pattern, with increased low-density-lipoprotein cholesterol and decreased high-density-lipoprotein cholesterol concentrations. The accumulation of a triglyceride-rich low-density lipoprotein in the majority of patients with renal disease may be the consequence of low hepatic triglyceride lipase activity (27). (Figures 2-4).

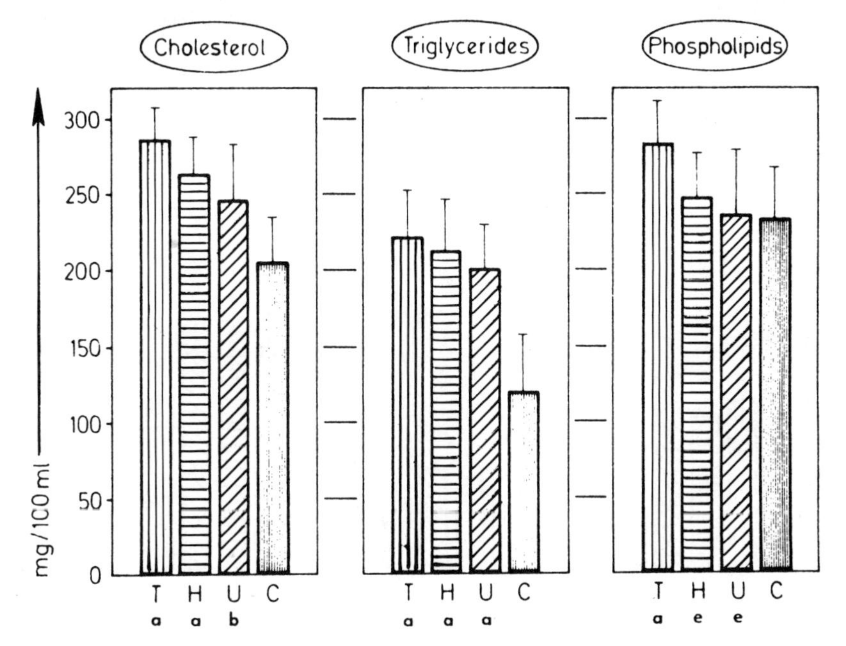

Figure 2. Plasma lipid concentrations in patients with renal disease as compared to normal controls. T: patients after renal transplantation, H: patients on maintenance hemodialysis, U: conservatively treated patients with chronic uremia, C: controls, a: $P < 0.005$, b: $P < 0.005$, c: $P < 0.01$, d: $P < 0.025$, e: difference not significant. (Reprinted by permission from the New England Journal of Medicine 297:1362, 1977.)

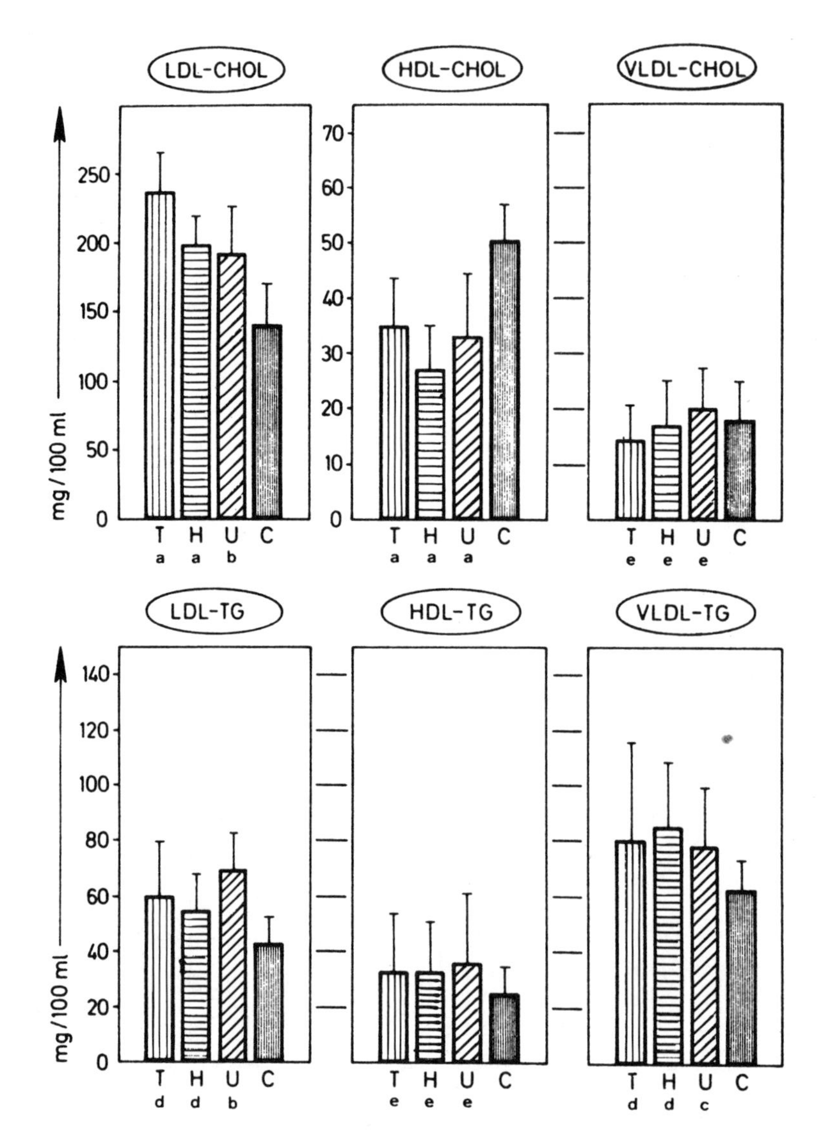

Figure 3. Cholesterol (CHOL) and triglyceride (TG) content of low-density (LDL), high-density (HDL) and very-low-density (VLDL) lipoprotein fractions of patients with renal disease and normal controls. For explanation of abbreviations, see Figure 2. (Reprinted by permission from the New England Journal of Medicine 297:1362, 1977).

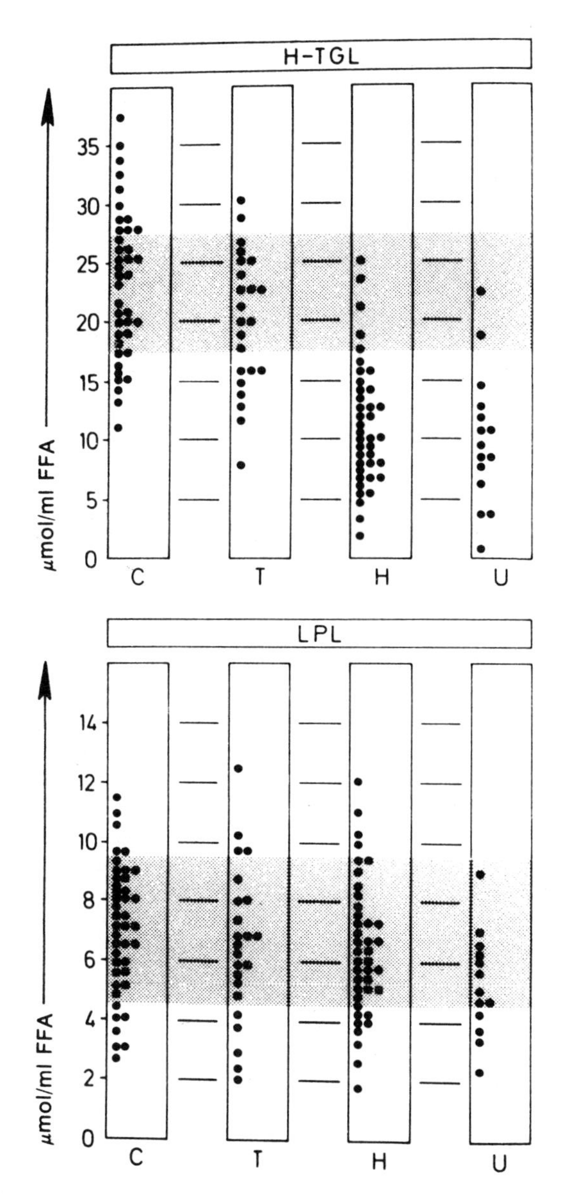

Figure 4. Hepatic triglyceride lipase (H-TGL) and lipoprotein lipase (LPL) in patients with renal disease and normal controls. The shaded area represents the normal range. For explanation of abbreviations, see Figure 2. (Reprinted by permission from the New England Journal of Medicine 297:1362, 1977.)

Significant hypertriglyceridemia, the most common lipid abnormality in renal failure, first occurs when the creatinine clearance falls to 50 ml/min. The prevalence of hypertriglyceridemia continues to rise as creatinine clearance falls further with the highest rate developing at a creatinine clearance of less than 10 ml/min. Hypertriglyceridemia is correlated with plasma glucagon levels but not with growth hormone or insulin levels. Plasma cholesterol values remain normal in the face of deteriorating renal function and show no correlation with the concentration of any of the hormones measured. Although the levels of all three hormones became elevated as renal function diminished, none was directly correlated with the GFR. There was a distinct decrease in the prevalence of hyperlipidemia after five years of maintenance hemodialysis therapy. Plasma growth hormone and glucagon through an effect on plasma triglyceride and plasma insulin by affecting plasma cholesterol may play a role in this decline in hyperlipidemia with duration of hemodialysis (28).

The plasma lipoprotein alteration and lipoprotein deposition in the kidney in patients with hepatorenal syndrome has been analyzed in an interesting paper. In addition to the renal lesion, an additional problem that contributed to the lipid abnormality was hepatic disease, which may frequently be associated with deranged lipid metabolism.

Four patients with advanced liver disease and progressive renal failure compatible with the diagnosis of hepatorenal syndrome were studied. All four patients had low lecithin/cholesterol acyltransferase activity in plasma, and the concentration of cholesterol esters was markedly reduced. The levels of the main classes of lipoprotein were abnormal, and the polar lipid content was increased. Electron microscopy of negatively stained lipoproteins from two of the patients revealed large particles with layered membranes (70-200 nm in diameter) corresponding to the large-molecular-weight fraction of the low-density lipoproteins. These structures were not present in the low-density lipoproteins from the other two patients. In both patients' renal biopsy specimens and in another patient's necropsy specimen, deposition of osmophilic material was found in the glomeruli (especially located subendothelially), in the basement membrane, and in the mesangial regions. The deposits were similar to those previously described in patients with familial lecithin/cholesterol acyltransferase deficiency and most probably represent cholesterol and phospholipids. It is suggested that the renal deposition of lipid may be related to the large-molecular-weight low-density lipoprotein fraction, and that the mechanisms involved in this lipid deposition are similar to those occurring in familial lecithin/cholesterol acyltransferase deficiency (29).

GLUCOSE ABNORMALITIES

Mechanism of Hyperglycemia

Carbohydrate intolerance in CRF is a common clinical biochemical abnormality. A number of important investigations attempting to analyze the hyperglycemic stage have been published. This abnormality most commonly includes previously unrecognized carbohydrate intolerance and even overt fasting hyperglycemia in more than 50% of azotemic patients. However, the development of spontaneous hypoglycemia has also been reported. On the other hand, an amelioration of the diabetic state with decreasing insulin requirements is a well-known clinical concomitant of advancing renal failure in patients with previously established diabetes mellitus. In a group of patients, the interrelation among glucose, insulin, and growth hormone was investigated and confirmed the presence of carbohydrate intolerance and hyperinsulinemia (30). In addition, the authors demonstrated alterations in growth hormone regulation, characterized by *(1)* the lack of suppression of growth hormone by orally induced hyperglycemia and paradoxical increase in serum levels of growth hormone after the administration of intravenous glucose or glucagon and *(2)* lack of release of growth hormone with induced hypoglycemia and an exaggerated response to levodopa administration. Furthermore, thyrotrophin-releasing hormone stimulated growth hormone release, a phenomenon not observed in the control population. Their studies show an impaired hypothalamic regulation of growth hormone secretion in patients with renal failure undergoing long-term hemodialysis (30).

Blood glucose, plasma nonesterified fatty acids, immunoreactive insulin, growth hormone, and immunoreactive glucagon responses to intravenous glucose were determined in 16 children on regular hemodialysis for CRF, and in 9 healthy children. In the 16 children, the fractional disappearance rate of glucose was significantly reduced; basal immunoreactive insulin was significantly raised; and, while the early immunoreactive insulin response to glucose was similar in patients and controls, the late response was increased. Basal growth hormone was elevated in the patients and rose paradoxically after glucose administration. Fasting immunoreactive glucagon response was significantly higher in the patients and was not suppressed by glucose. Plasma nonesterified fatty acid levels were lower in the patients and fell more markedly after glucose. Alanine levels, which were significantly raised in those with poor glucose tolerance, fell to normal after glucose administration and did not vary in those patients with more normal glucose tolerance. It is speculated that the metabolic and hormonal alterations may be interrelated and may result from failure of normal glucose utilization (31).

An investigation was carried out with glycosylated hemoglobin A (HbA) in control subjects, diabetics, and some patients on chronic hemodialysis. The authors found raised values in 18 patients on long-term dialysis (mean: 8.5%, SD: 1.3), compared with values in 108 adult control subjects (mean: 7.45%, SD: 0.73). They believe that this increase can be explained by coexisting hyperglycemia, and no other factor need be proposed. Figure 5 shows the relationship between fasting blood-glucose and total glycosylated HbA values in 20 normal subjects, 180 diabetics, and 15 patients on long-term dialysis. The best-fit line is linear when normoglycemia or moderate hyperglycemia exists but

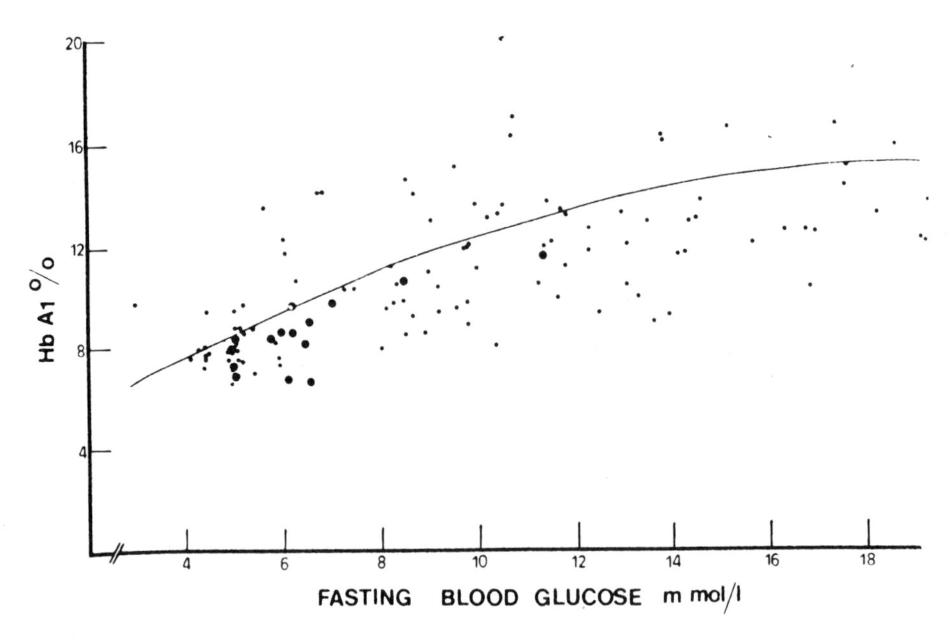

Figure 5. HbA1 in relation to blood-glucose. • = normals and diabetics. ● = chronic renal failure. (Reprinted by permission of authors and Lancet, ref. 32)

flattens out at higher blood-glucose levels. If one concentrates on the dialysis patients, whose results are represented by large dots, all their values fall within the expected range. Indeed the total glycosylated HbA values are lower than might be expected for the appropriate blood glucose, as we would expect in a situation of reduced blood-cell survival. Thus, hyperglycemia can be the sole explanation for the increase in total glycosylated HbA in CRF, and no other unknown factor involved in modifying this nonenzymatic reaction need be proposed. However, the findings do emphasize the importance of controlling glucose intolerance in renal failure. This reversible factor will increase glycosylation of HbA, diminishing tissue oxygen delivery already grossly impaired by anemia and acidosis (32).

To delineate the potential role of disordered glucose and glucose-precursor kinetics in the abnormal carbohydrate metabolism of CRF, alanine and glucose production and utilization and gluconeogenesis from alanine were studied in patients with chronic compensated renal insufficiency and in normal volunteers. With simultaneous primed injection and continuous infusions of radiolabeled alanine and glucose, turnover rates, metabolite and precursor-product interrelationships were calculated from the plateau portion of the appropriate specific activity curves. All subjects were studied in the postabsorptive state. In 13 patients with CRF (creatinine: 10.7 ± 1.2 mg/100 ml; mean ± SEM), glucose turnover was found to be 1,035 ± 9.93 μmol/ min. This rate was increased 56% (P = 0.003) over that observed in control subjects (664 ± 33 μmole/min). Alanine turnover was 474 ± 96 μmole/min in azotemic patients. This rate was 191% greater (P = 0.007) than the rate determined in control subjects (163 ± 19 μmole/min). Gluconeogenesis from alanine and the percentage of glucose production contributed by gluconeogenesis from alanine were increased in patients with CRF (192% and 169%, respectively) as compared with control subjects ($P < 0.05$ for each). Alanine utilization for gluconeogenesis was increased from 40.2 ± 3.9 μmole/min in control subjects to 143 ± 39 μmole/min in azotemic patients ($P < 0.05$). The percentage of alanine utilization accounted for by gluconeogenesis was not altered in chronic renal insufficiency. In nondiabetic azotemic subjects, mean fasting glucose and immunoreactive insulin levels were increased 24.3% (P = 0.005) and 130% (P = 0.046), respectively.

These results in patients with CRF demonstrate *(a)* increased glucose production and utilization, *(b)* increased gluconeogenesis from alanine, *(c)* increased alanine production and utilization, and *(d)* a relative impairment of glucose disposal. It is concluded that chronic azotemia is characterized by increased rates of glucose and glucose precursor flux and by a relative impairment of glucose disposal. These findings may suggest an underlying hepatic and peripheral insensitivity to the metabolic action of insulin in patients with chronic renal insufficiency (33).

PROTEIN METABOLISM ABNORMALITIES

Albumin

Important investigations relevant to protein metabolism have been published. The effects of CRF on albumin, amino acid, urea, and creatinine levels have been analyzed. A reduced serum albumin level frequently is observed in CRF. This abnormality has usually been ascribed to proteinuria. However, a recent study has suggested another possible etiology.

The pathophysiology of albumin metabolism in uremia was investigated by turnover measurements in a large series of uremic patients, either on conservative management or on dialysis therapy. A total of 62 turnover studies were performed in patients on dietary treatment. The patients were divided into two groups according to the duration of the low protein diet: 35 subjects from 6 to 30 days, 27 subjects from six months to five years. Albumin catabolism and distribution were measured by the two-tracer technique (^{131}I-albumin and ^{125}I-iodide, simultaneously injected intraveneously), while albumin synthesis was directly determined in 10 patients by the use of ^{14}C-carbonate and ^{131}I-albumin. Sixteen turnover studies were also performed in a group of endstage uremic patients on dialysis therapy by a two-tracer procedure especially designed to determine albumin catabolism in the course of a single peritoneal or hemodialysis treatment. The main features of albumin metabolism observed in the patients on dietary management were normal intravascular albumin mass and marked reduction of the extravascular and total albumin pools, with proportionally reduced catabolism. As to the uremic patients on dialysis therapy, the catabolic rate of albumin was increased threefold in three patients showing clinical features of "hypercatabolism" in the early phase of uremia or during relapse. Albumin turnover rates returned to normal when measured during clinical steady-state conditions. All these findings suggest that a marked body protein depletion exists in chronic uremia and that dietary treatment *per se* is not responsible for such a depleted state. Instead, the depletion of protein stores observed in the steady phase of chronic uremia may have originated from the exaggerated increased catabolism in the early phase of renal failure, which was not compensated by a proportional increase of the synthetic rate, due to both the state of uremic intoxication and to the reduced dietary protein intake during the early phase (34).

Amino Acids - General

There are many causes of altered amino acid and protein metabolism in uremia that may lead to impaired growth, wasting, malnutrition, and other aspects of the uremic syndrome. These causes have complex interrelationships, which are not well understood and include altered nutrition due to poor intake, losses of nutrients during dialysis, and abnormal metabolism of many nutrients. Uremic toxins, superimposed catabolic illnesses, elevated or reduced serum hormone levels, reduced capacity of the kidney to synthesize certain amino acids and to degrade other amino acids, peptides, and small proteins, and decreased excretion of certain amino acids and peptides may also contribute to altered amino acid and protein metabolism. The response of certain plasma amino acids to protein restriction appears to differ in uremic patients as compared with normal subjects. Increased plasma levels of many products of amino acids and proteins in renal failure are due primarily to decreased urinary clearance by the kidney. However, for some metabolites, increased synthesis or decreased degradation may also contribute to elevated levels. These latter compounds include guanidinosuccinic acid, methylguanidine, certain intermediate molecules, and, in some patients, phenylpyruvic acid (35).

Valine

Valine metabolism was investigated in five normal and three nondialyzed chronically uremic subjects eating 40 ± 1 (SEM) and 53 (range 40 to 80) gram protein diets, respectively, in a metabolic research unit. Subjects were injected intravenously with a tracer dose of L-valine-1-^{14}C while they fasted, and specific activity of plasma valine-^{14}C and expiration of $^{14}CO_2$ was not different from normal subjects. A two-pool model for valine metabolism was derived and indicated that in uremic patients there was a significant decrease in both valine pools and in the rate of irreversible loss, that is, valine incorporated into larger molecules, degraded, or excreted. Valine degradation was estimated to be decreased in the uremic patients (36).

Other Essential Amino Acids

The effectiveness of a mixture of five analogues of essential amino acids and the remaining four essential amino acids, as compared with preceding treatment period of the nine essential amino acids, was evaluated in 16 chronic uremic patients fed a low-protein diet. During amino acid analogue supplementation, there was a tendency for the blood urea nitrogen level to fall, whereas the creatinine level did not change. The serum phosphate concentration decreased in most patients, whereas the serum calcium concentration rose in some subjects. Protein metabolism, as judged by serum transferrin, Clq and C3c levels and total complement activity, was improved. Furthermore, the concentrations of prealbumin and retinol-binding protein, which are elevated in uremia, showed a further increase that might favor vitamin A intoxication (37).

Urea

The role of urea in CRF patients has been studied. The amount of urea nitrogen released and the amount reincorporated into albumin has been measured in healthy and uremic persons on both normal and low-protein diets. The albumin synthesis rate was measured simultaneously. Gut urea breakdown was only 50% higher in renal failure than in health, but the efficiency of utilization of the nitrogen thus released was increased more than sixfold in renal failure and was higher on a low-protein diet than on a normal-protein diet. The lower the albumin synthetic rate, the greater was the efficiency of incorporation of urea nitrogen into albumin. The rate of urea nitrogen incorporation into albumin increased on average 14-fold in CRF. The absolute rate of utilization (84 μmole/hr) was, however, small and comprised, on average, only 2.4% of the nitrogen used in albumin synthesis. These findings suggest that although some urea-derived nitrogen is incorporated into albumin, the amount is not nutritionally significant even under conditions of protein deprivation and high urea availability (38).

Gastrointestinal bacteria have been thought to be beneficial to uremic patients since they could provide an internal source of nitrogen by degrading urea and could function as an alternative means of clearing waste products. Data were presented to indicate that the bacteria of uremic patients do not clear significantly more urea than do the intestinal flora of normal subjects. Analysis of urea metabolism before and during oral administration of aminoglycoside antibiotics demonstrated that nitrogen derived from urea is not used by uremic patients for amino acid synthesis. In addition, it was found that nitrogen balance, on the average -0.98 ± 0.41, significantly improved to -0.18 ± 0.29 gm of nitrogen per day during the antibiotic period. The possible explanations for this were discussed. It was concluded that intestinal bacteria adversely affect uremic patients by promoting catabolism and by producing toxins that accumulate in body fluids (39).

Serum Creatinine

The ability to predict increasing renal insufficiency from serum creatinine measurement has been studied. Progression of renal insufficiency was estimated from serial serum creatinine (SCr) determinations in 45 pediatric patients with an SCr greater than 2 mg/dl. This was done by plotting the logarithm of SCr (log Scr) versus time or the reciprocal (1/SCr) versus time. Calculation of data by linear regression analysis was restricted to 38 of these patients where at least four sequential Scr determinations were available. In 24 patients, a similar straight line relationship was observed on both plots. The semilog plot appeared more suitable in two, and the reciprocal plot in seven instances. In 5 patients, progression of renal insufficiency was clearly nonlinear. In the majority (33 of 38) of the patients, loss of renal function thus occurred at an exponential rate (log SCr) or at a constant rate (1/Scr). The mean rate of progression of chronic renal insufficiency, expressed as SCr doubling time on the semi-log plot, was 2.4 months in 10 patients with acquired glomerular diseases, 17 months in 3 patients with cystinosis, 27.7 months in 5 patients with renal hypoplasia or dysplasia, and 21.8 months in 13 patients with urinary tract abnormalities (40).

Creatinine appearance, defined as the sum of daily creatinine excretion in urine (average over five days) plus accumulation in body water, measured over the same interval, was calculated in 27 patients with severe CRF (creatinine clearance less than 0.15 1/kg/day). Creatinine appearance per kilogram of body weight in patients with the lowest clearances decreased to values as low as one-third of those predicted from age and sex. The absolute value of measured creatinine accumulation was only 11 ± 2% of creatinine appearance and thus could not account for such deficits in appearance and therefore renal excretion. One explanation for these results is that extrarenal clearance (CM) remains constant, that is, that the quantity of creatinine degraded (M) is proportional to serum creatinine (S) according to the following: CM = M/S. When the values for extrarenal clearance necessary to account for the measured deficit in creatinine appearance were calculated, they were found to be quite constant: 0.042 ± 0.004 1/kg/day (SEM, n=13) in men and 0.041 ± 0.004 1/kg/day (SEM, n=14) in women. Renal creatinine clearance in these patients, predicted from age, sex, serum creatinine, and the assumed constant value for extrarenal clearance, corresponded closely to observed clearance (r = 0.93). From these calculations, decreased creatinine appearance (and excretion) of uremic patients may be explained by a constant extrarenal clearance, indicating degradation (41).

β-Melanocyte Stimulating Hormone (MSH) and Related Hormones

It is well known that patients with CRF who are being treated with hemodialysis develop increased pigmentation and darkening of their skin. The presence of darker skin, weakness, vomiting, and biochemical abnormalities of metabolic acidosis, hyponatremia, hyperkalemia, and increased BUN are found in adrenal insufficiency. These clinical findings and biochemical abnormalities are also characteristic of CRF and are known as pseudo-Addison syndrome. It was described by Thorn and is also commonly referred to as Thorn syndrome. An attempt to explain the increased melanin in the skin is the subject of an interesting investigation.

The plasma immunoreactive "β-MSH" was studied in hemodialysis patients to determine its basal level, plasma disappearance rate, gel filtration and immunological characteristics. All patients had increased plasma β-MSH (90-440 pg/ml; normal < 90 pg/ml). Plasma ACTH and cortisol values were within the normal range. Cortisol infusion over two hours induced almost no plasma β-MSH variation as compared with control subjects in whom β-MSH decreased rapidly (apparent half-life: 90 min); more prolonged administration of corticosteroids (dexamethasone, 0.5 mg every six hours for two days) caused a slight (20%) but significant ($P < 0.001$) decrease of β-MSH. On Sephadex G-50, endogenous β-MSH eluted in a molecular weight range of 6,000 to 10,000 daltons. In the radioimmunoassay used, dilution curves of endogenous β-MSH paralleled that of synthetic human β-MSH but not that of purified human β-lipotrophin (LPH). In conclusion, hemodialysis patients showed a clear dissociation between elevated β-MSH and normal ACTH plasma levels. β-MSH probably has a decreased plasma disappearance rate and seems related to a substance different from authentic human β-MSH (42).

Parathyroid Hormone (PTH)

The kinetics of parathyroid hormone metabolism have been studied In humans, the distribution and metabolic degradation of bovine PTH (bPTH) were measured after infusion of 400 units of bPTH and blood sampling up to two hours. Disappearance rates of intact 1-84 bPTH, carboxyl- and amino-regional peptides were calculated for healthy subjects (n-12); patients suffering from moderate (GFR 15-30 ml/min, n=4) and advanced (GFR $\leqslant$ 10 ml/min, n=30) CRF; and bilateral nephrectomized patients (n=3). Two components (distribution, metabolism) with a rapid and a slow disappearance rate could be separated. Half-lives were found to be in the range known from animal experiments with marked differences between intact 1-84 PTH and peptide fragments. The influence of impaired renal function on metabolic turnover rates of PTH was described (43).

RENIN ABNORMALITIES

The importance of renin in the pathogenesis of hypertension in CRF has been of considerable interest for many years. Recent investigations in man and animals have further elucidated the association of renin and hypertension in renal failure.

A study was initiated to examine the influence of different sodium loads on renin release in the hypertensive and normotensive state of CRF. Blood pressure (BP), plasma renin concentration (PRC), and exchangeable sodium (NaE) were measured in 18 patients with advanced CRF, in 9 hypertensive and 9 normotensive patients, and in 7 normal subjects *(a)* six days after a fixed sodium intake of 10 mmole/day and *(b)* six days after a fixed sodium intake of 150 mmole/day. Mean NaE was 14 to 19% higher in the hypertensive patients compared with the normotensive patients, and values of NaE correlated significantly to values of mean BP. No significant differences were present in PRC between the groups of patients and controls on either of the sodium regimens, and no correlation was found between BP and PRC. However, the average decreases of PRC (33 to 34%) in the hypertensive patients on high sodium intake were significantly lower than the corresponding values of 69 to 73% in the normotensive patients and 69 to 71% in the controls. Furthermore, the percentage changes of PRC on high-sodium intake correlated significantly with mean BP as well as with NaE. These results suggest that renin release is relatively unresponsive to different sodium intakes in hypertension after CRF. This alteration in renin release may contribute to the maintenance of hypertension in CRF; PRC is "inappropriately" increased in relation to the sodium excess (44).

Plasma renin activity was determined by bioassay before, during, and after a two-hour infusion of norepinephrine into the renal artery in unilaterally nephrectomized dogs to examine the role of the renin-angiotensin system in norepinephrine-induced acute renal failure (ARF) which was induced in five of eight dogs receiving 0.75 μg/kg/min of norepinephrine but not in the remaining three dogs, nor in two dogs infused with 0.6 and 0.4 μg/kg/min of norepinephrine. There was no difference in plasma renin activity in renal venous blood between the dogs with and without ARF when followed up to two hours after the discontinuation of the infusion. The same results were obtained when the plasma renin activity in the foreleg vein was followed at 24, 48, and 72 hours after the infusion. The renin-angiotensin system does not seem to contribute to the reduction of renal function in norepinephrine-induced ARF in dogs (45).

Serial changes in plasma of PRC, renin-substrate concentration (RSC) and renin activity (PRA) were followed in rats during glycerol-induced ARF. There was an early but transient increase in PRC and a more delayed and prolonged rise in RSC. PRA began to rise a few hours after glycerol administration with a concomitant increase in renin, but at the end of 24 hours the high activity was maintained largely by increases in RSC. No correlations between PSC, PRC or PRA changes and severity of renal failure could be demonstrated. These results suggest that these changes are not causally related to the development of kidney failure (46).

DRUG METABOLISM

Aluminum Magnesium Hydroxide

Aluminum magnesium hydroxide is usually used to decrease gastrointestinal absorption in patients with CRF. This therapeutic measure decreases the serum phosphorus concentration and results in a decreased incidence of bone disease. The rationale is based on the fact that increased phosphorus retention leads to hypocalcemia with consequent hyperparathyroidism and resulting osteomalacia.

Serum aluminum may increase because of the renal insufficiency and may accumulate in the brain, causing central nervous system toxicity. Thus, it may be important to be able to quantitate serum aluminum levels. An important paper offers a biochemical procedure to determine aluminum levels in biological samples by the use of atomic absorption spectrophotometry with a graphite furnace (47). No sample preparation is required, and the procedure is sensitive at the appropriate concentrations. A sample of serum and urine is pipetted into the interior of the graphite tube, where it is sequentially dried, charred, and atomized. Precautions for sample handling are discussed, and instrument settings are defined. Precision and accuracy of the method are evaluated, as are the effects of salts, protein content of serum, and specific gravity of urine. Serum aluminum concentrations of 23 persons who were not consuming aluminum-containing antacids were 28 ± 9 (SD) μg/liter (1.02 ± 0.33 μmol/l).

Calcium Carbonate

The effects of oral administration of calcium carbonate, aluminum hydroxide gel, dihydrotachysterol (DHT) and sodium bicarbonate on metabolic acidosis and plasma calcium and phosphate levels were studied in seven patients with CRF. The administration of calcium carbonate alone alleviated the acidosis and increased the urinary bicarbonate excretion. These effects were potentiated when aluminum hydroxide gel was administered in combination with calcium carbonate. The plasma calcium level was increased by this combination therapy. The effects of these two agents on acidosis and plasma calcium levels were further enhanced by the additional administration of DHT. Urinary bicarbonate excretion was less during the treatment with aluminum hydroxide gel and calcium carbonate than with aluminum hydroxide gel and sodium bicarbonate, when the excretions were compared at similar concentrations of plasma bicarbonate. Aluminum hydroxide gel and DHT are likely to enhance the effect of calcium carbonate, which works as an alkalinizing salt on acidosis, probably through increasing calcium absorption in the intestine. The three agents suppress the leak of bicarbonate into the urine, contributing to the improvement of acidosis (48).

Cimetidine

The metabolism of other drugs, including cimetidine, in CRF was extensively investigated in a series of reports. Serial blood samples for determination of drug levels were obtained after intravenous administration of 300 mg of cimetidine. Sixteen patients with varying degrees of renal failure were studied. There was a prolongation of drug half-life in patients with renal insufficiency compared with normal control subjects ($P < 0.001$). A significant inverse relationship between the half-life and the creatinine clearance was noted ($r = 0.69$, $P < 0.01$). The effect of hemodialysis was studied in 12 patients. Cimetidine was found to be dialyzable. This was demonstrated both by a shortening of the half-life of the drug during dialysis and by measurement of its concentration in the dialysate. This suggests that the dose schedule should be modified for patients with renal insufficiency and for those on hemodialysis. A single dose of cimetidine, administered intravenously, was well tolerated by the patients. One patient developed an urticarial skin rash, believed to be allergic in nature. There was a transient, mild (but significant) rise in blood urea nitrogen and serum creatinine concentrations in five patients with moderate renal failure (49) (Figure 6).

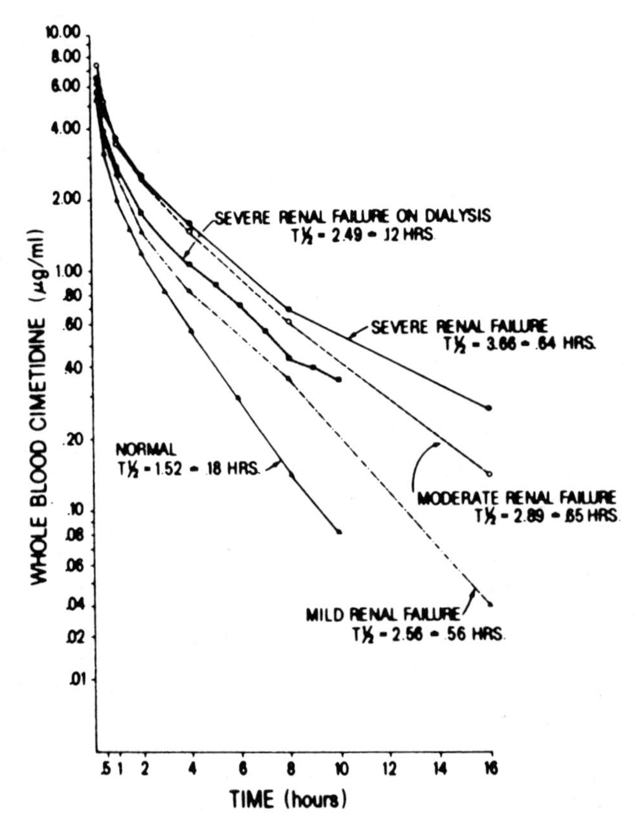

Figure 6. Blood cimetidine fall-off curves showing mean values for normal subjects and patients with varying degrees of renal failure including those on hemodialysis. The half-life ($T_{1/2}$) in hours is indicated for each group. There is a significant difference between normal controls and each group with renal insufficiency ($P < 0.001$), between those with mild and severe renal failure ($P < 0.01$) and between those with severe renal failure studied during and off hemodialysis ($P < 0.005$). (Reproduced with permission from the author and Gastroenterology, ref. 49.)

Furosemide

Furosemide kinetics were studied in normal volunteers and patients with renal failure. Comparison of results from intravenous bolus and intravenous infusion in normal subjects showed no significant model dependency of estimations of furosemide clearance, although the average clearance by fitting to a one-compartment model was 16% higher than that obtained by fitting to a two-compartment model. Normal subjects had a total body clearance of furosemide of 1.53 ± 0.11 (SE) ml/min/kg, a volume of the central compartment of 2.61 ± 0.37 liters, a volume of the peripheral compartment of 2.48 ± 0.24 liters, and a half-life of 0.8 ± 0.06 hour; they absorbed 68.9% ± 7.1% of a solution of furosemide given by mouth. The corresponding values in patients with renal failure were 0.27 ± 0.03 ml/min/kg, 8.02 ± 0.96 liters, 14.1 ± 3.57 liters, 14.2 ± 2.30 hours, and 43.4% ± 8.0%, all differing significantly from the normal subjects. The bioavailability of 500-mg tablets of furosemide in the patients with renal failure was 43.4% ± 7.4%, equivalent to the absorption of the same dose given to the same patients in the form of a solution (50).

Diazepam

Sex differences in the extent of binding of diazepam to plasma protein were assessed in a series of patients with renal insufficiency (51). Among identifiable independent variables, sex alone accounted for the greatest proportion of variability in percentage of unbound diazepam ($r = 0.33$), whereas age and serum creatinine concentration accounted for practically none. The mean (± SE) percentage of unbound diazepam in women (10.8 ± 2.8%) was larger (P ± 0.1) than in men (5.8 ± 1.3%). Since only the unbound fraction of diazepam in plasma is available for pharmacological activity, the intensity and duration of its clinical action in patients with renal insufficiency might differ between men and women.

Sulfonylureas

When the patient with diabetes mellitus develops renal failure due to Kimmelsteil-Wilson disease, he manifests more frequent hypoglycemic reactions, presumably due to decreased renal clearance of insulin. Consequently, his insulin level has to be reduced. Renal insufficiency is a factor that predisposes to hypoglycemic accidents in persons treated with hypoglycemic sulfonylureas. Renal insufficiency has little effect on the biotransformation of glibenclamide (glyburide). To determine to what extent the retention of the metabolites intervenes in such hypoglycemic accidents, rats with ligatured ureters received intraperitoneal injections of 1 mg/kg glibenclamide or hydroxy-glibenclamide (the main metabolite) or of saline. For each animal there was a control, which had undergone a simulated operation. For six rats with renal insufficiency, glibenclamide caused hypoglycemia of the same intensity as in the control group, but the hypoglycemia was more prolonged. With hydroxy-glibenclamide, the blood glucose level was significantly lower than in the control group. Hydroxy-glibenclamide has an obvious hypoglycemic activity, which represents one sixth that of the parent drug but 50 to 100 times that of tolbutamide. Its retention contributed to the intensification and prolongation of the hypoglycemic effect of glibenclamide in rats with renal insufficiency (52).

ABNORMALITIES OF SERUM AND SWEAT ELECTROLYTES

A number of investigations have been published analyzing the serum and sweat electrolytes in CRF. Studies were carried out in 10 normal persons and in 16 patients with moderately severe CRF to determine the quantities of potassium (K), sodium (Na), and water in muscle tissue obtained by needle biopsy and in white blood cells (WBC) from peripheral venous blood. Depletion of intracellular K with high levels of Na and normal water were found in patients with CRF. Therefore, the cellular electrolyte pattern was not substantially different from that reported by others in patients with advanced uremia although there was no increase in intracellular water (as can occur in endstage CRF). These data suggest that in endstage CRF, accumulation of intracellular water could be relatively independent of intracellular electrolyte balance (53).

Sweat collected from the forearms of patients with CRF and control patients, after iontophoretic stimulation with pilocarpine, was analyzed for Na, K, Cl, Mg, phosphate, and urea. Concentrations of Ca, Mg, and phosphate in sweat from patients with CRF were significantly elevated as compared with controls. Tentatively, it was concluded that the increase in Ca, Mg, and phosphate in uremic sweat is due to an increase in the secretion of these electrolytes in the secretory portion of the sweat gland, while the reabsorptive duct is normal (54).

OTHER MISCELLANEOUS ABNORMALITIES

Uric Acid

The patient with CRF usually manifests an increase in serum phosphorus and uric acid concentrations as renal failure becomes more advanced. However, hyperuricemia secondary to various disease states, such as leukemia, may cause renal failure; recent investigations studied this phenomenon. Acute uric acid nephropathy is a reversible type of renal failure that results from the deposition of uric acid crystals in the collecting tubules. The results of a number of laboratory tests were compared in 5 patients with a clinical diagnosis of this disorder and in 27 patients with acute renal failure due to other causes. Neither the serum creatinine, BUN, serum urate concentrations, nor the ratio of serum urate to BUN concentration differentiated between these two groups of patients. However, the ratio of uric acid to creatinine concentration on a random urine specimen did differentiate between these two patient populations. All patients with uric acid nephropathy had a ratio greater than 1.0, while all patients with other types of acute renal failure had ratios of less than 1.0 (55,56).

Phosphorus Metabolism-Further Considerations

It is important to reduce the elevated serum phosphorus level associated with CRF since hypocalcemia occurs and induces secondary hyperparathyroidism with resultant bone disease. Oral aluminum magnesium hydroxide antacid therapy reduces phosphorus absorption and lowers serum phosphorus. If the serum phosphorus concentration falls below 0.5 mg/dl, as can happen with aggressive antacid therapy and glucose hyperalimentation, erythrocyte ATP is markedly reduced and profound hemolysis may occur. The levels of 2,3 DPG in the erythrocytes may also be low, resulting in a shift of the oxygen dissociation curve to the left. Both of these important problems are reviewed in a recently published paper (57).

The control of hypophosphatemia is emphasized in a recent investigation with a detailed description of the therapeutic regime. Hypophosphatemia is a much more common condition than is generally recognized and may cause ill-defined disturbances in the course of illness in patients with renal insufficiency. The addition of 5 to 10 mmole of phosphate (155-310 mg of phosphorus) per 100 kcal of the nutrient solution at the start of parenteral nutrition in patients with chronic renal insufficiency is recommended. The phosphate dosage must be further increased, at least temporarily, in hypophosphatemic patients with acute renal insufficiency. Serum phosphate determination should be obligatory in patients with renal insufficiency when the patient is first seen. It should also be performed at least every second day during the first week of parenteral nutrition. Thereafter, the determination of serum phosphate twice weekly should suffice to control the dosage of phosphate required by the patient (58).

Magnification Phenomenon

The magnification phenomenon in CRF is analyzed especially as it relates to phosphorus and sodium balance. If a normal person with a GFR of 120 ml/min ingests 7 gm of salt (120 m*E*q of sodium), a nephron must excrete one-two hundredth of the sodium that it filters. By contrast, in a patient with severe uremia (2 ml/min) on the same salt intake, each nephron must excrete almost one third of its filtered sodium. In response to an identical perturbation of the extracellular fluid, the fraction of filtered sodium excreted must be 64 times greater by the uremic nephron than by the normal nephron. This transformation in nephron function in uremia constitutes the central and pivotal feature of the magnification phenomenon, which is defined as follows: *the addition or loss from extracellular fluid of any given amount of a substance that is actively regulated by the kidneys will evoke an excretory response per nephron that varies inversely with the number of surviving nephrons.*

The biological basis of the magnification phenomenon for phosphorus is as follows. Owing, at least in part, to preexisting chronic secondary hyperparathyroidism, phosphate reabsorption is substantially inhibited in the fasting state. Consequently, there is a sustained basal hyperphosphaturia per nephron. After ingestion and absorption of a fixed quantity of phosphorus, serum phosphate levels rise more in a uremic than in a normal person. Consequently, there is a greater reciprocal fall in ionized calcium, and PTH levels, although already supernormal, rise at least as much in uremia as in health for a given decrease in calcium. There also appears to be a distinct possibility that the sensitivity of the nephron to the phosphaturic effect of PTH is increased in uremia. Thus, the degree to which the inhibition of tubular reabsorption of phosphate is augmented from the control or fasting state will depend on the interplay between the rise in PTH, the number of PTH-receptor sites that are available, and the degree of responsitivity of the tubule to the hormone. The rise in the filtered load of phosphate per nephron that results from the sharp increase in serum phosphate levels contributes decisively to the magnified phosphaturia, for, owing to the hormonally mediated reduction in the capacity of the tubular epithelia to reabsorb phosphate, a high percentage of the additional filtered phosphate will escape reabsorption and enter the urine (59) (Figure 7).

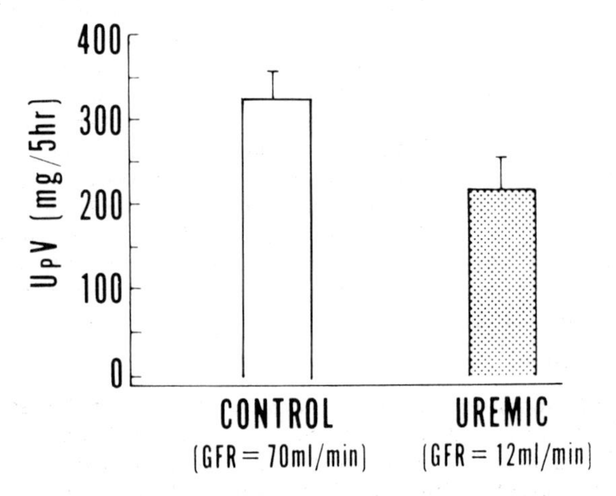

Figure 7. Urinary excretion of phosphate (UpV) in relation to glomerular filtration rate (GFR). (Reproduced with permission from the New England Journal of Medicine 299: 1287, 1978.

Renal Tubular Injury

Finally, important investigations relating to diagnostic tests to identify renal tubular injury have been published. Ligandin, an intracellular organic anion-binding protein, having glutathione-S-transferase activity, was detected in concentrated perfusing solutions from 8 of 13 kidneys preserved for homotransplantation. The presence of ligandin in the perfusate correlated well with oliguric acute renal failure following transplantation. Testing the perfusate for ligandin may be useful in predicting tubular damage in renal transplants (60).

A human proximal renal tubular epithelial antigen (designated HRTE-1) was isolated and purified from a crude tubular preparation by a process of salt fractionation, DEAE anion-exchange chromatography, and Sephadex G-200 gel filtration. Using

^{125}I-HRTE-1 and a rabbit antiserum specific for the proximal tubular brush border, as determined by immunofluorescent microscopy, a radioimmunoassay by competitive protein-binding was developed. HRTE-1 was demonstrated in serum and urine and in extracts of a variety of body organs. A range of concentrations for normal random urine samples and 24-hour urine excretion rates were determined. Random urine samples from 36 patients with a variety of functional and pathological renal disorders were assayed for the HRTE-1 antigen. Twenty-three of 24 patients with either chronic nephropathy or prerenal azotemia had normal urinary antigen concentrations, despite wide differences in urine flow rates, the degree of existing renal function, and the amount of proteinuria. Ten of 12 patients with acute tubular necrosis, however, had statistically abnormal HRTE-1 concentrations (high in 8 patients, undetectable in 2). These findings suggest that HRTE-1 antigen can be detected in both normal and pathological urines, that altered antigen concentrations can be documented in at least one renal disorder, and that quantitation of HRTE-1 in urine may have clinical value as a marker of acute tubular injury (61).

REFERENCES

1. Coolen RB, Herbstman R, Hermann P: *Clin Chem* 24:1636, 1978.
2. Aleyassine H, Tonks DB, Kay M: *Clin Chem* 24:492, 1978.
3. Vladutiu AO, Cunningham EE, Walshe J: *Clin Chem* 24:1084, 1978.
4. Chuga DJ, Bachner P: *Clin Chem* 24:1286, 1978.
5. Stepan J, Pilarova T, Melicharova D: *Acta Univ Carol (Med Monogr) (Praha)* 78(pt 2):75, 1977.
6. Kopple JD, et al: *Kidney Int (suppl 8):*S20, 1978.
7. Adler SN: *Arch Intern Med* 138:816, 1978.
8. Rabin EZ, et al: *Proc Eur Dial Transplant Assoc* 14:528, 1977.
9. Hofmann V, et al: *Schweiz Med Wochenschr* 108:1835, 1978.
10. Hemmelof Andersen KE: *Clin Nephrol* 8:510, 1977.
11. Muller-Wiefel DE, et al: *Eur J Pediatr* 128:103, 1978.
12. Wallner SF, Vautrin RM: *J Lab Clin Med* 92:363, 1978.
13. Wallner SF, et al: *J Lab Clin Med* 92:370, 1978.
14. Raich PC, Korst DR: *Arch Pathol Lab Med* 102:73, 1978.

15. Schulz E, Modder B, Fisher JW: *Contrib Nephrol* 13:69, 1978.

16. Vincent PC, et al: *Lancet* 2:864, 1978.

17. Ellis HA, Peart KM, Pierides AM: *J Clin Pathol* 30:960, 1977.

18. Bank N, Su WS, Aynedjian HS: *J Clin Invest* 61:884, 1978.

19. Stanbury SW: *Contrib Nephrol* 13:132, 1978.

20. Mennes P, et al: *Ann Intern Med* 88:206, 1978.

21. Christiansen C, et al: *Lancet* 2:700, 1978.

22. Berl T, et al: *Ann Intern Med* 88:774, 1978.

23. Massry SG, Ritz E: *Arch Intern Med* 138:853, 1978.

24. Lins LE: *Acta Med Scand* 203:309, 1978.

25. Graben N, et al: *Dtsch Med Wochenschr* 102:1903, 1977.

26. Verschoor L, Lammers R, Birkenhager JC: *Metabolism* 27:879, 1978.

27. Mordasini R, et al: *N Engl J Med* 297:1362, 1977.

28. Frank WM, et al: *Am J Clin Nutr* 31:1886, 1978.

29. Hovig T, et al: *Lab Invest* 38:540, 1978.

30. Ramirez G, et al: *Arch Intern Med* 138:267, 1978.

31. El-Bishti MM: *Am J Clin Nutr* 31:1865, 1978.

32. Stanton KG, Davis R, Richmond J: *Lancet* 1:100, 1978.

33. Rubenfeld S, Garber AJ: *J Clin Invest* 62:20, 1978.

34. Bianchi R, et al: *Am J Clin Nutr* 31:1615, 1978.

35. Kopple JD, et al: *Am J Clin Nutr* 31:1532, 1978.

36. Jones Mr, Kopple JD: *Am J Clin Nutr* 31:1660, 1978.

37. Heidland A, Kult J, Rockel A: *Am J Clin Nutr* 31:1784, 1978.

38. Varcoe AR, et al: *Am J Clin Nutr* 31:1601, 1978.

39. Mitch WE: *Am J Clin Nutr* 31:1594, 1978.

40. Leumann EP: *Helv Paediatr Acta* 33:25, 1978.

41. Mitch WE, Walser M: *Nephron* 21:248, 1978.

42. Vertagna X, et al: *J Clin Endocrinol Metab* 45:1179, 1977.

43. Lustenberger N, et al: *Contrib Nephrol* 13:115, 1978.

44. Kornerup HJ: *Scand J Clin Lab Invest* 38:147, 1978.

45. Sasaki Y, et al: *Contrib Nephrol* 9:35, 1978.

46. Carvalho JS, Page LB: *Nephron* 20:47, 1978.

47. Gorsky JE, Dietz AA: *Clin Chem* 24:1485, 1978.

48. Ueda H, et al: *Tohoku J Exp Med* 124:1, 1978.

49. Ma KW: *Gastroenterology* 74:473, 1978.

50. Tilstone WJ, Fine A: *Clin Pharmacol Ther* 23:644, 1978.

51. Greenblatt DJ, Harmatz JS, Shader RI: *Pharmacology* 16:26, 1978.

52. Loutan L: *Schweiz Med Wochenschr* 108:1782, 1978.

53. Montanari A, et al: *Clin Nephrol* 9:200, 1978.

54. Prompt CA, Quinton PM, Leeman CR: *Nephron* 20:4, 1978.

55. Kelton J, Kelley WN, Holmes EW: *Arch Intern Med* 138:612, 1978.

56. Ettinger DS, et al: *JAMA* 239:2472, 1978.

57. Laiken S, Beck Jr CH, Timms RM: *West J Med* 128:343, 1978.

58. Kleinberger G: *Wien Klin Wochenschr* 90:169, 1978.

59. Bricker NS, et al: *N Engl J Med* 299:1287, 1978.

60. Feinfeld DA, et al: *Nephron* 21:38, 1978.

61. Zager RA, Carpenter CB: *Kidney Int* 13:505, 1978.

4
Hepatobiliary Disease

Norman B. Javitt

The recognition, differential diagnosis, and management of diseases of the liver and biliary tract - - perhaps more than any other organ - - have traditionally been dependent on biochemical tests. Numerous tests have been proposed, and only occasionally does one find critical appraisals of their value. These too tend to be of limited scope, usually comparing the relative value of several tests. Perhaps the major criticism is that we fail to ask sufficiently precise questions. It is important to distinguish among tests that *(1)* determine if liver disease exists, *(2)* differentiate the type of liver disease, *(3)* evaluate the severity of the disease, and *(4)* may permit a statement about progression of the disease.

By using this classification, it is possible to focus more precisely on the true value(s) of a particular procedure. For instance, serum bile acid determinations, which are becoming quite popular, are much more valuable for the detection of liver disease than for classifying the type or extent. In contrast, determination of the serum albumin level and of the prothrombin time indicates the severity of the disease, but these tests are too insensitive to use for its detection.

This initial chapter on the biological methods for evaluating liver status will give some background for each of the specific tests that are reviewed and will attempt to put them in the perspective of their use as classified above and also in caring for patients. As the series continues, I hope to review other tests as they relate to these categories, keeping as the goal the minimum number of tests providing the greatest amount of information.

HEPATITIS A VIRUS (HAV-INFECTIOUS HEPATITIS)

Isolation of virus particles from the feces of a person with acute disease has led to the identification of an RNA enterovirus of the family Picornaveridae, namely, hepatitis B virus. This virus can infect the liver of chimpanzees and marmosets, and infection is followed by virus shedding in the stool in five to seven days and by a transient viremia. These events occur typically before there are clinical symptoms or elevations in the levels of serum alanine aminotransferase (GPT). Specific anti-HAV immunoglobulin M levels begin to rise early in the course of the disease and reach peak levels much earlier than the increase in anti-HAV immunoglobulin G levels. Recovery from illness is associated with a rapid decline in IgM levels and with a continuing increase in IgG levels (Fig. 1).

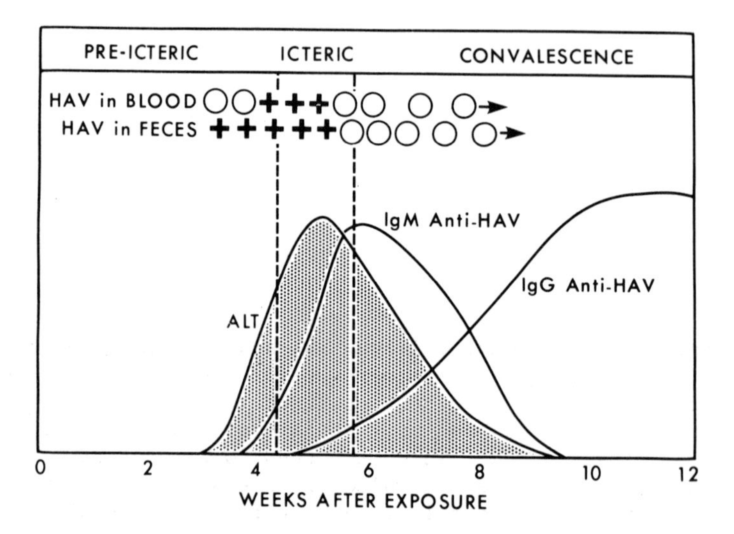

Figure 1. Time course of changes in serum ALT activity and HAV-specific immunoglobins in a typical case of uncomplicated type A viral hepatitis. (Reproduced with permission from Koff, ref. 42.)

Several radioimmunoassay techniques using HAV have been developed. In one procedure (1), polystyrene beads are coated with HAV virus obtained from primate liver and are added to an ^{125}I-IgG anti-HAV serum standard. Sera of patients to be tested are admixed with the ^{125}I-IgG serum before the addition of the polystyrene coated beads. The presence of IgG or IgM antibodies to HAV in the patient's serum will lower the binding of radioactive IgG, giving positive results.

IgG antibody can be distinguished from IgM antibody by using protein A, a membrane component of Staphyloccocus aureus cells that has a very high binding affinity for IgG. Prior absorption of the patient's serum on a suspension of staphyloccoccus aureus cells (Newman D_2C strain) will remove anti-HAV IgG but not IgM. Thus, a fall in titer after absorption indicates that mostly IgM antibody is present and that the illness is acute. The use of the differential absorption test permits the specific diagnosis of all instances of acute hepatitis A.

HEPATITIS B VIRUS

A number of antigenic components related specifically to the activity of the hepatitis B virus have been identified and are designated as: *(1)* surface antigen (Hb_sAg), *(2)* core antigen (Hb_cAg), and *(3)* E antigen (Hb_eAg). In addition, the virus possesses a DNA polymerase enzyme that can also be found in the serum of infected people. Tests for various components of the complete virus (Dane particle) relate to diagnosis and also to prognosis. In the typical acute illness with complete recovery, Hb_sAg appears transiently in serum, together with antibodies to the core antigen (anti-Hb_c). This stage is followed by a rising titer of antibody to the surface antigen (anti-Hb_s), which remains elevated for a variable period of time. Failure to make a complete immune response is associated with the persistence in serum of Hb_sAg, anti-Hb_c, and DNA polymerase activity. The existence of the carrier state is verified by, and the screening of donor blood is based on, sensitive tests for the presence of surface antigen. Some investigators have reported on people who have persistent disease and test negative for Hb_sAg but who are positive for anti-Hb_c. These findings are worrisome and have raised a question about the need for an anti-Hb_c screening test (2).

The detection of Hb_eAg in patients with persisting Hb_sAg in their serum is thought to correlate with progressive liver disease and, more likely, with transmission during pregnancy. Purification and partial characterization of HB_eAg indicates that it behaves like an IgG molecule in that it binds to protein A, has a molecular weight of 320,000 daltons, and contains two major polypeptides with molecular weights of 22,000 and 55,000 daltons, respectively (3).

The appearance of anti-Hb_e in the serum correlates with the disappearance of the carrier state and the achievement of a complete immune response. The sequence of antibody and antigen changes in a typical case of hepatitis B infection is shown in Figure 2.

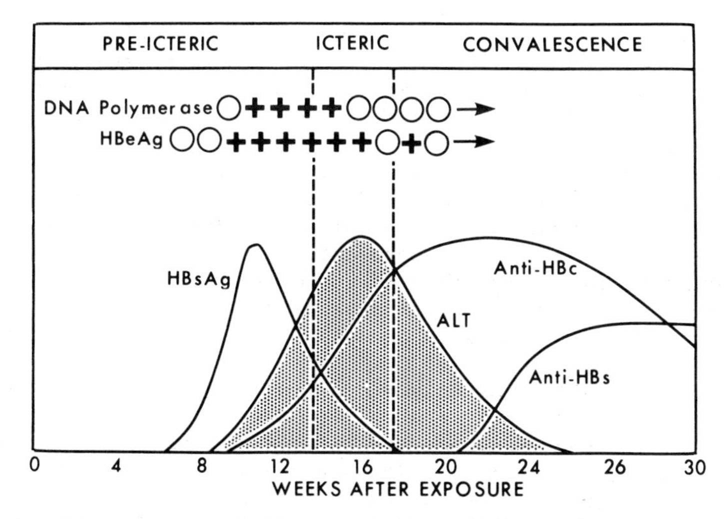

Figure 2. Time course of changes in hepatitis B viral components and antibodies compared with serum ALT activity in a typical case of uncomplicated type B viral hepatitis. (Reproduced with permission from Koff, ref. 42.)

γ-GLUTAMYL TRANSFERASE (GGT)

In the last several years a considerable amount of information has accumulated on parameters for evaluating the state of the liver; these data indicate that the estimation of the GGT activity in serum is one of the most useful for such evaluations. Indeed, it has been suggested that a knowledge of the serum activity of GPT, GGT, and cholinesterase can help to characterize virtually all types of liver disease (4). The enzyme, a glycoprotein, is found in high concentration in the human kidney and liver (5). Both organs yield enzymes with very similar kinetic properties, but these enzymes may differ considerably in their carbohydrate residues, since they exhibit very different binding properties to various lectins (6). Based on this analysis, the enzyme found in serum is considered to be derived mostly from liver.

Histochemically, the enzyme can be localized to the cell membrane. In the liver, for the most part, it is found in membranes bordering the biliary canaliculi and bile ductules. A number of studies indicate that the activity of the enzyme, as judged histochemically, increases in the liver when there is mechanical obstruction to bile flow or in response to the administration of drugs known to affect the generation of bile flow. Presumably, the increased amounts in the liver are reflected by increases in the amount in serum. This argument can be further supported by the inhibition of activity in both liver and serum by treatment with cycloheximide, a presumed inhibitor of protein synthesis (translation), before bile duct ligation.

Less is known about the location of the enzyme within the membrane and about how it is chemically and physically related to other membrane constituents. The need to use proteolytic enzymes and/or detergents in the preparation of a "solubilized" enzyme indicates that it exists in the membrane as a much larger complex. Part of this complex may represent a hydrophobic binding between the enzyme and lipid constituents of the membrane (5). Because of the methods used in isolating and characterizing the enzyme in tissues and biological fluids, some difficulty has been encountered in deciding whether the different electrophoretic mobilities and other properties represent evidence for isoenzymes or for different forms of the same enzyme. Huseby (6) has concluded that the multiple forms found in normal human serum, liver, and bile represent aggregates of the enzyme with lipids and proteins. From this interpretation it follows that the heterogeneity found in pathological sera may also represent the association of the enzyme with amounts or types of lipoproteins not normally found in serum (7).

Diagnostically, the enzyme appears most useful in the evaluation of patients with cholestasis, although it does not code for this problem as specifically as the 5'-nucleotidase or liver-specific hepatic alkaline phosphatase. Its advantage in recognizing cholestasis lies in the observation (4) that cholestasis can be present when the levels of the latter enzymes are still within the normal range, particularly in extrahepatic biliary tract disease and infiltrative liver disease (8,9). Of practical advantage is the ease of estimating GGT by automated methods, and the fact that in children older than two months (1) its level in the serum is not significantly different from the adult value implies that neither GGT nor 5'-nucleotidase reflect metabolic bone activity.

An important exception to the generalization that GGT is more sensitive in detecting cholestasis than either alkaline phosphatase or 5'-nucleotidase exists in relation to natural and synthetic estrogens (11,12). During pregnancy or oral contraceptive therapy, the serum enzyme level may not reflect cholestasis. In one study (12) 10 patients (men and women) with known extrahepatic obstruction and jaundice and with elevated serum levels of alkaline phosphatase and GGT were given 15 mg of esoestrol per day for seven days. In all patients, GGT activity in plasma diminished without significant change in other parameters. In two patients with initial GGT values of 48 and 90 mU/ml, respectively, the activity fell into the normal range (< 20 mU/ml).

Although serum levels of GGT are closely associated with cholestasis, it is important to recognize that elevated levels also indicate noncholestatic conditions, particularly alcohol consumption and other forms of continuing or chronic liver injury (13-16). It is not possible at present to provide a unified hypothesis to account for the elevated levels associated with cholestasis, moderate alcohol consumption, and chronic hepatitis. Based on histochemcial studies showing intense staining of fibroblasts in areas where connective tissue is forming (17), it is attractive to consider that persistently elevated GGT levels reflect the slow process by which the liver becomes cirrhotic in the presence of continuing inflammation, whether chemically induced or related to persistent viral infection.

Elevations in GGT levels are also known to occur in pancreatitis, diabetes, myocardial infarction, and congestive heart failure. Perhaps a final common denominator to account for these findings relates to coexistent biliary tract obstruction in instances of pancreatitis and diabetes and to mild chronic passive congestion of the liver in instances of heart disease.

Much more information needs to be delineated about the biological functions of the enzyme. Its usefulness in the detection of hepatobiliary disease should accelerate the rate at which such knowledge is accumulated.

SERUM BILE ACIDS

In 1957 Rudman and Kendall (18) indicated that elevations in the levels of the serum bile acids occurred in most types of liver disease. Because of the difficulty encountered in analyzing bile acids, further exploration of this observation did not occur for approximately 20 years. In the interim more has been learned about the enterohepatic circulation of bile acids, and simplified methods of enzymic and immunological analysis have become available (19-24).

A number of studies have confirmed the observation (25) that the test for the postprandial bile acid concentration in serum is more sensitive than that for the fasting bile acid concentration for the detection of liver disease (26-33), as shown in Figure 3. In concept, the test depends on the normal ileal transport of a normal-sized bile acid pool, which represents an endogenous infusion into the portal vein of 2,000-4,000 mg of bile acids during a brief period of time. Factors affecting these underlying assumptions, such as ileitis, can

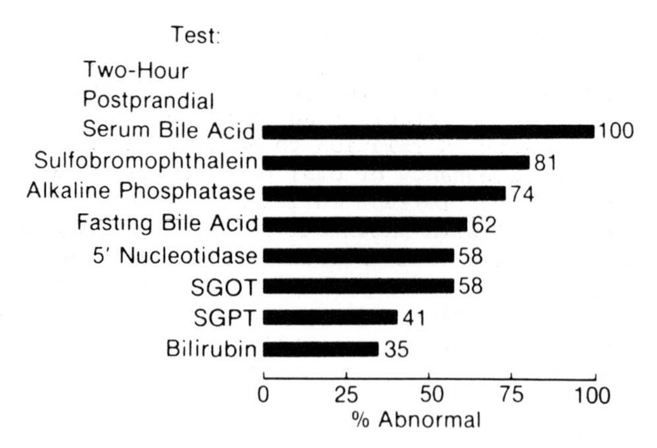

Figure 3. Frequency of abnormal biochemical tests in 26 patients with hepatobiliary diseases. Bar graph indicates the two-hour postprandial serum bile acid test was the only test abnormal in the entire group. When not tested for in a given patient, the sulfobromophthalein level was assumed to be abnormal (Reproduced with permission from Kaplowitz et al, ref. 25.)

give rise to false-negative results. In patients who have had a cholecystectomy, it is thought that the bile acid pool between meals is sequestered in the jejunum. Under these circumstances peak ileal absorption will occur earlier than two hours. Since inpatients with a portacaval shunt, even with a normally functioning liver, the bile acids must enter directly into the systemic circulation, the very high postprandial levels of bile acids in the serum do not reflect the state of liver function. However, barring these unusual circumstances, the postprandial bile-acid levels in serum reflect the extraction capacity of the liver for bile acids, which correlates well with the presence of liver disease.

From a clinical viewpoint, estimation of the total serum bile acid concentration will be useful much more often than estimation of the levels of the individual bile acid components. For this reason, immunoassay of individual bile acids is unnecessarily expensive. Also unless the immunoassay includes an estimation of serum chenodeoxycholate, the results may be misleading. The normal serum bile acid pattern shows a predominance of chenodeoxycholate compared with cholate (34). Minimal liver dysfunction may be associated with elevation in the chenodeoxycholate level, while the cholate level remains within the normal range, and, therefore, immunoassay of cholylglycine alone will not reflect hepatic deompensation. At these slightly elevated levels the chenodeoxycholate/cholate ratio does not necessarily have the same diagnostic implications as the reported ratios, which refer to much higher levels (35).

Patients with advanced cirrhosis are known to have chenodeoxycholate predominantly circulating in plasma. Typically, a patient with chronic active hepatitis that progresses to cirrhosis will show elevated postprandial serum bile-acid levels, in which the proportion of chenodeoxycholate to cholate progressively increases. Usually by the time the chenodeoxycholate level is greater than 70% of the total, there are other clinical, biochemical, or biopsy findings indicating liver disease. A knowledge of the proportions of bile acids present in serum does not usually provide any unique diagnostic or prognostic information.

Patients with cholestatic liver disease, caused by either benign or malignant obstruction of the bile ducts or by drugs in which hepatic cell necrosis is minimal or absent, have elevated serum bile-acid levels in which cholic acid is predominant. In instances where there is uncertainty about hepatic cell function, a predominant cholic acid pattern implies no severe underlying parenchymal cell disease. Some of these points are evident from consideration of the trihydroxy/dihydroxy bile-salt concentration ratio, shown in Figure 4.

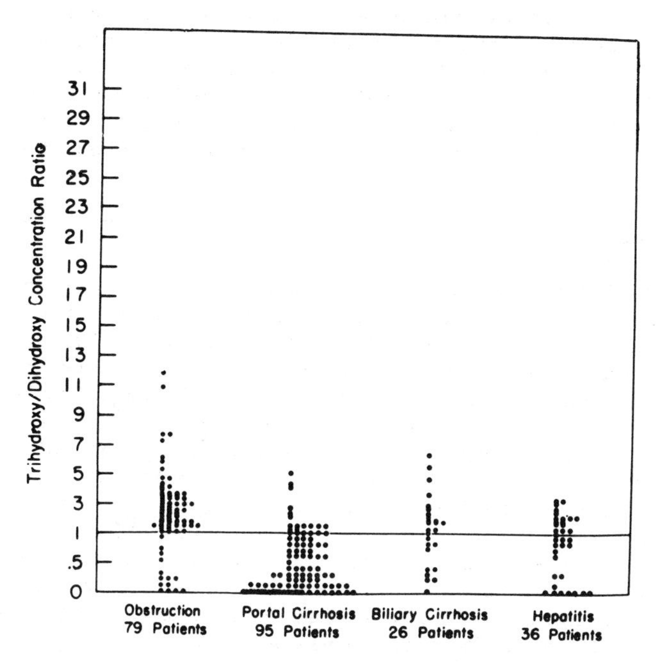

Figure 4. Distribution of values for the trihydroxy-dihydroxy bile-salt concentration ratios in serum for four major types of hepatobiliary disease. (Reproduced with permission from Javitt, ref. 35.)

As more data on serum bile-acid patterns in liver disease accumulate, it may be possible to relate different bile acid patterns more precisely to different diseases, but, most likely, knowledge of the total value will be most useful to the physician. Total serum levels estimated enzymatically represent the least expensive, most widely applicable analytical approach.

Enzymatic methods of analysis are still in the process of evolution; however, a standard kit is available (20). Other procedures for using 3α-steroid dehydrogenase have been described (19), and from the limited data available, it appears that the upper limits of normal for the postprandial test may be higher using enzymatic methods than more specific isotope dilution techniques (34). One problem is that too few patients have been studied to consider the results applicable to a large population. Using the isotope dilution method in five patients, a mean of 5.1 μM was obtained (24). One enzyme study using 3α-steroid dehydrogenase in seven patients gave an upper limit of normal of 20 μM (21). Since the isotope dilution technique estimated chenodeoxycholate, cholate, and deoxycholate levels, part of the difference can represent small amounts of ursodeoxycholate and perhaps other bile acids and hormones in serum that would be estimated by the enzyme. However, it is unlikely that these constituents can account for all the difference. Another possible source of error is that the 3α-steroid dehydrogenase enzyme may also have 7α-dehydrogenase activity. In this event each molecule of chenodeoxycholate and cholate may generate two molecules of NADH and give values twice as high as the actual molarity. Until all these differences are studied more carefully, some uncertainty will continue to exist about the upper limit of normal. Nevertheless, serum bile-acid levels occurring with hepatocytic disease are generally sufficiently higher than normal for the abnormal population to be easily distinguished, for the most part.

MISCELLANEOUS

The results of two studies, one on serum enzyme tests, and the other on serum immunoglobulins and autoantibody titer, have been published (36,37). Each study included more than 1,000 patients, and therefore each contained suitable sample size for reliable statistics. Determinations of 5'nucleotidase activity were again found to be more specific and sensitive than those of alkaline phosphatase in detecting hepatobiliary disease. A comparison with GGT was not made in this study; however, in another recent study, GGT activity was slightly more sensitive than the 5'nucleotidase activity in detecting known liver metastases (9), but only if one is willing to accept a higher incidence of false-positive results. In regard to the problem of liver metastases, perhaps, it is most accurate to say that if both the 5'nucleotidase and GGT levels are normal, it is unlikely (less than 4 per cent) that the patient has a metastasis. Other enzyme tests, that is, those for isocitrate dehydrogenase, guanase, and glutamate dehydrogenase, had no advantages over the GOT (aspartate-aminotransferase) and GPT tests. Adenosine deaminase determination may warrant further evaluation, since it is rarely elevated in cholangitis and cholecystitis but is elevated in hepatitis and, therefore, can be supportive of one diagnosis or the other in instances when the transaminase levels are increased but not enough to distinguish between hepatitis and biliary tract disease. Tables 1 and 2 summarize some of these findings.

In regard to immunoglobulins and immunological phenomena, it is important to note that patients without liver disease but with neoplastic disease and chronic respiratory disease can have a variety of serum antibodies, including those to mitochondria, as well as elevated levels of all the immunoglobulin classes. Diabetes mellitus can also be associated with similar abnormalities,except for the absence of antibodies to mitochondria. It seems clear that these types of analyses are not useful for the detection of liver disease but are

Table 1. Percentages of Patients with Diseases of the Liver and Biliary System with Abnormally High Serum Enzymatic Activities[a,b]

Disease Group[c]	*GOT*	*GPT*	*GOT/GPT > 1.0*	*ICDH*	*GDH*	*Guanase*	*AD*	*5NT*	*AP*
Gallstones	95	89	14	74	89	76	16	94	97
Carcinoma of head of pancreas	100	100	53	78	89	89	64	94	100
Carcinoma of bile duct	100	95	64	86	82	95	32	100	100
Intrahepatic cholestasis	86	76	35	34	31	69	34	66	79
Biliary inflammation	73	77	15	55	76	64	14	84	85
Portal cirrhosis	90	64	90	68	66	75	90	71	67
Primary biliary cirrhosis	100	96	72	78	100	96	87	100	100
Infectious hepatitis (early)	100	100	12	100	100	100	94	100	94
Infectious hepatitis (late)	93	100	10	82	86	89	93	68	82
Chronic active hepatitis	87	85	65	85	87	87	83	74	74
Alcoholic hepatic disease	78	75	79	70	81	56	24	41	48
Hepatic metastases	86	74	80	76	81	81	41	91	85

[a] Reproduced with permission from Ellis et al(36).

[b] GOT: aspartate aminotransferase, GPT: alanine aminotransferase, GOT/GPT: aspartate aminotransferase/ alanine aminotransferase ratio, ICDH: isocitrate dehydrogenase, GDH: glutamate dehydrogenase, AD: adenosine deaminase, 5NT: 5'nucleotidase, AP: alkaline phosphatase.

[c] Number of cases in each disease group listed in Table 3.

Table 2. Mean Values for Serum Enzymatic Activities in Patients with Diseases of the Liver and Biliary System[a,b,c]

Disease Group[d]	*GOT*	*GPT*	*ICDH*	*GDH*	*Guanase*	*AD*	*5NT*	*AP*
Gallstones	104	158	19	39	8.1	28	64	44
Carcinoma of head of pancreas	114	128	17	30	9.9	39	72	54
Carcinoma of bile duct	86	82	11	15	6.5	33	63	52
Intrahepatic cholestasis	75	74	12	14	4.1	32	17	27
Biliary inflammation	71	106	16	17	5.2	25	40	30
Portal cirrhosis	112	91	13	31	7.0	48	126	56
Infectious hepatitis (early)	1561	1869	130	45	87.3	66	33	28
Infectious hepatitis (late)	192	505	22	26	10.7	49	30	28
Chronic active hepatitis	301	245	25	13	15.7	58	34	26
Alcoholic hepatic disease	65	54	13	14	4.0	30	14	18
Hepatic metastases	111	80	23	22	8.3	37	66	45

[a] Reproduced with permission from Ellis et al (36).

[b] Mean values given as IU/l at 37°C, except for alkaline phosphatase, which is given as King Armstrong Units/dl.

[c] For definitions of abbreviations, see footnotes to Table 1.

[d] Number of cases in each disease group listed in Table 3.

helpful for further categorizing it. In this regard, abnormal results were found in every category of hepatobiliary disease from gallstones to alcoholic hepatic disease. The presence of an elevated value *per se* therefore, may not be very helpful in differential diagnosis. However, the absence of antibodies to mitochondria and a normal IgM level would make the diagnosis of primary biliary cirrhosis unlikely, since abnormal values occur in 83% and 96% of instances, respectively. Similarly, one would expect an elevated IgA level in 68% of the patients with portal cirrhosis. Data from this paper (37) appear in Tables 3 and 4.

A simplified method for the quantitative estimation of lipoprotein-X (LP-X) concentration has been published (38). It is based on the extractability into ether of the phospholipids and bilirubin associated with LP-X, in contrast to the nonextractability of phospholipids of other lipoproteins. Cetyltrimethylammonium bromide is added to the aqueous phase to enhance extraction into the ether phase. The extracted material can then be quantitated spectrophotometrically by analyzing for phosphorus or bilirubin. Although the estimation of LP-X concentration was initially considered reliable for distinguishing between intrahepatic and extrahepatic cholestasis, enough data have now accumulated to indicate that too much overlap occurs to permit a decision for surgery to be based on LP-X concentration in serum. However, the structure and morphological characteristics of the abnormal serum lipoprotein and the mechanism of its formation, release,

Table 3. Distribution of Patients with Diseases of the Liver and Biliary System and Percentages with Abnormally High Serum Immunoglobulin Concentrations and Antibody Titers[a,b]

Disease Group	*No. Cases*	*Percentage with Abnormal Values*					
		IgG	*IgA*	*IgM*	*ANF*	*AMA*	*SMA*
Gallstones	38	16	16	11	8	5	5
Carcinoma of head of pancreas	37	8	28	3	17	11	17
Carcinoma of bile duct	20	18	36	18	9	5	14
Intrahepatic cholestasis	27	7	14	28	10	0	7
Cholecystitis	25	3	7	7	7	7	10
Cholangitis	13	38	38	46	8	23	8
Portal cirrhosis	116	49	68	44	13	3	21
Primary biliary cirrhosis	24	61	35	96	0	83	26
Infectious hepatitis (early)	20	47	12	76	0	6	59
Infectious hepatitis (late)	39	11	0	64	0	7	46
Chronic active hepatitis	38	78	26	54	43	7	46
Alcoholic hepatic disease	59	16	27	22	5	5	6
Hepatic metastases	79	16	23	15	16	6	18

[a] Reproduced with permission from Ward et al (37).

[b] ANF: antinuclear factor, AMA: antimitochondrial antibodies, SMA: smooth-muscle antibodies.

Table 4. Mean Concentrations of Serum Immunoglobulins in Patients with Diseases of the Liver and Biliary System[a]

Disease Group (Reference Ranges)	*IgG (mg/dl) 500-1,600*	*IgA (mg/dl) 125-425*	*IgM (mg/dl) 47-170*
Gallstones	1,190	315	112
Carcinoma of head of pancreas	1,060	337	104
Carcinoma of bile duct	1,269	401	147
Intrahepatic cholestasis	1,156	269	145
Cholecystitis	1,103	285	99
Cholangitis	1,546	388	181
Portal cirrhosis	1,711	605	174
Primary biliary cirrhosis	1,987	399	667
Early infectious hepatitis	1,644	292	344
Late infectious hepatitis	1,279	207	361
Chronic active hepatitis	2,384	391	221
Alcoholic hepatic disease	1,204	364	127
Hepatic metastases	1,194	321	118

[a] Reproduced with permission from Ward et al (37).

[b] Number of cases in each disease group listed in Table 3.

and reshaping in plasma is an area of continuing interest (39-41).

REFERENCES

1. Bradley DW, et al: *J Clin Microbiol* 5:521, 1977.

2. Lander JJ, et al: *Vox Sanguinis* 34:77, 1978.

3. Fields HA, et al: *Infect Immunol* 20:792, 1978.

4. Schmidt E: In: *Evaluation of Liver Function* Demers LM, Shaw LM (eds): Urban & Schwarzenberg, Baltimore, pp 79-101, 1978.

5. Shaw L: In: *Evaluation of Liver Function* Demers LM, Shaw LM (eds) : Urban & Schwarzenberg, Baltimore, pp 103-121, 1978.

6. Huseby NE: *Biochim Biophys Acta* 522:354, 1978.

7. Freise J: *Clin Chim Acta* 73:267, 1976.

8. Kryszewski AJ, et al: *Clin Chim Acta* 47:175, 1973.

9. Kim NK, et al: *Clin Chem* 23:2034, 1977.

10. Richterich R, Cantz B: *Enzyme* 13:257, 1972.

11. Combes B, et al: *Gastroenterology* 72:271, 1977.

12. Frezza M, et al: *Gastroenterology* 74:800, 1978.

13. Jacobs WLW: *Clin Chim Acta* 38:419, 1972.

14. Rosalki SB, Rau D: *Clin Chim Acta* 39:41, 1972.

15. Lum G, Gambino RS: *Clin Chem* 18:358, 1972.

16. Miyazaki S, Okumura M: *Clin Chim Acta* 40:193, 1972.

17. Tamaka M: *Acta Path Jap* 24:641, 1974.

18. Rudman D, Kendall FE: *J Clin Invest* 36:530, 1957.

19. Siskos PA, Cahill PT, Javitt NB: *J Lipid Res* 18:666, 1977.

20. Fausa O: *Scand J Gastroenterol* 10:747, 1975.

21. Osuga T, et al: *Clin Chim Acta* 75:81, 1977.

22. Demers LM, Hepner G: *Clin Chem* 22:602, 1976.

23. Schwartz HP, Von Bergmann K, Paumgartner G: *Clin Chim Acta* 50:197, 1974.

24. Barnes PJ: *American Laboratory* May 1976, p 67.

25. Kaplowitz N, Kok E, Javitt NB: *Clin Chim Acta* 50:197, 1974.

26. Angelico M, Attile AF, Capocaccia L: *Amer J Dig Dis* 22:941, 1977.

27. Pennington CR, Ross PE, Bouchier IAD: *Gut* 18:903, 1977.

28. Barnes S, Gallo GA, Trash DB: *J Clin Path* 28:506, 1975.

29. Barbara L, et al: *Rendic Gastroenterol* 8:194, 1976.

30. Fausa O, Gjone E: *Scand J Gastroenterol* 11:537, 1976.

31. Stiehl L, et al: *Inn Med* 5:14, 1978.

32. Poupon RY, et al: *Gastroenterol Clin Biol* 2:475, 1978.

33. Morita T: *Gastroenterol Japonica* 13:491, 1978.

34. Angelin B, Bjorkheim I: *Gut* 18:606, 1977.

35. Javitt NB: In: *Diseases of the Liver* Schiff L (ed), Philadelphia, JB Lippincott, chapt 4, pp 111-137, 1975.

36. Ellis G, et al: *Amer J Clin Path* 70:248, 1978.

37. Ward AM, Ellis G, Goldberg DM: *Amer J Clin Path* 70:352, 1978.

38. Talafant E, Tovarek J: *Clin Chim Acta* 88:215, 1978.

39. Hauser H, et al: *Biochim Biophys Acta* 489:247, 1977.

40. Sauar J, et al: *Clin Chim Acta* 88:461, 1978.

41. Fellin R, et al: *Clin Chim Acta* 85:41, 1978.

42. Koff RS: *Viral Hepatitis*. New York, John Wiley, 1978.

5 Diseases of the Gastrointestinal System

Gordon G. Forstner

THE STOMACH AND DUODENUM

Peptic Ulcer

Gastric Analysis

Gastric acid secretion is usually assessed in response to a meal or to one of a number of potent secretagogues such as histamine, gastrin, or insulin. Of the two approaches, measurement in response to a meal presents the most difficulties, not only because there is no definition of the ideal meal but also because the variable rate of gastric emptying alters the duration and strength of the stimulus and creates the opportunity for unmeasured acid loss into the duodenum. Nevertheless, food is the physiological stimulus for acid secretion. The integrated response that it provokes may be of much greater use in distinguishing between patients than the response to a single exogenous hormone. Taylor et al (1) have made an interesting contribution to this field by comparing acid response to a standard Oxo meal with the maximal acid output in response to pentagastrin in eight healthy volunteers and in nine patients with chronic duodenal ulceration. After an overnight fast and following instillation of the Oxo meal by means of a double-lumen gastric tube, total acid output was measured over a period of one hour at three-minute intervals by titration to pH 5.5 with sodium bicarbonate. Since this technique, first introduced by Fordtran and Walsh (2), relies on achieving adequate neutralization before acid can escape into the duodenum, patients had to be rotated frequently from side to side, and the left hypochondrium had to be compressed intermittently throughout the procedure to ensure adequate mixing. In the healthy volunteers the acid response to the Oxo meal was greater than the response to pentagastrin, but there were no differences in the duodenal ulcer patients. These findings are at variance with the response to the steak, bread, and butter meal of Fordtran and Walsh (2), possibly because of the different stimulus, but they are more consistent, in the authors' point of view, with an increased endogenous secretory drive in duodenal ulcer patients. The liquid Oxo meal is probably manipulated with greater ease than the solid meal and, therefore, is likely to be used more widely, but it is worth stressing that the stimulus may provoke a different response.

Duodenal ulcer patients as a group appear to have greater acid secretion than normal subjects, both at rest and after secretory stimulation (3). Gross et al (4) examined the possibility that an impairment of fat-induced inhibition of meal-stimulated gastric acid secretion could account for the poststimulatory increment in the patients with duodenal ulcer. Ten normal controls and 10 patients with chronic but nonsymptomatic duodenal ulcer disease were studied. The subjects received both a carbohydrate meal and a fat meal, each containing 50 gm of peptone, on separate, randomly ordered days. In contrast to the studies of Taylor and coworkers (1) a phenol-red marker was added immediately after the instillation of the meal and was mixed rapidly by tubal suction and expulsion. Gastric residual volume was calculated at 20-minute intervals by a serial dilution method (5), and these calculations, along with the results of acid titration and analysis of gastric pepsin, were used to determine gastric acid and pepsin secretion and duodenal acid and peptic loads (6) during a 140-minute period. Although the fat meal produced significant inhibition of both acid and pepsin secretion and duodenal load, results were the same in both subject groups, even though, in separate studies, the duodenal ulcer patients had greater responses to betazole and pentagastrin. Interestingly, the serum gastrin response was reduced by fat in the normal subjects but not in the patients with duodenal ulcer. The authors suggest that the inhibited secretion of acid and pepsin seen with the fat meal in both groups probably occurs independently of gastrin release, possibly by means of the release of gastric inhibitory polypeptide (GIP). These two papers (1,4) do not exhaust the considerable ingenuity displayed by investigators studying gastric secretion in patients (see also ref. 11). In an equally interesting study Schoon et al (7) examined the response of six normal persons and six duodenal ulcer patients to graded gastric antrum distension by 50-, 100- and 150-ml balloons, in the presence of a submaximal stimulus of continuously infused pentagastrin. In the normal subjects acid response was inhibited by 20% with the 150-ml balloon, but this stimulus had no effect in the duodenal ulcer patients. The results of these and similar studies are not as straight forward as one might wish since antral distension alone produces an increase in acid output in duodenal ulcer patients, which is independent of gastrin release (8), but they suggest that an inhibitory mechanism, affecting gastrin-stimulated secretion, predominates in healthy subjects and is defective in the duodenal ulcer patients.

From the point of view of routine gastric secretory analysis in clinical laboratories, these studies remind us that gastric secretion in response to food is a net response to a number of complex reflex and hormonal adjustments. These can only be dissected by specifically designed tests of function. The histolog- or pentagastrin-secretion techniques most commonly in use produce only one of these responses, and perhaps not the one of greatest clinical value.

Bile Reflux Post Surgery

After gastric surgery many patients experience abdominal discomfort, which is associated with regurgitation and is relieved by the vomiting of bile. According to Hoare et al (9,10) measurement of the fasting bile reflux (FBR) accurately distinguishes between symptomatic and asymptomatic postgastrectomy patients and is of value in predicting which patients with dyspepsia will respond to bile diversion surgery. To perform the test a nasogastric tube was passed into the stomach of fasting patients, and its position was checked fluoroscopically and by a water recovery test. Stomach contents were aspirated and discarded. Subsequently, three consecutive 10-minute aspirates were collected, and their bile-salt content was estimated. If all three samples were satisfactory, the results were expressed cumulatively as μmoles of bile salt refluxed per hour. Initially, 14 patients were studied, seven of whom were asymptomatic and seven of whom had dyspepsia relieved by vomiting bile. Fasting bile acid reflux was compared with bile acid reflux during stimulation with pentagastrin (60 minutes), after stimulation by a test meal (60 minutes), and during continuous overnight suction. Only the FBR distinguished between the symptomatic and asymptomatic groups. In a larger group of 20 asymptomatic patients all of whom had an FBR test, FBR values less than 120 μmole/hr were invariably found. In 17 of 22 symptomatic patients, FBR values were greater than 120 μmole/hr. All patients with high FBR values complained of bile regurgitation and vomiting. Endoscopic and radiological evidence of reflux was a less useful discriminant since reflux was seen in a number of asymptomatic patients. After an FBR test 16 patients underwent a bile diversion procedure consisting of an interposed Roux-en-Y loop (10). Eleven of the patients had FBR values of greater than 120 μmole/hr. In these, epigastric pain was abolished significantly more often than in the patients with lower values. Bile regurgitation was invariably improved in patients for whom it had been a problem before surgery. These papers are not totally satisfactory because of the relatively subjective nature of abdominal pain and discomfort. Patients with mild aching pain or continuous pain relieved by vomiting did best, while patients with other types of pain, particularly colicky lower abdominal pain, did poorly. Nevertheless, the FBR test introduces a quantitative, objective note into a very subjective field. If others have success with it, many patients would benefit, either from earlier surgery or from the avoidance of surgery where no benefit will result.

Zollinger-Ellison Syndrome

Most patients with Zollinger-Ellison syndrome (ZES) have continuously elevated serum gastrin levels, and their basal gastric acid secretory rate (BAO) is just about equal to their maximal secretory rate in response to histamine or pentagastrin. None of these facts is surprising, since the gastrinomas responsible for the syndrome might be expected to secrete gastrin at an autonomous and high rate at all times. Most surprising, however, is the observation that a mixed meal of steak, bread, ice cream, chocolate syrup, and butter should rapidly, if only transiently, reduce gastric acid output. Malagelada (11) measured gastric acid and pepsin output in six patients with ZES after insertion of double-lumen gastric and duodenal tubes. A ^{14}C-polyethylene glycol solution was constantly perfused via the proximal duodenal orifice and was collected at the distal orifice near the angle of Treitz to facilitate quantitative estimation of acid dumped into the duodenum. The homogenized mixed meal (pH 6.0) was delivered intragastrically over a 10-minute period, and gastric and duodenal samples were aspirated over a period of three hours. Blood samples were taken regularly for serum gastrin estimations, which remained unaltered during the procedure. In four patients with high BAOs, the gastric secretory output decreased below fasting levels, returning to basal levels over a one- to two-hour period. In two patients with relatively low BAOs, secretory output increased moderately. Two patients were restudied after removal of a duodenal gastrinoma and both exhibited a normal gastric secretory rise. Apparently there are inhibitory regulatory mechanisms triggered by food (antral distention, GIP), which take precedence over gastric drive in this pathological situation.

Recently, it has become apparent that cimetidine, the H_2-receptor antagonist, can be used to treat the ZES in lieu of gastrectomy. The effectiveness of this agent in peptic ulcer disease has been monitored principally by relief of symptoms or by the rate of ulcer healing. Straus et al (12) demonstrated that plasma secretin levels can be used to provide objective evidence of drug effectiveness. A 51-year-old man with ZES was studied. Basal plasma gastrin averaged 350 ± 20 pg/ml and plasma secretin, 300 ± 50 pg/ml (normal $<$ 20 pg/ml). After a standard test meal the plasma secretin level dropped to less than 20 pg/ml and remained suppressed for 80 minutes. BAO and plasma secretin levels were determined two and eight hours after oral doses of 300 mg, 600 mg, and 900 mg of cimetidine. At two hours, the results indicated that the suppression of acid secretion and the fall in plasma secretin concentration were dose dependent, requiring 900 mg of cimetidine before secretin reached normal levels. Plasma secretin concentration fell linearly with the fall in acid secretion (Fig. 1). The test depends on the fact that secretin release is proportional to the amount of gastric acid reaching the duodenum. The exciting hope is that the test may be applied to patients with duodenal ulcer as well as those with the relatively rare ZES. We must await the availability of secretin assays with the sensitivity to detect physiological levels of the hormone. Fortunately, these are beginning to appear (13,14).

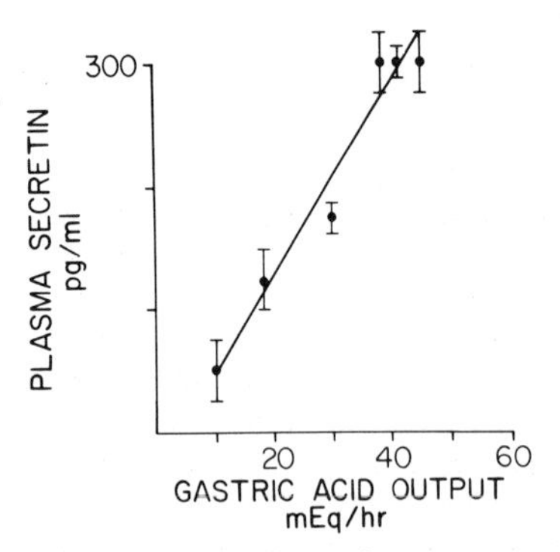

Figure 1. Mean (± SEM) plasma-secretin as a function of gastric acid output in patient with Zollinger-Ellison syndrome. (Reproduced with permission from Straus et al, ref. 12)

Plasma and Serum Gastrin

Serum Gastrin in Duodenal Ulcer

Serum gastrin emanates principally from the G cells of the gastric antrum, although the duodenum may also make a significant contribution (15). G cells can be estimated immunohistochemically in surgical specimens. Although antral G-cell density is increased in pernicious anemia (16), acromegaly (17), and hyperparathyroidism (17), this does not seem invariably to be the case for duodenal ulcer disease (18). However, two groups of investigators have recently provided evidence that suggests that the total G-cell population is increased in some patients with duodenal ulcer, correlating reasonably well with peak acid output to pentagastrin (19) and gastrin response to a meal (20). The fasting serum gastrin level also correlated with antral G-cell mass in patients with duodenal ulcer but not if accompanied by uremia (20). The results may have been influenced by the fact that the kidneys are a major site of gastrin degradation (21).

Gastrin in Uremia

Hallgren et al (22) suggested that there is a dependence of serum gastrin levels on glomerular filtration rate (GFR). They studied 70 patients with chronically reduced GFR as estimated by ^{51}Cr-EDTA clearance, creatinine clearance and β_2-microglobulin values. Figure 2 illustrates some of their data, demonstrating that, although scatter was large, basal serum gastrin values tended to increase as GFR fell. In 10% of a group of 31 patients on maintenance dialysis serum gastrin rose to levels commonly associated with the ZES.

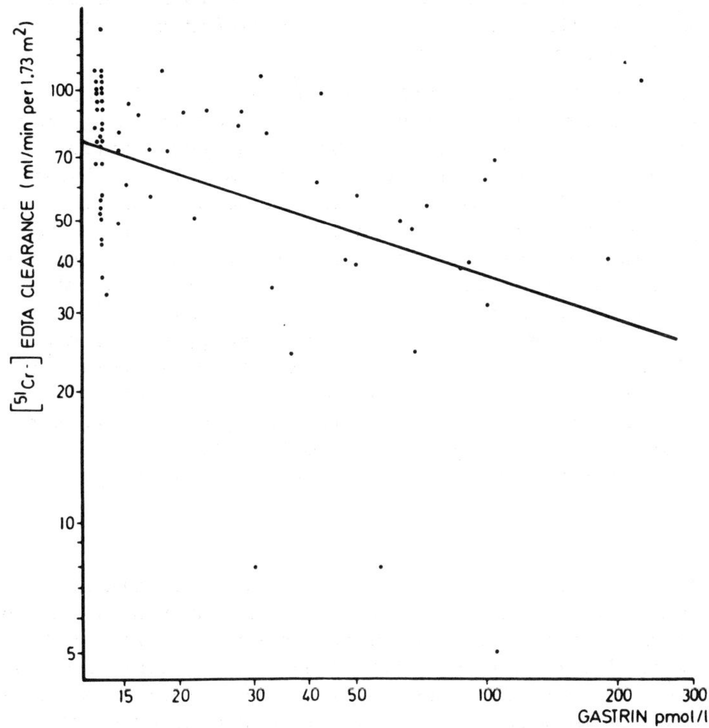

Figure 2. Basal serum gastrin in patients with chronically reduced GFR. (Reproduced with permission from Hallgren R, Karlsson FA, Lundqvist G, ref. 22, with permission of the Editor of Gut.)

Gastrin in Hypercortisonism

Plasma gastrin levels may also increase in response to long-term exposure to excess glucocorticoids according to the work of investigators at Kobe University in Japan (23). Fasting plasma gastrin levels were determined in 45 normal subjects, in 11 normal subjects given oral prednisolone (30 mg/day) for three days, in 10 patients who had received more than 500 mg of prednisolone or the equivalent over a period of at least three weeks, and in 6 patients with Cushing syndrome. All subjects were adults and the age range (17 to 58 years) was approximately the same for the control subjects and patients. Fasting gastrin levels were the same in the control subjects and in those receiving a short-term course of prednisolone, but both the long-term therapy group and the patients with Cushing syndrome had significantly elevated levels. Response to a test meal was determined in six control subjects, in six patients on short-term therapy, and in six patients on long-term therapy. In all three groups plasma gastrin levels rose approximately threefold at 30 and 60 minutes. Basal and stimulated levels in the control and short-term therapy groups were not significantly different, but both values were significantly higher in the patients on long-term therapy. In normal subjects infusion of 500 mg of hydrocortisone for one hour had no effect on plasma gastrin levels. At the moment there is no certain explanation for these results, but they may be of significance in the causation of steroid ulcers.

Gastrin in Idiopathic Hypertrophic Pyloric Stenosis (IHPS)

The interest in serum gastrin levels in this condition stems from the fact that injecting pregnant bitches with pentagastrin stimulates the development of hypertrophic pyloric stenosis in the offspring (24). Bleicher et al (25) obtained fasting serum from 20 infants with IHPS preoperatively, from 5 vomiting infants without IHPS, from 18 normal infants hospitalized for elective procedures, and from 20 normal adults. The mean gastric level in the normal infants was 103 ± 9 pg/ml (± SEM), similar to that of the vomiting infants (93 ± 9 pg/ml), whereas levels in the infants with pyloric stenosis were significantly higher (256 ± 26 pg/ml). Adult levels were considerably lower (28 ± 5 pg/ml). Previous workers had reported both normal and raised gastrin levels in IHPS. Bleicher and his coworkers suggested that some of the differences may reflect the length of the fast, which was relatively long (12 hours) in this study.

MICROASSAY FOR PEPSIN

Furihata et al (26) have described a very sensitive fluorescent assay for pepsin based on the use of a nonquenching substrate, succinylated albumin. Primary amines, liberated by proteolysis, were detected by complexing with fluorescamine. The assay was 200 times more sensitive than the micromethod of Anson (27). The principal applications at the moment lie in measuring the pepsinogen content of the fetal stomach and of stomach tumors or in purifying minor pepsinogen isoenzymes.

THE PANCREAS

Assessment of Exocrine Pancreatic Function

The singular peculiarity of the pancreas is its extreme reserve capacity. More than 90% of the gland must be destroyed before digestion is sufficiently impaired to cause the malabsorption of fat, carbohydrate, or protein. The only reliable way to probe the gland's reserve capacity is to stimulate it maximally with secretin and cholecystokinin, or with food, while collecting pancreatic secretions via a tube placed in the duodenum. Preferably collection should be absolute, and the ideal method for ensuring this is to perfuse the duodenal area with a solution containing a nonabsorbable marker to correct for losses at a distal collection site. Gastric secretions must be kept from the duodenum by gastric suction during the collection. The technique is not easy, nor is the patient particularly comfortable. In practice one avoids the procedure except when malabsorption or the presence of a disease such as chronic pancreatitis or cystic fibrosis suggests a relatively high likelihood of functional impairment. The result is that the true incidence of moderate pancreatic impairment is poorly understood. A simple screening procedure, providing a more sensitive index of pancreatic failure than the appearance of fat in the stool or diarrhea, is badly needed.

The Chymotrypsin-Labile Peptide

Several recent studies (28-30) continue to suggest that the oral administration of a peptide, N-benzoyl-L-tyrosyl-p-aminobenzoic acid (Bz-Ty-PABA), which is cleaved by chymotrypsin liberating p-aminobenzoic acid (PABA), might provide a useful basis for a suitable screening test. The principle, first enunciated by Imondi et al (31) is quite simple. After cleavage, PABA is absorbed, conjugated in the liver, and excreted almost quantitatively in the urine where it can be measured conveniently as an aromatic amine (32). If, as the result of inadequate delivery of chymotrypsin, Bz-Ty-PABA is not cleaved, PABA excretion will be low. Excretion does not appear to be appreciably affected by hepatic and renal disease (33). Certain drugs, such as sulfonamides, thiazides, and chloramphenicol, and certain foods, such as prunes and cranberries, may interfere with the assay in urine (34). Impaired absorption as a result of rapid intestinal transit, or accelerated digestion by bacterial chymotrypsin-like proteases in the intestinal lumen, might possibly alter the excretory pattern, but these are not likely to affect most studies.

Immamura et al (18) compared the results of a pancreozymin-secretin test with the Bz-Ty-PABA test in a large number of adults, including 4 healthy volunteers, 10 diabetic patients, 9 patients with assorted gastrointestinal problems, and 15 patients with suspected or proven chronic pancreatitis. The dose of Bz-Ty-PABA was 0.5 gm, containing 169 mg of PABA. The PABA excreted in eight hours was expressed as a percentage of the administered dose. PABA excretion correlated significantly ($P < 0.01$) with the volume, maximal bicarbonate concentration, and total amylase output of the stimulated pancreatic juice. In a larger group of 14 healthy control subjects PABA excretion always exceeded 69%, whereas 12 of the patients in the chronic pancreatitis group excreted less than 66%, with a range of 21 to 66%. Of a group of 12 patients with either a subnormal maximum bicarbonate concentration or low amylase output, 10 excreted less than 66% PABA, whereas of 26 patients with apparently normal pancreatic function, 4 fell within this range. Three of the 4 patients giving apparently false-positive results had diabetes mellitus. These studies are particularly important because they suggest that the Bz-Ty-PABA test will detect most patients with pancreatic insufficiency, even of moderate degree. Unfortunately, fat balance data were not accumulated and, therefore, it is not clear how many of the group with pancreatic insufficiency actually represented patients with normal fat excretion but diminished pancreatic reserve.

Sacher et al (29) examined the usefulness of the oral test in infants and children. A dose equivalent to 15 mg of PABA per kg of body weight was administered with breakfast or with a morning bottle, and six hours later the percentage of PABA recovered in the urine was determined. In 20 children, aged 5 months to 16 years, with various extraintestinal and extrarenal problems, mean PABA excretion was 58.5% ± 11.2% (± SD), with a lower limit of normal (-2 SD) of 36%. In children with cystic fibrosis (CF) and pancreatic insufficiency, aged seven weeks to 15 years, the six-hour excretion was 5.5 ± 5.3%. In nine healthy newborn infants, however, the mean excretion was 23.4% ± 17.7%, and in eight of the infants the value was less than 36% In four instances the value was sufficiently low to overlap with that of the CF infants. In all the newborn infants fecal chymotrypsin activity was within the normal range, suggesting that fecal chymotrypsin determinations may be more useful in delineating pancreatic insufficiency in this age group. These results obviously indicate that caution must be exercised in interpreting the results of the Bz-Ty-PABA test in infants. In a group of older children consisting of 24 patients with CF, aged 2 to 19 years, and 14 control subjects aged 3 to 18 years Nousia-Arvanitakis and colleagues (30) found that low PABA excretion sharply delineated the CF patients with pancreatic insufficiency from the control subjects. Therefore, the test appears to be reasonably reliable in children over the age of 3 years. Thus far, only Sacher et al (29) appear to have given Bz-Ty-PABA with a meal, thereby providing an additional stimulus for pancreatic secretion. Whether the routine addition of an intraluminal stimulus would extend the useful range of the test is an issue that deserves investigation.

Fecal Chymotrypsin

We all use this test. Its usefulness in screening for conditions such as CF with major impairment of pancreatic function is well established and, as noted above, it may still be the best screening procedure in infants (29). Durr et al (35) address themselves to the crucial question of its diagnostic value as a screening test in patients with minor impairment of pancreatic function. They compared results of chymotrypsin activity expressed as IU-gm in randomly collected fecal specimens with the results of a secretin-cholecystokinin test. Four groups of patients were delineated on the basis of the secretin-cholecystokinin test: *(a)* established insufficiency, all enzymes low (18 patients); *(b)* borderline cases, at least one enzyme low (7 patients); *(c)* low normal, all enzymes between -1 and -2 SD below normal (5 patients); and *(d)* normal function (50 patients). Fecal chymotrypsin activity was normal in all borderline cases, in all but one of low-normal cases, and in 5 of the 18 established cases of insufficiency. Seven percent of the patients with normal function had low chymotrypsin activity. On the other hand, as shown in Figure 3, stool chymotrypsin correlated well with decidedly low pancreatic output. The

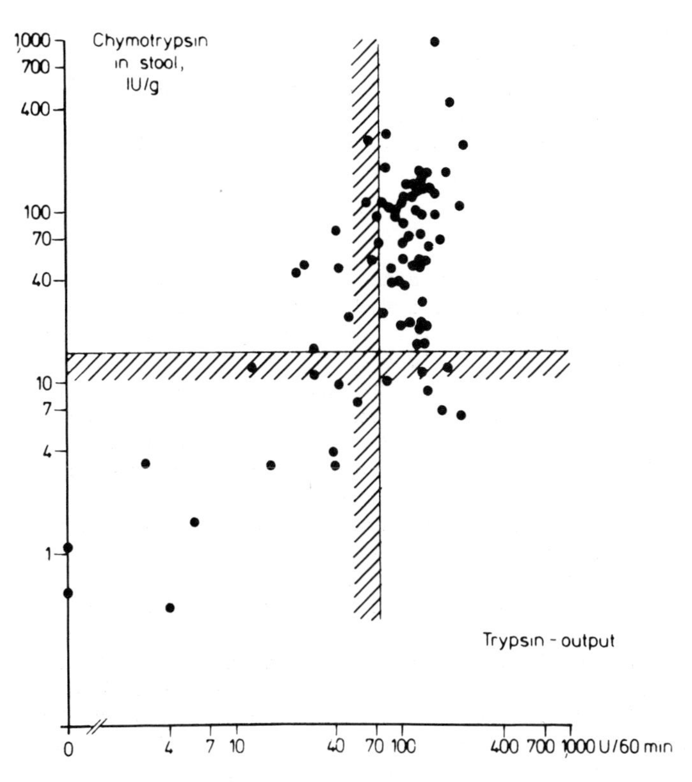

Figure 3. Correlation between trypsin outputs in duodenal juice and chymotrypsin activities in fecal specimens. Hatched lines indicate lower limits of normal. (Reproduced from Dürr HK, Otte M, Forell MM, Bode JC: *Digestion* 17:404-409, 1978, by permission of S. Karger AG, Basel.)

conclusion seems to be quite clear. The fecal chymotrypsin test is a simple procedure that will detect most cases of very poor pancreatic function, but it is not very sensitive.

Chronic Pancreatitis

Lactoferrin Test

Fedail et al (36) presented a novel observation which may ultimately be of use in the diagnosis of chronic pancreatitis. Forty patients, 19 with chronic pancreatitis as judged by retrograde pancreatography, secretin or Lundh meal stimulation test, 10 with carcinoma of the pancreas, and 11 control patients with other problems were studied by retrograde cannulation of the pancreatic duct. The concentration of lactoferrin in pancreatic juice was determined by a radioimmunoassay using pure human milk lactoferrin as the antigen. The control subjects and all but one patient with carcinoma had lactoferrin levels of less than 500 ng/ml, while all patients with chronic pancreatitis had higher values ranging from 800 to 12,000 ng/ml. Lactoferrin is an iron-binding protein found in milk and a wide variety of exocrine secretions. It is also found in leukocytes, which might explain its presence in an inflammatory condition. At any rate the fact that in chronic pancreatitis, the lactoferrin level is elevated, whereas the levels of most elements of pancreatic juice are reduced, could make its estimation of considerable confirmatory value in the diagnosis. This is an initial report, and one hopes that the observation will be quickly examined by other investigators.

Cystic Fibrosis (CF)

Sweat Test

The sweat test is the only practical and reliable laboratory test for CF. The standard version, first described by Gibson and Cook (37), depends on the stimulation of sweat using pilocarpine iontophoresis and on the analysis of a minimum of 50 mg of sweat, carefully collected on a sealed gauze pad, by microchemical testing for sodium or chloride. The test requires considerable technical experience and is somewhat time consuming, so there is a great incentive to develop alternative, simpler procedures for wider use. Instrumentation for measuring the chloride concentration of sweat by direct application of a chloride ion electrode to sweating skin or for measuring the electrical conductivity of sweat is used in many laboratories, particularly where sweat testing is seldom performed. Each year papers appear that support the diagnostic or screening value of one of these instruments, and 1978 was no exception (38-40). In experienced hands and under controlled laboratory conditions there is little doubt that they perform well. Rosenstein et al (41), from the John Hopkins School of Medicine add an important cautionary warning, however. Among 234 patients referred to their laboratory for testing by the Gibson-Cooke method over a period of 21 months, 62 had had a previous test; the results of the test in 29 were reported as negative and in 33 as positive. Repeat testing led to a change of diagnosis in 43.5% of these cases, of which there were 25 false-positive diagnoses and 2 false-negative diagnoses. Of these 27 patients, a chloride electrode had been used previously on 17 patients and electrical conductivity on 1. This number of false diagnoses over a short period of time is truly alarming. Instrumental problems may be responsible in part, but Rosenstein and his coworkers noted that, in laboratories where few of these tests are carried out, failure to confirm positive test results or to repeat tests with borderline results and mistakes in interpretation are probably of equal importance. The apparent conclusion that one can draw from this study is that sweat testing must be carried out in a central laboratory where technicians are experienced, and where quality control is continually maintained, and where results are interpeted by physicians experienced with CF.

Meconium Screening

Meconium from newborn infants with CF contains a high concentration of albumin (42), which offers the possibility for country-wide newborn screening. The advantages of such a program are obvious, since CF is an exceedingly common genetic disease, and early detection would pay substantial dividends in terms of genetic counselling and early care. Widespread screening is now accomplished by using albumin-sensitive test strips or sensitive immunochemical assays for albumin. As with sweat chloride analyses, sensitivity and precision seem to deteriorate when testing is done in a number of hospitals in a community (43). These problems are circumvented to a large extent by centralized testing of meconium samples transported from surrounding nurseries. Of the two methods currently available for estimating albumin, immunochemical approaches appear to be preferable, since they are more accurate and less subjective than are the test strip methods. Schattinger and Zinterhofer (44) have reported on their experience with a nephelometric quantitative, automated, immunoprecipitation assay for meconium albumin, which seems to be particularly well suited to the accurate and relatively inexpensive processing of large numbers of samples. Instrumentation consisted of a sampler, proportioning pump, a fluoronephelometer, and a strip chart recorder, all of which can be used for other purposes. Antihuman albumin prepared from goats and diluted 1/40 in a solution of polyethylene glycol was interacted with meconium supernate. The light scattered by the immunoprecipitate was measured and compared with an albumin standard curve. Sample blanks can be used as required but were used routinely only when initial results were elevated. The detection limit was 0.2 mg of albumin/per ml of meconium, with reasonable linearity to 60 mg of albumin per ml of meconium. An imprecision of approximately 8.5% was found for day-to-day variation at albumin concentrations of 4.6-14.9 mg/ml. The upper limit of the normal range was 3.8 mg of albumin per ml. The major sources of false-positive results were blood in the meconium, which was easily tested by checking each elevated sample for occult blood, and prematurity, which was present in 9 of 11 cases with albumin elevations of undetermined origin.

Circulating Pancreatic Enzymes

Amylase

In 1976 Pierre et al (45) described the analysis of α-amylase activity using the following reaction sequence:

$$\text{soluble starch} + H_2O \xrightarrow{\alpha\text{-amylase}} \text{maltose} + \text{malto-oligosaccharides}$$

$$\text{maltose} + \text{orthophosphate} \xrightarrow{\text{maltosephosphorylase}} \text{glucose} \pm \beta\text{-D-glucose-1-P}$$

$$\beta\text{-D-glucose-1-P} \xrightarrow{\beta\text{-phosphoglucomutase}} \text{glucose-6-P}$$

$$\text{glucose-6-P} + \text{NAD} \xrightarrow{\text{glucose-6-P dehydrogenase}} \text{6-phosphogluconate} + \text{NADH}$$

This coupling system is used in the Beckman kit for α-amylase (Enzymatic Amylase Reagent, Beckman Instruments Inc.), with the addition of a further conversion of 6-phosphogluconate to ribulose-5-phosphate to yield another mole of NADH. Every new assay introduces new problems, it seems. Hansen and Yasmineh (46) confirmed the results of Alexander (47) and of Van Wersch et al (48) that the reaction rate produced by the kit is not linear, having an initial lag phase of 10 minutes. By substituting glycogen for soluble starch, the lag phase was reduced to 3 minutes, and the subsequent reaction rate was linear. Linearity was also obtained when glycogen was added to the starch reagent, so the kit reagents could be used without modification. In one patient a dialyzable inhibitor was found that affected the Beckman assay, but not when amylase was assayed with the Phadebas reagent. The authors suggested that endogenous keto and diketo acids might be responsible. This technique has the great advantage that endogenous glucose does not interfere with it. It is quite sensitive and easy to perform, and it is probable that most of its problems will be easily ironed out. Nevertheless, one hesitates to recommend it as a replacement for the currently used chromogenic and dye-starch methods until further trials are carried out.

Trypsinogen
Human trypsin exists in two forms, anionic and cationic, of which the anionic form accounts for approximately one third of the total activatable trypsinogen in pancreatic secretions. A paper by Largman et al (49) describes a specific radioimmunoassay for the anionic enzyme and demonstrates its presence in normal serum. Binding by α_1-antitrypsin and α_2-macroglobulin was prevented by inactivating the radioactive trypsin tracer used in the assay by tosyl-L-lysine chloromethylketone. Almost all the measurable circulating antigen appeared to be anionic trypsinogen. The normal concentration was 5.45 ± 1.3 ng/ml ($\pm$ SD).

Acute Pancreatitis

Amylase-Creatinine Clearance Ratios
Farrar and Calkins (50) studied 29 patients in whom a diagnosis of acute pancreatitis was made by the criteria of serum amylase level above 200 Somogyi units per ml, clinical presentation and course, and exclusion of other causes of abdominal pain. The serum amylase level, the one-hour urinary amylase excretion and the amylase-creatinine clearance (C_{am}/C_{cr}) ratio were determined on several occasions from the time of admission until at least eight days post admission. The one-hour urinary amylase excretion was elevated in 21 patients initially and the C_{am}/C_{cr} in 17. On the days when the patients experienced abdominal pain, 69% of the serum amylase levels and 74% of the one-hour urinary amylase excretions were abnormal, whereas only 33% of the C_{am}/C_{cr} were elevated. The C_{am}/C_{cr} was abnormal on only two of 34 occasions when a one-hour urinary amylase excretion was normal, whereas the one-hour urinary amylase excretion was abnormal on 15 of 47 occasions when the C_{am}/C_{cr} was normal. The mean serum amylase level and the one-hour urinary amylase excretion were elevated for a more prolonged period than the C_{am}/C_{cr} after hospitalization. The authors concluded that the C_{am}/C_{cr} estimation lacks the sensitivity of the serum amylase and one-hour urinary amylase determinations in patients with acute pancreatitis.

The C_{am}/C_{cr} has been promoted as an extremely sensitive and accurate indicator of pancreatitis, but reservations such as those of Farrar and Calkins (50) are beginning to appear. Specificity is also at issue as Levitt and Johnson pointed out in an excellent editorial (51). The problem is that it is extremely difficult to make the case for diagnostic specificity when a reliable, independent method for confirming the diagnosis does not exist. Since most patients with pancreatitis improve, histological confirmation is rarely available, and there is unfortunately no clinical sign that provides unequivocal evidence of the disease. As was pointed out in a thoughtful case conference from Washington University (52) there are many other causes of hyperamylasemia. At times the serum amylase level may be normal at the height of an attack, particularly in patients with hyperlipemia.

Pancreatic Neoplasia

Diagnostic Tests

Di Magno and colleagues (53) from the Mayo Clinic have made an important contribution to the field of diagnostic testing for gastrointestinal diseases by studying 70 patients with suspected pancreatic cancer on a prospective basis to compare the predictive accuracy of a battery of tests commonly used in clarifying pancreatic function. The tests included a cholecystokinin stimulation test, duodenal cytological studies, ultrasonography, pancreatic scan, thermography, arteriography, and endoscopic retrograde pancreatography. All tests were performed in more than 60 of the 70 cases, except for retrograde pancreatography, which was performed in 52. Thirty of the patients were found to have pancreatic cancer and seven had pancreatitis. The cholecystokinin stimulation test and ultrasonography were the most sensitive and specific tests for detecting pancreatic disease. Retrograde pancreatography and arteriography were much superior to cytological studies in diagnosing cancer. Pancreatic scanning and thermography were not of great value in either area. The authors suggested that ultrasonography should be performed on initial suspicion. If the results of the ultrasonogram are negative, a pancreatic function test should be performed. If the results of either test are positive, retrograde pancreatography should be carried out. This combination of tests resulted in about 90% accuracy for pancreatic disease and cancer These results are particularly interesting because they reinforce the importance of the pancreatic function test in the early screening of patients with suspected pancreatic tumors. The procedure will often alert one to the presence of small neoplasms, which will not be evident on the ultrasonogram but which are nevertheless capable of blocking the pancreatic duct.

Somatostatinoma

These D-cell tumors are not common but initial reports of proven cases suggest that they produce an interesting constellation of biochemical findings because of the ability of somatostatin to inhibit the release of so many hormones. Ganda and colleagues (54) have described the case of a 46-year-old woman who showed clinical evidence of hyperglycemia and a pancreatic tumor. An oral glucose tolerance test and an arginine infusion test produced serum insulin, plasma glucagon, and growth hormone responses typical of diabetes mellitus. After removal of the tumor, which contained large amounts of somatostatin, the clinical evidence of diabetes disappeared, suggesting that the hyperglycemia was probably initiated by impaired release of insulin. Elsewhere, Larsson et al (55) have described a patient with achloryhydria, steatorrhea, and a somatostatin-producing tumor. Schlegel et al (56) have demonstrated very nicely that the steatorrhea could easily be due to inhibition of cholecystokinin release. In five healthy volunteers, aged 19 to 29, 40 ml of olive oil was introduced into the duodenum on two occasions, and the serum CCK-PZ levels were determined by radioimmunoassay. An intravenous infusion of somatostatin on the second occasion completely abolished the rise in the serum CCK-PZ concentration induced by the olive oil (Fig. 4).

Two good reviews of the use of laboratory procedures in the diagnosis of pancreatic disease have appeared during 1978 (57,58).

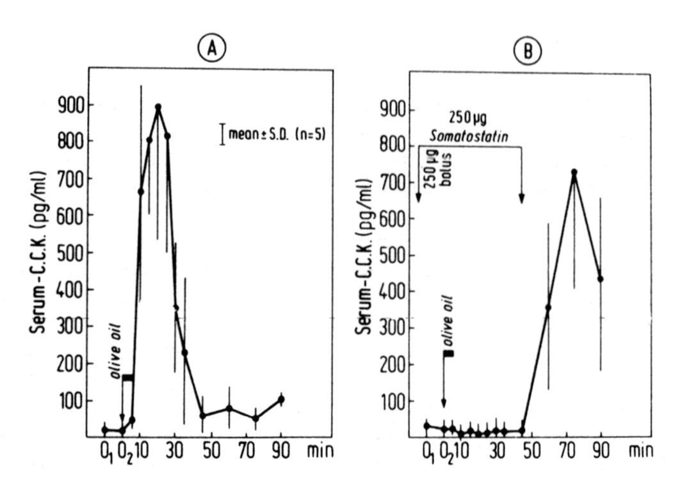

Figure 4. Effect of intraduodenal olive oil on serum-CCK *(A)*, and complete abolition of this response by intravenous infusion of somatostatin *(B)* (Reproduced with permission from Schlegel et al, ref. 56.)

SMALL INTESTINE AND COLON

Coeliac Disease

Leukocyte-Migration-Inhibition-Factor (LIF) Assay
Cases of coeliac disease continue to be overlooked in spite of the great interest in the problem and the generally high level of physician awareness, because the degree of malabsorption is not always such as to promote attention, and the clinical signs of wasting, anemia, and osteomalacia, appear rather late. A screening test that is more specific for the disease would be of great clinical benefit since it might allow clinicians to select those patients with minimal clinical signs who would not otherwise be candidates for a duodenal biopsy. Ashkenazi and colleagues from Jerusalem have presented initial evidence that a LIF assay could fulfill this function (59). The assay depends on the observation that gluten fractions stimulate the lymphocytes of patients with coeliac disease to produce an LIF. Frazer's fractions B_2 and B_3 from gluten were prepared and mixed with a cell preparation consisting of measured aliquots of lymphocytes and polymorphonuclear leukocytes from patients and controls. Leukocyte migration was measured by an agarose droplet technique, in which migration could be assessed over 24 hours by means of an inverted microscope with an ocular scale. Fifty-five patients with biopsy-proven coeliac disease, of whom 18 were on a normal diet and 37 on a gluten-free diet, and 32 control patients were studied. The gluten fractions B_2 and B_3 both produced significantly greater inhibition of leukocyte migration in the cells from the patients with coeliac disease, but for each there was considerable overlap with control values. However, when the highest inhibitory values for either fraction were used, discrimination was improved, so that overlapping values were confined to two or three cases. The mean inhibition for patients was 24 ± 6% (± 1 SD) and for controls 5 ± 6% (± SD). LIF assays probably reflect the development of delayed hypersensitivity. In most of the patients

tested in this study, the diagnosis of coeliac disease had been established previously, and the control subjects were free of chronic intestinal disease. It is thus not clear how well the test might perform early in the development of the disease or in distinguishing between coeliac disease and other gastrointestinal problems. The early results are nevertheless quite impressive, and it is hoped that others will soon seek to verify them.

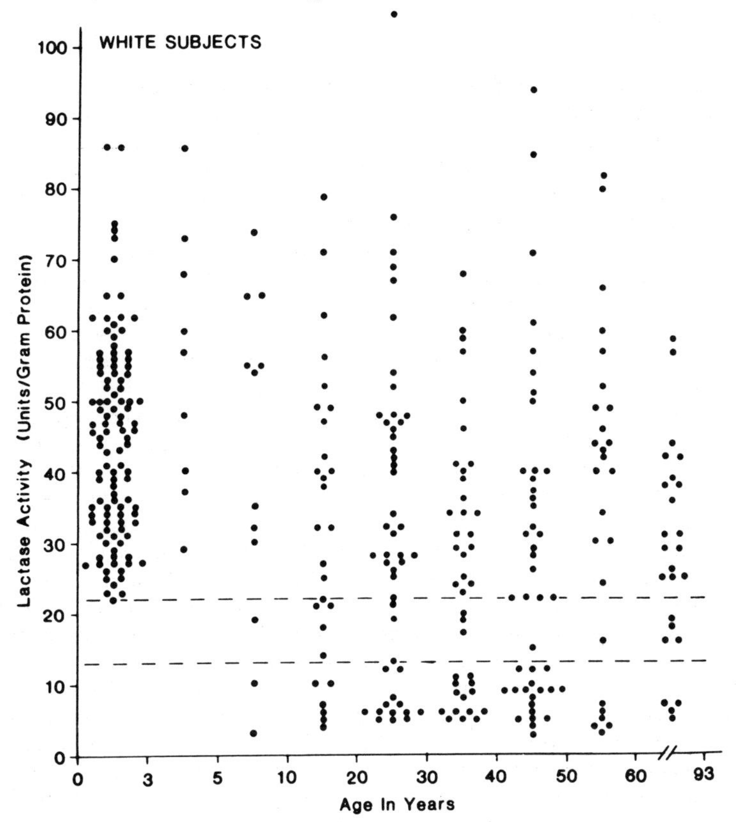

Figure 5. Lactase activity in normal small intestinal mucosa from 339 white subjects between the ages of 2 months and 93 years. The dotted lines designate values considered low (< 13 U) and intermediate (< 21.9 U). (Reproduced with permission from Welsh JD, Poley JR, Bhatia M, Stevenson DE: *Gastroenterology* 75:847-855, © 1978, American Gastroenterological Society.)

Disaccharidase Deficiency

Lactase Deficiency

Figure 5 illustrates the range of lactase activity in 339 white subjects, aged 2 months to 93 years, as reported by Welsh and colleagues from Oklahoma and New Jersey (60). The conclusions seem quite clear: lactase deficiency does not develop in whites before 5 years of age. Similar data were obtained from 55 biopsy specimens from blacks. Again lactase levels were normal at birth and, somewhere between the ages of 3 and 10 years the levels fell in those patients who later developed lactose intolerance. In this study there were no age-related changes in the activities of other disaccharidases or alkaline phosphatase. In 13 patients with secondary disaccharidase deficiency who had repeated biopsies, the emergence of primary low intestinal lactase activity was related to age and race rather than mucosal damage. The results imply that acquired deficiency is probably not related to initial mucosal damage.

Breath Tests

Intestinal biopsies are difficult to perform and not without risk. The assay of disaccharidase activities in biopsy specimens is also relatively time consuming. The possibility that a simple measurement of expired hydrogen after a disaccharide meal might identify patients with disaccharidase deficiencies is therefore attractive. Intestinal bacteria are the principal, if not the only, sources of respiratory hydrogen (61). When fed carbohydrate, enterobacteria respond by producing large quantities of H_2 through fermentative processes. Of course, since the bulk of these bacteria normally reside in the colon and most of the dietary carbohydrate is absorbed before the chyme passes the ileocecal valve, respiratory H_2 production after meals is relatively low. In patients with disaccharidase deficiency or other forms of sugar intolerance, this happy state of affairs ceases to exist and, as a result, sugar entering the colon causes the increased production of H_2, which makes breath testing feasible. Clearly, other variables must be considered in evaluating results, such as the possibility of bacterial overgrowth in the upper intestine where unabsorbed sugar is normally present or alteration of colonic flora by therapy (see below), but these issues do not invalidate the potential usefulness of the breath tests in large-scale population screening. The major problems to date have been technical. Gas-liquid chromatographic (GLC) detectors are not very sensitive, and investigators have had to resort to determining the concentration of H_2 expired by rebreathing in a closed circuit or by repetitious end-respiratory sampling. These collection techniques pose problems for young children, who, except for patients with acquired lactase deficiency, constitute the population that is most likely to develop disaccharide intolerance. Several recent papers have suggested solutions to this problem. Douwes et al (62) from Rotterdam have described a sampling device consisting of a curved perspex tube with an anesthetic mask at the proximal end. The mask is fitted with a low-resistance outlet valve, and the tube is fitted with a one-way valve for collecting expired air, in addition to two gas-tight stopcocks to close off the expired volume at 25 ml. The mask is placed over the child's nose and mouth, and collections are usually complete at the end of two to four expirations. Whether all children will accept this indignity is open to question, but the authors stated that reasonable results were obtained when two collections were made, and the highest result was selected. In five lactose-intolerant children, aged 4 to 11 years, results correlated well with a rebreathing collection method. Perman et al (63) have advanced an alternative method, which seems to have the benefits of extreme simplicity and probable acceptability. Breath samples were collected by means of a nasal prong consisting of 3-5 cm of silastic tubing, attached by a longer 9-cm segment of tubing to a syringe connector. While the patient breathed normally, the nostril without the tubing was occluded by finger pressure, and 3- to 5-ml expiratory fractions from the intubated nostril were aspirated during late expiration into a 50-ml plastic syringe. Results were obtained with sucrose in six children with sucrase-isomaltase deficiency, aged 6 to 11 years, and in 16 sucrose-tolerant control patients, aged four months to 16 years. The sucrose-intolerant patients achieved peak H_2 excretion values of 114 ± 6.3 (± SD) ppm above baseline compared with 2.4 ± 3.6 ppm above baseline for the control patients. Best discrimination occurred 90 minutes after the sucrose feeding (Fig. 6). Perhaps the most significant advance in this area, at least in potential, has been presented however in a paper by Payne-Bose and associates (64). A technique is described in which a GLC thermal conductivity detector, an electronic peak integrator, and a sample pressurizing pump were combined to increase the sensitivity of H_2 detection to the level where it could be used on whole human breath collected in a single bag.

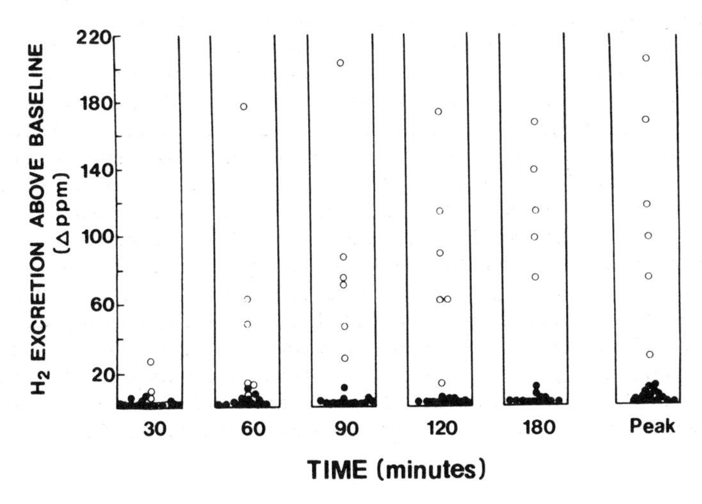

Figure 6. Breath hydrogen excretion following a 2 gm/kg (maximum 50 gm) sucrose load. ●: tolerant patients, o: intolerant patients. (Reproduced with permission from Perman JA, Barr RG, Watkins JB: *J Pediatr* 93:17-22, 1978.)

The only disturbing note in this scenario has been provided by the work of Gilat and colleagues (65), who studied end-expiratory air after feeding the poorly absorbed sugar lactulose to 18 patients prepared for colonoscopy by laxatives and enemas, to 12 patients prepared for colonic surgery with neomycin and enemas, to 17 patients on a variety of antibiotics, and to 15 control subjects. The laxatives and the antibiotic routines both greatly reduced H_2 production. When control values and post-treatment values obtained one to six weeks after cessation of therapy were evaluated, 11 of the 55 subjects failed to produce significant amounts of H_2. Although nonhydrogen producers are known, the incidence has been thought to be less than 4% of the population (66). The 20% incidence reported in this study suggests that a great deal more work will be needed to assess the recovery of colonic flora after bouts of diarrhea, or after the use of laxatives and antibiotics, before general population surveys will have value.

Stool Disaccharides

In children with suspected disaccharide intolerance, the best approach to the diagnosis is to demonstrate the presence of excess sugar in the stools. At the bedside, reducing substances can be detected with a Clinitest tablet, before and after acid hydrolysis. In most instances this test is sufficient for the purposes of initial treatment, since dietary sugar can be denied until diarrhea subsides, and a trial of lactose, sucrose, or even glucose-free formulas can be instituted during recovery. Even though this procedure might strike one as perfectly satisfactory from the point of view of management, it is nevertheless true that a diagnosis could be made with much greater precision if there were a simple, good technique for measuring individual disaccharides in stool. In some instances clinical management would be greatly improved by this knowledge. In susceptible populations with a high incidence of intolerance, stool disaccharide analyses could also provide useful screening data. The contribution of Thompson and coworkers (67) in developing a relatively simple GLC technique for measuring stool disaccharides is therefore significant. Stool supernatants are mixed with a trehalose standard and are then dried and permethylated under nitrogen. The products are then extracted in chloroform and applied to GLC. Inadequate permethylation occurred in samples with large amounts of disaccharide but was remedied by repeated permethylation of diluted samples. The lowest level of detectability for disaccharides was 5-10 mg/100 gm of feces, which is sufficient to detect sucrose and occasionally lactose in healthy controls. Sucrose, maltose, lactose, and cellobiose were determined directly without interference from extraneous peaks. The technique seems accurate, sensitive, and relatively quick. Clinical laboratories in hospitals in which a relatively large number of cases of childhood diarrhea are seen might find this technique quite useful.

Bacterial Overgrowth

Clinical Value of Bile Acid Breath Tests

In humans, the amide bond of glycine-conjugated bile acids is cleaved exclusively by bacterial enzymes, and the glycine moiety, after its liberation, is promptly converted to CO_2, which is expired. Normally most of the conjugated bile acid in the small intestine is absorbed in the ileum without cleavage. When small intestine is contaminated by bacteria, however, conjugates are split off in significant quantities. In such a situation, after administration of ^{14}C-cholyl glycine, the specific activity of the $^{14}CO_2$ expired will increase greatly. This is the principle of the bile acid breath test, and since its introduction in 1971 (68,69), many reports have supported its clinical usefulness. Investigators at the Mayo Clinic (70), where one of the initial studies was performed, reviewed their retrospective experience with 200 bile acid breath tests. When results were compared with cultures from small intestinal aspirates in 43 patients who had had both procedures, the test exhibited a sensitivity of 0.7 and a specificity of 0.9. According to analyses using Bayes' theorem, the test appeared to double the probability with which the clinician could be certain of the presence or absence of bacterial overgrowth. On the other hand the test was insensitive and rarely useful in patients suspected of malabsorbing bile acids. A 30% false-negative and a 10% false-positive record is not very impressive, but this probably represents the worst case for the test, since bacterial overgrowth is not always easily verified by intestinal cultures. Certainly, in comparison with duodenal or jejunal intubation, the test is quite simple, and it will continue to be useful in guiding clinical wisdom. A negative result should never be used to exclude the possibility of bacterial overgrowth, however.

Protein-Calorie Malnutrition

Somatomedin and Growth Hormone

It is a striking fact that patients with malnutrition do not respond to correction of caloric and protein intake by an immediate increase in growth and weight. Continuing malabsorption due to slow recovery of pancreatic and intestinal functions may be one issue and redistribution of body water another, but it seems likely that there are a number of hormonal adjustments that also lag. A study undertaken in Thailand (71) explores the possibility of a malfunctioning growth hormone-somatomedin axis. Twenty-seven patients with marasmus, marasmus-kwashiorkor, and kwashiorkor, ranging in age from 8 months to 60 months, were studied during the treatment and recovery phase for a period of 50 days; 21 well-nourished children served as controls. Plasma growth hormone was assayed by immunoassay and plasma somatomedin by a bioassay that depends on its stimulation of $^{35}SO_4$ incorporation by porcine costal cartilage (72). Fasting heparinized blood was taken at 2, 8, 29, and 50 days. The somatomedin level was significantly depressed on days 2 and 8 but was not different from that of the controls on days 29 and 50, whereas growth hormone levels were high on days 2 and 8 and subsequently fell. These results suggest a functional block in the synthesis and/or release of somatomedin in cases of malnutrition. Somatomedin is a peptide that appears to mediate many of the activities of growth hormone. It is probably released in response to growth hormone, and the levels in plasma are generally parallel to those of growth hormone. The inverse relationship found in the early recovery stage in these infants with malnutrition certainly suggests that the growth hormone response may be impaired.

Vegetarians

Nutritional Deficiency in a Breast-Fed Infant

Higginbottom and colleagues (73) have described the case of a four-month-old boy who was admitted to the hospital in coma. The mother was an apparently healthy 26-year-old vegan who had eaten no animal products for eight years and had taken no supplemental vitamins. The baby was exclusively breast-fed and appeared to be developing normally until the age of four months, when loss of head control, decreased vocalization, and lethargy were noted. Over the month before admission he fed poorly, was often irritable, and bruised easily. Scattered ecchymoses were noted on his legs and buttocks, and there was hyperpigmentation of the extremities. His hemoglobin level was 5.4 gm/dl, and his white cell, reticulocyte and platelet counts were

3,800/cumm, 0.1%, and 45,000/cu mm, respectively. A bone marrow examination revealed frank megaloblastic changes. An EEG was markedly abnormal. Additional investigations indicated methylmalonic aciduria (79 μmole/mg of creatinine), and homocystinuria (0.85 μmole/gm of creatinine), cystathioninuria, glycinuria, methylcitric aciduria, 3-hydroxypropionic aciduria and formic aciduria. Serum vitamin B_{12} concentration was 20 pg/ml (in contrast to a normal level of 150-1,000 pg/ml). Both the biochemical and clinical abnormalities rapidly reverted to normal on treatment with vitamin B_{12}. The mother had low normal serum vitamin B_{12} levels and excess methylmalonic acid in urine. After treatment the baby continued to breast feed, and during a two-month follow-up period, his serum vitamin B_{12} level fell from 1,260 to 190 pg/ml.

This case is a striking demonstration of the importance of vitamin B_{12} in infant feeding. The deficiency manifested by mimicking diseases associated with an inborn error in vitamin B_{12} metabolism and by chemically exhibited evidence of impaired activity of two enzymes that require B_{12} metabolites as cofactors, methylmalonyl-CoA mutase and N^5-methyltetrahydrofolate-homocysteine methyltransferase. Megaloblastic anemia is rarely seen as part of an inborn error. Its early appearance in this case, presumably as a result of the failure to build up vitamin B_{12} stores *in utero*, suggests that it might be a useful clue to the presence of maternal nutritional deprivation.

Tests of Absorptive Function

One-Hour Blood D-Xylose Level

The quantitative determination of D-xylose levels in blood and urine after an oral load is widely used as a test of mucosal absorptive capacity, but its usefulness has been challenged repeatedly (74,75). However, two careful studies appeared in 1978 that support the diagnostic value of determining the one-hour blood D-xylose level, particularly when allowance is made for differences in surface area (76, 77). Haeney and his colleagues (76) gave a 5-gm dose of D-xylose to 216 subjects, aged 14 to 92 years, of whom 146 were normal controls and 70 had proven malabsorption. In 88 control subjects and in 34 patients with malabsorption blood values were determined at intervals over a three-hour period. Optimum discrimination between normal and abnormal subjects was provided by determining the one-hour blood xylose level corrected to a constant body surface area. A reference range for one-hour values was established by linear extrapolation at the 2.3 and 97.7 percentiles of the cumulative probability distribution. On the basis of these results the authors obtain a corrected one-hour blood xylose level in 139 patients, aged 15 to 65 years, who could be divided into four groups on the basis of the results of jejunal biopsies, culture of intestinal contents, and ancillary investigation: *(1)* 32 patients with abnormal jejunal biopsy specimens, including 21 patients with coeliac sprue; *(2)* 16 patients with contaminated bowel syndrome; *(3)* 53 patients with normal biopsy specimens and sterile aspirates but with definite evidence of gastrointestinal dysfunction; and *(4)* 38 patients with normal biopsy specimens, sterile aspirates, and no evidence of gastrointestinal dysfunction. The incidence of false-positive results was 2.2%. False-negative results in groups 1 and 2 amounted to 10.4%, but in the 21 patients with coeliac disease this incidence fell to 4.8%.

Buts and associates (77) studied 435 children over a five-year period from 1970 until 1975. Each child was given 14.5 gm/m^2 of D-xylose to a maximum of 25 gm, and blood xylose levels were determined at 30, 60, 90, 120, and 180 minutes. The children were divided into four groups on the basis of retrospective information: *(1)* 126 children with normal gastrointestinal function; *(2)* 47 children with coeliac disease; *(3)* 48 children with CF, and *(4)* 214 children with various digestive disorders. All but two of the patients with coeliac disease had a one-hour blood xylose value under 25 mg/dl, whereas all 126 children in group 1 had values equal to or greater than this value. All 48 patients with CF had normal one-hour levels. Interestingly, at times beyond 90 minutes, mean values for this group were significantly higher than in the normal children. In group 4, 63 children had levels under 25 mg/dl. In this group acute enteritis, cow's milk allergy, and intractable diarrhea were the most common diagnoses (Fig. 7). These studies suggest that the one-hour D-xylose level is highly discriminating. For a dissenting opinion one might read the paper by Christie (78), which evaluates the one-hour value in 33 infants and children who had diarrhea for more than two weeks. A single 5-gm dose of D-xylose was given. A very high incidence of false-negative and false-positive results was obtained. The results are so much at variance with those in the preceding articles that one is tempted to blame the fixed dose. This must have led to a very uneven distribution of xylose in body fluids, since the weight of the patients ranged from 4 to 47 kg.

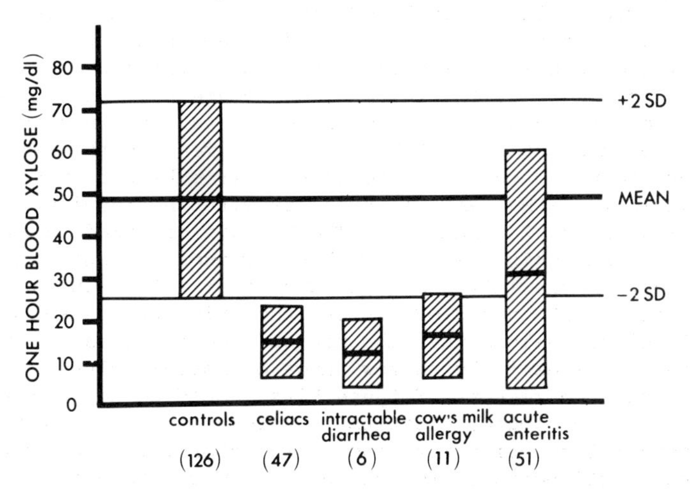

Figure 7. One-hour blood xylose in childhood illnesses. The number of patients per group appears in brackets. (Reproduced with permission from Buts J-P, Morin CL, Roy CC, et al: *J Pediatr* 92:729-733, 1978.)

Although diminished D-xylose absorption is generally considered to reflect mucosal damage, this may not be true in the blind loop syndrome. In an interesting study with a newly developed D-(^{14}C)xylose breath test, Toskes and his associates (79) demonstrate quite convincingly that rats with jejunal self-filling blind loops expire much more of an orogastric dose of the sugar as $^{14}CO_2$ than nonoperated rats or rats with self-emptying blind loops. Urinary xylose secretion fell by an amount correlating with the expiration of $^{14}CO_2$. In this experimental situation, which represents a somewhat extreme bacterial overgrowth syndrome, poor xylose absorption appears to be attributable chiefly to intraluminal catabolism of D-xylose by bacteria.

Folic Acid Radioassay

Serum folic acid determination is a useful and relatively simple procedure for screening patients suspected of having small intestinal mucosal dysfunction, and of course it also provides a useful indication of malnutrition. Serum levels are most commonly used, but, as an indication of actual folate deficiency, erythrocyte levels might be preferable, since they are subject to less short-term fluctuation with diet. Unfortunately, although the primary form of the vitamin in serum, methyltetrahydro-folic acid, is readily used by the organisms in the standard microbiological assays, the polyglutamate derivatives in the red cell require preliminary hydrolysis before they are measurable. Radioassays are not subject to this limitation, since they are not seriously affected by glutamate residues and therefore can be used to measure red cell folic acid directly. In addition they are somewhat faster and easier to perform than the microbiological assays and are not affected by antibiotics or other antibacterial factors in serum that interfere with microbiological determinations. Radioassays, therefore, appear to represent the wave of the future in folate measurement, and a number of manufacturers have introduced easily manipulated kits to facilitate transition to them. The kits generally rely on the folate-binding properties of either β-lactoglobulin from milk or a porcine serum protein and use tracers of folate I^{125}, folate Se^{75}, or folate H^3. Unfortunately, as McGowan and his associates (80) have pointed out in a very careful study, all of the kits cannot be counted to measure the same thing. The authors compared serum folate results from kits from BIORAD, New England Nuclear, (NEN), and RIA Products with the results of a microbiological assay of 200 samples. The NEN kit compared unfavourably with the microbiological assay, although reasonable agreement was obtained between the latter and the two other kits. On whole blood samples both the NEN and BIORAD kits (the only two tested) gave significantly higher values than the microbiological assay, and the NEN kit yielded an excessively narrow range of normal values. The authors concluded that the results of these kits should be correlated carefully with microbiological results in reference sets of sera or whole blood before introduction into one's own laboratory. During the study, the constituents of two of the kits were changed with considerable effect on results. One hopes that manufacturers would accept responsibility for providing a constant product, but perhaps that is too much to expect at this stage. In a second study, Baril and Carmel (81) compared results of a microbiological assay in 415 sera with a fourth kit (Diagnostic Products Corp.) not examined by McGowan et al (80), which uses ^{3}H-pteroyl glutamic acid (^{3}H-PGA) as a tracer. A correlation coefficient of 0.87 was obtained, but the scatter was large, with 25% of the cases giving discrepant results as judged by the fact that values fell into different diagnostic categories (high, normal, indeterminate, or low). By retrospective chart review of 85 of these cases, 74% of the discrepancies appeared to be due to errors in the microbiological assay, principally because of a presumptive suppression by antibiotics. The radioassay gave falsely high values in patients with liver disease and falsely low ones in patients who had received isotopes for scanning purposes. In an addendum the authors introduce the issue of a serum supernate control for endogenous serum binding. This factor appeared to be unimportant with the ^{3}H-PGA assay but might be necessary in other instances.

Ileal Resection and Circulating Bile Salts

Schalm and his coworkers (82) compared measurements of serum conjugates of chenodeoxycholic and cholic acid made at 15- to 20-minute intervals during the day and hourly during the night throughout a 24-hour period, which included three standardized liquid meals and an overnight fast. Bile salts were assayed by radioimmunoassays that were specific for either conjugate. Four subject groups were studied: *(1)* five healthy volunteers; *(2)* five patients with a previous cholecystectomy; *(3)* five patients with documented bile-salt malabsorption due to ileal resection; and *(4)* four pregnant women in the sixth to ninth month of gestation. In healthy subjects the fasting levels of the chenodeoxycholic conjugates were higher than those of the cholic acid conjugates, and they rose more rapidly to a higher peak. In cholecystectomized patients the fasting levels of the chenodeoxycholic conjugates were higher than in the healthy controls and the percentage rise postprandially was slightly higher, although the integrated postprandial rises were not significantly different. Similar results were obtained with the cholic acid conjugates, except that fasting levels were not elevated. In the patients with ileal resection the postprandial rises of the chenodeoxycholic conjugate levels remained relatively constant throughout the day, although they were somewhat smaller than those of the healthy controls, whereas the postprandial rise in the cholic acid conjugate levels fell with each successive meal. In three of the four pregnant women the postprandial rise of the chenodeoxycholic conjugate levels was lower than normal, although cholic acid conjugates behaved normally.

This is not the first paper of its genre (see Hepner and Demers, ref. 83, and Barbara et al, ref. 84), but it is agreeably complete. The principal message seems to be that the major circulating bile acids are chenodeoxycholic acid conjugates, and that the chenodeoxycholic conjugate pool is reasonably well maintained in patients with ileal resection in contrast to the cholic acid pool, which becomes depleted by the demands of successive meals. The results in the pregnant women are interesting, suggesting that pregnancy affects the chenodeoxycholic conjugate pool in a relatively specific manner.

Vitamin E

De Leenheer and his colleagues (85) have described a sensitive, accurate, and rapid assay for the naturally occurring vitamin E compounds in serum that employs high-performance liquid chromatography. Two micrograms of tocol were added to 200 µl of serum as an internal standard, and the tocopherols were extracted with n-hexane. After drying under nitrogen, the residue was dissolved in methanol and injected directly on to a reversed-phase column packed with 10-µm RP18. The column was operated in the isocratic mode and was eluted with 100% methanol at a flow rate of 2 ml/min. The eluate was monitored at the absorption maximum for α-tocopherol (292 nm). About 88% of the total vitamin E in serum is in the alpha form, 2% is in the beta form, and 10% is in the gamma form. β- and γ-tocopherol give a single peak in this system, eluting between tocol and α-tocopherol (Fig. 8). Within-day imprecision for α-tocopherol levels was 2.3% (CV) for 24 samples, and day-to-day imprecision was 3.2% for 20 days. Analytic recovery with fortified sera varied from 89 to 100%. The lower limit of detection was 0.6 mg/l, well below the normal range in adolescents of 3-15 mg/l. These results are excellent. The technique seems to be extremely simple, and the total analysis time is less than one hour. Vitamin E deficiency is probably recognized much less frequently than it occurs; and one hopes that this, or a similar, method will soon become a routine test, particularly in laboratories in hospitals with infant populations, where it might provide the explanation for many otherwise rather vague episodes of anemia and hemolysis that complicate malabsorption.

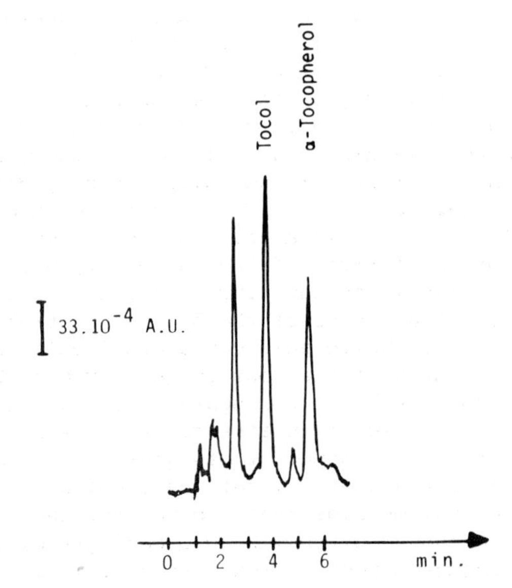

Figure 8. Typical HPLC chromatogram of an extract of normal serum to which tocol was added as an internal standard. (From De Leenheer et al: *Clinical Chemistry* 24:585-590, 1978, by permission.)

Inflammatory Bowel Disease

Serum and Fecal Lysozyme

In 1975 Falchuck et al (86) suggested that serum lysozyme activity might be an ideal marker for Crohn's disease, since it appeared to be markedly elevated in this condition and minimally affected by other inflammatory diseases of the gut. In a subsequent paper in the same year Falchuck and Perroto (87) suggested that serum lysozyme activity might also be helpful in assessing the severity of this disease, since levels tended to rise with exacerbations and fall with remissions. Concurrently and subsequently, a large number of investigators have studied serum lysozyme activity in inflammatory bowel disease without providing strong support for these expectations. To this list one must add at least three publications that appeared in 1978 (88-90). Klass and Neale (88) measured lysozyme activity, by a lysoplate method and by a turbidimetric method, in both the sera and stools of patients with inflammatory bowel disease, of patients with other gastrointestinal diseases, of patients without gastrointestinal disease, and of normal control subjects. The turbidimetric method gave serum lysozyme values of 9.2 ± 2.7 (mean SD) μg/ml in patients with active Crohn's disease and 4.4 ± 2.0 μg/ml in the sera from healthy controls, but the results of this technique could not be used to distinguish patients with Crohn's disease from those with ulcerative colitis or a variety of other gastrointestinal conditions. The lysoplate method produced higher values, with considerable overlap between the results for the normal control subjects and those for the patients with Crohn's disease. The correlation between the two methods was poor (r - 0.56). Fecal lysozyme activity was much higher in patients with inflammatory bowel disease and patients with other causes of diarrhea than in the normal control subjects or in diseased patients without diarrhea, but the results could not be used to distinguish patients with Crohn's disease from other patients with diarrhea. El-Khateb et al (89) found that, although the mean serum lysozyme activity as determined by the lysoplate method was higher in patients with Crohn's disease (10.5 ± 6.8 μg/ml) and with ulcerative colitis (9.6 ± 4.1 μg/ml) than in normal controls, there was no correlation with the diagnosis or the degree of disease activity. Using a turbidimetric method, Feldman and Stern (90) found no differences between serum levels in the majority of their patients with Crohn's disease or with ulcerative colitis and those in the normal control subjects and no evidence of a change in lysozyme activity in two patients with Crohn's disease whose progress was followed during periods of exacerbation and remission. These results seem to be quite representative of the literature after 1975. Some of the variation in results may be methodological, since the substrates for lysozyme are not particularly standardized nor are the two favorite assay methods equivalent, but it seems clear that the measurement of serum lysozyme activity is not helpful in determining either the cause of inflammatory bowel disease or the degree of disease activity in a single patient.

Is there a place for fecal lysozyme determinations in inflammatory bowel conditions? Krawczuk and colleagues (91) think there is. They studied 152 infants with acute diarrhea and 50 healthy infants. Fecal lysozyme was assayed by the lysoplate method in stools dried at 37°C and extracted with phosphate buffer and acetone. Normal values varied from 14.9 to 77.0 units/gm of dry feces. In 72.4% of the cases of acute diarrhea, values exceeded this range and were highest in patients judged to have the most severe disease. In general, values fell with the disappearance of clinical symptoms, and they persisted at high levels if diarrhea was prolonged or if other evidence of mucosal abnormality, such as malabsorption, was present. The authors feel that this simple determination may have prognostic value in infantile diarrhea, where it probably reflects inflammatory activity in the intestinal wall.

Protein-Losing Enteropathy

Diagnosis by α_1 Antitrypsin Clearance

Every once in a while a diagnostic technique appears that is so simple that one cannot understand why anything so obvious had not been tried before. The use of an endogenously circulating glycoprotein like α_1 antitrypsin to determine protein clearance by the intestine certainly fits into this category. For many years the precise diagnosis of protein-losing enteropathy seemed to demand the use of a variety of radioactively labelled macromolecules, such as albumin I^{131}, albumin Cr^{51}, ceruloplasmin Cu^{67}, PVP I^{131}, and so on. In two fascinating articles Crossley and Elliott (92) and Bernier et al (93) indicate first that α_1 antitrypsin fecal/serum ratios derived from random samples may provide a useful screening test for protein-losing enteropathy (92), and second that clearance data obtained for α_1 antitrypsin using prolonged stool collections may be as reliable as those for ^{51}Cr clearance (93). In 10 control children the fecal/serum ratio in 17 samples varied from 0 to 0.66, whereas in two patients with protein-losing enteropathy, 9 samples displayed ratios of 1.8 to 5.2 (92). Bernier and colleagues compared 10 patients with protein-losing enteropathy, including patients with coeliac disease, Crohn's disease, lymphonodular hyperplasia, post-gastrectomy syndrome, and gastric ulcer with 13 control subjects. Blood samples were taken on days 1, 5, and 10, and all stools were collected for 10 days. Feces were stored at -20°C, thawed, diluted in isotonic NaCl with mixing, centrifuged at 1500 g for 20 minutes, and the supernatant was assayed for α_1 antitrypsin. On the controls clearance ranged from 0.5 ml/day to 7.0 ml/day, and in the patients with protein-losing enteropathy it varied from 16.5 ml/day to 218 ml/day. Stools kept at 37°C for three days lost less than 5% of their α_1 antitrypsin. The secret seems to lie in the extraordinary stability of α_1 antitrypsin deriving no doubt from its ability to inhibit many proteases. The serum level is reasonably stable, and amounts appearing in the stool are easily measurable In this study results compared favorably with those of ^{51}Cr clearance, although α_1 antitrypsin clearances were lower than ^{51}Cr clearance in both study groups. Tests were performed on stools collected for 10 days, but, as the authors state, 3-day collections should suffice. Crossley and Elliot and Bernier and his coworkers are currently investigating the simplest conditions needed for reliable diagnosis.

Gastrointestinal Malignancy and Tumor Markers

Carcinoembryonic Antigen (CEA)

The serum CEA determination has had a checkered career as a screening test for malignancy, particularly of the gastrointestinal tract. Initial investigations suggested that elevated levels were likely to be both sensitive and accurate in diagnosing malignancy (94), but as time has passed it has become obvious that there are many nonmalignant causes of false-positive results (95,96), and the claim for sensitivity has steadily receded. Few would claim now that there is a screening role for the test in any but high-risk populations. However, since the test is readily available to most clinicians, it is appropriate to ask whether its routine use even in highly selected populations is likely to have any effect on the accuracy and speed of diagnosis or on the efficacy of care in the individual patient. Hine and coworkers (97) have addressed themselves to one aspect of this problem in a fascinating, careful, prospective, long-term study of the usefulness of the test performed in the initial stages of the investigation of 381 patients with clinical problems commonly associated with gastrointestinal malignancy. The total group could be divided into four categories: *(1)* 176 patients with upper gastrointestinal symptoms; *(2)* 97 patients with lower gastrointestinal symptoms; *(3)* 34 patients with unexplained iron deficiency anemia; and *(4)* 74 patients with disordered liver function. CEA results were compared against the final diagnosis, either of malignancy established by histological findings or by evidence at laparotomy or in patients initially judged not to have a malignancy by verification at a follow-up visit four years later. A clear indication of a final diagnosis was established at this time in all but eight patients. Malignancy was diagnosed in 26 patients with upper gastrointestinal symptoms, in 38 with lower gastrointestinal symptoms, in 7 with iron deficiency anemia, and in 38 with disordered liver function. In groups 1, 2, and 3 many patients with or without malignancy had moderately elevated CEA levels (20-50 ng/ml). Levels exclusively associated with malignancy were high: > 58 ng/ml in group 1, > 47 ng/ml in group 2, and > 38 ng/ml in group 3. Overall, such high values were seen in only 16 of the 71 patients with malignancy making up these three groups. Unfortunately, when the role of the CEA test was evaluated in these 16 patients, only three cases were found in which the diagnosis was materially influenced by the results. In the patients with disordered liver function the CEA test fared somewhat better. Sixteen of the 38 diagnoses of malignancy were associated with CEA levels higher than found in other diagnoses (> 45 ng/ml), and in 9 of these, the CEA was, or could have been, helpful in establishing the diagnosis, largely by shortening the time required for biopsy or laparotomy. The authors have concluded that CEA testing is not of much use in detecting cancer in patients with gastric or colonic symptoms or iron deficiency. The results of CEA testing seemed to be useful in establishing the diagnosis in patients with liver involvement, particularly if values exceeded 45 ng/ml. For those who are troubled by the various units used in the different CEA assays, 50 ng/ml in the assay used by Hine et al (91) is roughly equivalent to 10 ng/ml by the Z-gel assay. The problem, of course, is that the tumors in patients with disordered liver function have usually metastasized. Not surprisingly Hine and his colleagues did not feel that determining the CEA level led to earlier or more successful treatment in cases they studied. Jubert et al (98) suggested, however, that preoperative CEA values may be of some prognostic significance. Of 97 patients with colorectal carcinoma, 42% had normal preoperative CEA levels (< 2.5 ng/ml). When compared with the remainder, the low level group had more Duke A lesions, were less likely to suffer recurrences or to have elevated CEA levels at the time of recurrence, and had a longer disease-free interval. There are many possible explanations for these results, but the simplest is that the serum CEA level depends both on the rate of production by the tumor, which is probably a function of metabolic activity and size, and the accessibility of the antigen to the circulation, which is a function of the invasion of draining lymphatics and blood vessels. Low levels of the antigen in the circulation are therefore likely to indicate a rather somnolent neoplasm with minimal invasion. Unfortunately, this does not help one find a tumor any faster or to treat it any differently.

Although the early concept held that tumors produce an antigen that is unique to themselves or to embryonic tissues, it has gradually become apparent that normal adult tissues produce CEA, and that the major difference between neoplastic and normal tissues is a quantitative one. Shively and his coworkers (99) have reported, for example, that they were able to isolate a glycoprotein from the colons of healthy people by lavage, which was not only immunologically identical to CEA but showed homology in 23 of the first 24 amino acids in the NH_2-terminal sequence. If true, the lack of specificity displayed by the CEA assay is not surprising.

Alkaline Phosphatase Isoenzymes

Three electrophoretic variants in addition to the common hepatic, osseous, and intestinal alkaline phosphatases have been described in association with tumors. The Regan isoenzyme resembles placental alkaline phosphatase in every respect except that it is found in the absence of pregnancy. The Nagao isoenzyme is somewhat similar, but although heat stable, it is inhibited to a greater extent by L-leucine and EDTA. The Regan variant differs from both in its greater electrophoretic mobility and heat sensitivity, but it resembles the Nagao isoenzyme in its susceptibility to L-leucine and EDTA. Crofton and Smith (100) have described six cases of tumor associated with the Regan variant. The importance of their study derives from the fact that only one of the six patients had hepatocellular carcinoma, whereas three lacked evidence of liver metastases. The nonhepatic tumors derived from the stomach, larynx, and colon. This is not the first paper to report cases of metastasizing gastrointestinal tumors in association with the Regan variant, but it runs counter to the prevailing view that the isoenzyme most commonly indicates the presence of a hepatocellular carcinoma (101). Most reported cases have been associated, however, with hepatic metastases. This may still be true, because although three of the patients in Crofton and Smith's study displayed no evidence of metastases at the time the variant was discovered, all three were dead within a matter of months. However, Nishimura et al (102) have extracted an alkaline phosphatase isoenzyme resembling the Regan variant from a gastric adenocarcinoma suggesting that the isoenzyme can be derived from the tumor *per se*.

A more comprehensive listing of biochemical tests proposed in the diagnosis and monitoring of patients with gastrointestinal cancer is found in the Chapter by Lawrence and Neville in this volume.

REFERENCES

1. Taylor TV, et al: *Gut* 19:865, 1978.

2. Fordtran JS, Walsh JH: *J Clin Invest* 52:645, 1973.

3. Baron JH: *Scand J Gastroent* 6(suppl):9, 1970.

4. Gross RA, et al: *Gastroenterology* 75:357, 1978.

5. George J: *Gut* 9:237, 1968.

6. Hunt JN: *Guys Hosp Rep* 103:161, 1954.

7. Schöön IM, et al: *Gastroenterology* 75:363, 1978.

8. Bergegardh S, Nilsson G, Olbe L: *Scand J Gastroent* 11:475, 1976.

9. Hoare AM, et al: *Gut* 19:166, 1978.

10. Hoare AM, et al: *Gut* 19:163, 1978.

11. Malagelada JR: *Gut* 19:284, 1978.

12. Straus E, Greenstein R, Yalow R: *Lancet* 2:73, 1978.

13. Greenberg G, et al: *Gut* 18:A981, 1977.

14. Tai H, Chey W: *Anal Biochem* 87:376, 1978.

15. Royston CM, et al: *Brit J Surg* 62:664, 1975.

16. Rubin W: *Gastroenterology* 57:641, 1969.

17. Creutzfeldt W, et al: *Europ J Clin Invest* 1:461, 1971.

18. Creutzfeldt W, et al: *Gut* 17:745, 1976.

19. Royston CMS, et al: *Gut* 19:689, 1978.

20. Stave R, Gedde-Dahl D, Gjone E: *Scand J Gastroent* 13:845, 1978.

21. Newton WT, Jaffe BM: *Surgery* 69:34, 1971.

22. Hallgren R, Karlsson FA, Luncqvist G: *Gut* 19:207, 1978.

23. Seino S, et al: *Gut* 19:10, 1978.

24. Dodge JA, Karim AA: *Gut* 17:280, 1976.

25. Bleicher NA, et al: *Gut* 17:794, 1978.

26. Furihata C, et al: *Anal Biochem* 84:479, 1978.

27. Anson ML: *J Gen Phys* 22:79, 1939.

28. Imamura K, et al: *Amer J Gastroent* 69:572, 1978.

29. Sacher M, Kobsa A, Shmerling D: *Arch Dis Child* 58:639, 1978.

30. Nousia-Arvanitakis, et al: *J Ped* 92:734, 1978.

31. Imondi AR, Stradley RP, Wohlgemuth R: *Gut* 13:726, 1972.

32. Smith H, et al: *J Clin Invest* 24:388, 1945.

33. Gyr K, et al: *Gut* 17:27, 1976.

34. Arvanitakis C, Cooke A: *Gastroenterology* 74:932, 1978.

35. Dürr H, et al: *Digestion* 17:404, 1978.

36. Fedail S, et al: *Lancet* 1:181, 1978.

37. Gibson L, Cook R: *Pediatrics* 23:545, 1959.

38. Bray PT, et al: *Arch Dis Child* 53:483, 1978.

39. Warwick W, Hansen L: *Clin Chem* 24:381, 1978.

40. Finley PR, et al: *Amer J Clin Pathol* 69:615, 1978.

41. Rosenstein BJ, et al: *JAMA* 240:1987, 1978.

42. Schutt WH, Isles TE: *Arch Dis Child* 43:178, 1968.

43. Holsclaw D, et al: *Cystic Fibrosis Club Abst No 54*. 18th Annual Meeting, 1977.

44. Schuttinger G, Zinterhofer L: *Clin Chim Acta* 83:109, 1978.

45. Pierre J, Tung KK, Nadj H: *Clin Chem* 22:1219, 1976.

46. Hansen NQ, Yasmineh WG: *Clin Chem* 24:762, 1978.

47. Alexander R: *Clin Chem* 23:1369, 1977.

48. van Wersch J, Quast O, Kleesiek K: *Fresenius Z Anal Chem* 290: 173, 1978.

49. Largman C, et al: *Biochim Biophys Acta* 543:450, 1978.

50. Farrar W, Calkins W: *Arch Intern Med* 138:1958, 1978.

51. Levitt MD, Johnson SG: *Gastroenterology* 75:118, 1978.

52. Ladenson J, McDonald J: *Clin Chem* 24:815, 1978.

53. di Magno E, et al: *N Engl J Med* 297:737, 1977.

54. Ganda O, et al: *N Engl J Med* 296:963, 1977.

55. Larsson L, et al: *Lancet* 1:66, 1977.

56. Schlegel W, et al: *Lancet* 2:166, 1977.

57. Arvanitakis C, Cooke AR, Greenberger NJ: *Med Clin N Amer* 62: 107, 1978.

58. Arvanitakis C, Cooke AR: *Gastroenterology* 74:932, 1978.

59. Ashkenazi A, et al: *Lancet* 1:627, 1978.

60. Welsh J, et al: *Gastroenterology* 75:847, 1978.

61. Levitt MD: *N Engl J Med* 281:122, 1969.

62. Douwes A, Fernandes J, Rietveld W: *Clin Chim Acta* 82:293, 1978.

63. Perman J, Barr R, Watkins JB: *J Ped* 93:17, 1978.

64. Payne-Bose D, et al: *Anal Biochem* 88:659, 1978.

65. Gilat T, et al: *Gut* 19:602, 1978.

66. Levitt M, Donaldson RJ: *Lab Clin Med* 75:937, 1970.

67. Thompson R, et al: *Clin Chim Acta* 84:185, 1978.

68. Fromm H, Hofmann A: *Lancet* 2:621, 1971.

69. Sherr H, et al: *N Engl J Med* 285:656, 1971.

70. Lauterburg B, Newcomer A, Hofmann A: *Mayo Clin Proc* 53:227, 1978.

71. Hintz R, et al: *Trop Ped* 92:153, 1978.

72. Van den Brande J, Du Caju V: *Excerpta Med Int Cong Ser* 256:41, 1972.

73. Higginbottom NC, Sweetman L, Nyhan WL: *N Engl J Med* 299:317, 1978.

74. Sladen GE, Kumar PJ: *Brit Med J* 3:223, 1973.

75. Lamabadusuriya S, Packer S, Harries J: *Arch Dis Child* 50:34, 1975.

76. Haeney M, et al: *Gastroenterology* 75:393, 1978.

77. Buts JP, et al: *J Ped* 90:729, 1978.

78. Christie DL: *J Ped* 92:725, 1978.

79. Toskes P, et al: *Gastroenterology* 74:691, 1978.

80. McGown EL, et al: *Clin Chem* 24:2186, 1978.

81. Baril L, Carmel R: *Clin Chem* 24:2192, 1978.

82. Schalm SW, et al: *Gut* 19:1006, 1978.

83. Hepner GW, Demers LM: *Gastroenterology* 72:499, 1977.

84. Barbara L, et al: *Rendiconti di Gastro-enterologia* 8:194, 1976.

85. DeLeenheer AP, et al: *Clin Chem* 24:585, 1978.

86. Falchuk KR, Perroto JL, Isselbacher KJ: *N Engl J Med* 292:395, 1975.

87. Falchuk KR, Perroto JL: *Gastroenterology*

88. Klass HJ, Neale G: *Gut* 19:233, 1978.

89. El-Khatib O, et al: *Dig Dis* 23:297, 1978.

90. Feldman C, Stern R: *Israel J Med Sc* 14:278, 1978.

91. Krawczuk J, Sawicki A, Krawczynski J: *J Clin Chem Clin Biochem* 16:343, 1978.

92. Crossley J, Elliott R: *Brit Med J* 1:428, 1977.

93. Bernier J, et al: *Lancet* 2:763, 1978.

94. Thompson DM, et al: *Proc Natl Acad Sci* 64:161, 1969.

95. Moore T, et al: *Gastroenterology* 63:88, 1972.

96. Hansen HJ, et al: *Human Path* 5:139, 1974.

97. Hine K, et al: *Lancet* 2:1337, 1978.

98. Jubert AV, Talbott TM, Maycroft TM: *Cancer* 42:635, 1978.

99. Shively J, et al: *Cancer Res* 38:503, 1978.

100. Crofton MM, Smith AF: *Clin Chim Acta* 86:81, 1978.

101. Higashino K, et al: *Clin Chim Acta* 40:67, 1972.

102. Nishimura H, et al: *Gastroenterology* 71:497, 1976.

6
The Plasma Lipoprotein System

Barry Lewis

Normal and abnormal lipoprotein metabolism has been reviewed extensively in the past three years, in a monograph (1) and other books (2-4) and in comprehensive articles (5-9). In this Chapter emphasis will be placed on major advances that have taken place since these literature surveys were completed. Inevitably this emphasis reflects a subjective assessment of the more substantial contributions to the subject. Although this Review will focus predominantly upon the 1978 literature, earlier work will be referred to where this is necessary in order that the reader will be provided with the background to place the newer literature in a proper perspective.

LIPOPROTEIN METABOLISM

Chylomicrons and Very-Low-Density Lipoproteins (VLDL)

Chylomicrons and VLDL, the largest lipid-bearing particles in plasma, are the major transport forms of dietary and of endogenous triglyceride, respectively. Chylomicrons, by definition, are synthesized in the small-intestinal mucosa. The diameter of most chylomicrons is in the range 80-600 nm. During fat absorption, these particles contribute largely to alimentary lipemia; but smaller triglyceride-rich particles are produced by the intestine, with diameters and lipid composition overlapping those of VLDL. The latter, which also play a role in alimentary lipemia (10), are variously termed *intestinal VLDL* or *small chylomicrons*.

The physiological relationships between the triglyceride-rich lipoproteins and high-density lipoproteins (HDL) are becoming clearer; several recent papers have shed light on this topic. Patients with gross hypertriglyceridemia due to lipoprotein lipase deficiency have characteristically low plasma HDL-cholesterol concentration (11); and a reciprocal relationship between HDL levels and plasma triglyceride or VLDL concentrations is evident from population studies (12,13). Hypertriglyceridemia is also a feature of Tangier disease, in which normal HDL is absent from plasma (14). Several interventions lead to a reciprocal increase in VLDL levels with a fall in HDL levels, for example, a high-carbohydrate diet (15), while clofibrate and weight reduction may produce an increase in HDL and a fall in VLDL concentration (16). It is chiefly the HDL_2 subclass that participates in these changes (19,20). The relationship is not universal: estrogen treatment (1,5,17) and alcohol (18) lead to an increase in plasma levels of both VLDL and HDL.

Nascent Lipoproteins

The small protein moiety of chylomicrons (about 1%) and of VLDL (5 to 12%) (1) is involved in mass transfers to and from other lipoprotein classes. Nascent hepatic and intestinal triglyceride-rich lipoproteins are relatively deficient in apolipoprotein (21,22), acquiring their full complement after secretion, by transfer of apo C from HDL (23,24). As these particles are catabolized, apo C transfers back to HDL (23,24). The acquisition of apo C by chylomicrons and VLDL is of great physiological importance; one of the C peptides, apo C-II, activates the enzyme lipoprotein lipase (25), which by hydrolyzing the glycerolipids of these particles is largely responsible for the intravascular phase of their catabolism and for clearance of circulating triglyceride fatty acids into muscle, adipose, lung, and other tissues.

Transfer of apo A Peptides

More recently, evidence has been accumulating that the major apoproteins of HDL (apo A-I and A-II) may be acquired by these lipoproteins, at least in part, after their secretion. The nascent HDL produced by the perfused rat liver is a flattened discoid particle in which apo E is the main peptide, apo A-I being a relatively minor component (26,27); by contrast peripheral plasma HDL, in the rat and in man, has a preponderance of apo A-I. Rat intestinal HDL (obtained from the thoracic duct) is also a discoid particle, but, unlike the HDL of hepatic origin, it is rich in apo A-I (28). It is probable that "maturation" of HDL involves the partial loss of apo E with gain of apo A-I (29).

The source of the A peptide of HDL is of great interest, not only physiologically but because of the epidemiological evidence that plasma HDL-cholesterol concentration is inversely predictive of ischemic heart disease (IHD) risk (30). Apo A-I and apo A-II are present in thoracic duct chylomicrons, both in rats (29) and in humans (31); apo A-I has been shown to be synthesized in the small intestine and is not, to an important extent, acquired from plasma (25,32-34). Therefore, it is possible that chylomicron-borne A peptides are an important source of HDL-apo A. Furthermore, the flux of A peptides is enhanced by fat feeding (32). Although experimental evidence for this hypothesis is as yet incomplete, it has been reported that the apo A-I and apo A-II of human thoracic duct chylomicrons can transfer to HDL (35). Recently it has been observed that "small chylomicrons", that is, VLDL-like particles produced during alimentary lipemia, also contain apo A-I and apo A-II, unlike the VLDL present in plasma in the fasted state (36); such particles are also a potential source of HDL peptides. The origins of high-density lipoproteins are discussed further in a later section.

VLDL Kinetics in Man

The production and metabolic fate of VLDL have been studied extensively, after a long period of methodological uncertainty. Radio-iodination of the apoproteins, followed by analysis of the specific activity/time curve of the main structural peptide apo B has been valuable (37-42); precursor labeling with ^{14}C-leucine has also been used (43). The older procedure, in which the triglyceride of VLDL is precursor-labeled with radioglycerol (44,45), has regained popularity in the past few years; its simplicity is attractive, but cogent reasons have been put forward for caution in interpretation of the complex specific activity/time curve of the labeled product (46,47).

The Lipoprotein Cascade

VLDL isolated from peripheral plasma ranges from 30 to 80 nm in diameter, and its composition also varies considerably (1). Its initial catabolism is intravascular, occurring in a series of steps during which its composition is modified and its particle size decreases progressively (48). Hence, it is possible that the largest, least dense, most triglyceride-rich particles represent the primarily secreted VLDL, while the smaller particles are catabolic products or VLDL remnants. This open catenary model is an incomplete description, however, for there is evidence that the VLDL entering plasma is not confined entirely to the least dense particles, but includes some in the denser subclasses (49).

The catabolism of VLDL and chylomicrons is initiated by hydrolysis of their triglyceride moiety by lipoprotein lipase. The enzyme is bound at the luminal surface of capillary endothelium, particularly in skeletal and cardiac muscle and in adipose tissue. The released fatty acids are taken up largely by the muscle cells and adipocytes. The fate of the intravascular products of lipolysis has been studied extensively. The structural peptide of VLDL, apo B, is retained as the lipoprotein particle is catabolized through a range of products including smaller VLDL, then intermediate-density lipoprotein (IDL, density range: 1.006-1.019 gm/ml) and finally low-density lipoprotein (LDL, density range: 1.019-1.063 gm/ml) (50). The steps in this conversion have been amply confirmed (51, 52).

VLDL-LDL Conversion: Quantitative Relationship

Hence, precursor-product relationships exist between VLDL and IDL, and between IDL and LDL. Not only the apo B but also cholesteryl ester are stable components of this series of lipoproteins. Quantitative aspects of this sequence of conversions have been studied (39,41,52,53). Normally, LDL-apo B and IDL-apo B are derived entirely from VLDL-apo B (39,53). About 90% of VLDL-apo B is converted to the denser lipoprotein classes, IDL and LDL. Although some studies indicate that most VLDL-apo B is normally converted to LDL-apo B (41,42), a large series of observations suggest that this conversion is only partial (53,54); it appears that most or all apo B is transferred from the VLDL density class to IDL, but that there are two subsequent pathways for IDL and its structural peptide. One is conversion to LDL, and the other is extravascular catabolism; hence, only about 50% of VLDL B peptide is converted to LDL B peptide. Presumably this is one reason for the relative independence of VLDL and LDL concentrations under most circumstances, despite their precursor-product relationship. The steady-state synthetic rates of VLDL-apo B and of LDL-apo B show a modest but significant correlation, $r = 0.41$ (54).

Under certain conditions, an inverse relationship is apparent between VLDL and LDL concentrations, as in some familial hypertriglyceridemic states and during high carbohydrate intake. This appears to be due in genetic hypertriglyceridemias to an increased fractional catabolism of LDL rather than to subnormal LDL production (55).

The most common kinetic cause of endogenous hypertriglyceridemia is increased VLDL production, as assessed by apo B turnover (56,54). Under such circumstances the fraction of the rapidly turning-over VLDL-apo B, which is converted to IDL, is much reduced, from about 90% to about 50% (39,55,56).

In the rat, it is clear that conversion of VLDL to LDL is very incomplete, most VLDL-apo B radioactivity being recoverable in the liver (7). If, as suggested, VLDL is not completely converted to LDL in humans, it is possible that the proportion of IDL and VLDL that is not converted to LDL also has the liver as its main extravascular site of clearance. The role of the liver in chylomicron metabolism is well documented. After lipolysis by lipoprotein lipase, chylomicron remnants are avidly taken up by the liver in the rabbit (57) and rat (58,59), suggesting a specific recognition mechanism for these remnant particles. In the rat the uptake of the cholesterol of chylomicron remnants has an important regulating role in hepatic cholesterol synthesis (58,59). Although small VLDL, and IDL, may be regarded as the analogues of chylomicron remnants, with which they seem to have physical similarities (60), their uptake and catabolism by the liver and the site of their conversion to LDL is still partly conjectural.

However, evidence for the uptake of IDL and small VLDL in the human splanchnic bed has recently been reported (61). By measuring arterial-hepatic venous concentration differences, and using radioiodinated lipoproteins, net uptake of these particles has been demonstrated (by contrast larger VLDLs show net secretion by the splanchnic organ).

In four of five patients with remnant (type III) hyperlipoproteinemia, a study of VLDL-apo B kinetics showed far lower catabolic rates than in patients with hypertriglyceridemia; the authors concluded that a defect in VLDL catabolism was responsible for the hyperlipidemia in type III (62).

Most of the foregoing studies have been based on relatively simple models; and in one series of observations a noncompartmental method was used, which avoided assuming any particular model (39, 52-54). Another approach, multicompartmental analysis, has been used to develop a detailed model of the behavior of apo B and apo C during the lipoprotein cascade (41). It was necessary to propose a highly complex model to fit the observed data on five patients with type III hyperlipoproteinemia, on five with other hyperlipoproteinemias, and on four normal subjects.

Metabolic Fate of VLDL Surface Components

Hence, a reasonably clear picture has emerged of the metabolism of the apo B and of the core lipids (triglyceride and cholesteryl ester) of VLDL. Elegant and extensive studies of the fate of the surface components of VLDL have also been reported by Eisenberg and his colleagues (63-67). These components - - apo C, cholesterol, and phospholipids - - form a polar or amphipathic shell surrounding the hydrophobic domain of the peptide. During lipolysis, depletion of the triglyceride content of the core renders some of the surface material redundant. Disposition of the surface components has been studied by *in vitro* procedures; one method involved the incubation of VLDL with a purified bovine lipoprotein lipase; in the other, VLDL was exposed to the enzyme by perfusing the quadruply labeled lipoprotein through the beating rat heart in a recirculating system. About one third of the phosphatidylcholine is hydrolyzed to lysolecithin. Progressive removal of surface constituents occurs, even in the absence of HDL. In its presence, apo C transfers rapidly between VLDL and HDL, reaching equilibrium apparently within a few minutes at 37°C. In the absence of HDL, apo C, free cholesterol, and phospholipids accumulate in the density range of 1.04 to 1.21 gm/ml, which largely corresponds with that of HDL, this material appears as discoid lipoprotein-like particles under the electron microsope. With the *in vitro* incubation system, particles appear in the density range of LDL and have a composition similar to but not identical with authentic LDL. In the presence of the HDL_3 subclass of HDL, surface material released during VLDL lipolysis *in vitro* combines with the HDL_3 to form lipoprotein that resembles HDL_2.

These findings shed light on the complex mechanisms by which "mature" HDL is produced. The physiology of LDL synthesis is almost certainly more complex than the action of lipoprotein lipase on VLDL, although this action normally initiates the process. Firstly, LDL is formed from labeled VLDL in patients with inborn deficiency of this enzyme (68); and secondly LDL is not produced when labeled triglyceride-rich lipoproteins are injected into the functionally hepatectomized rat, the product being the larger "remnant" particle (69).

Lipoprotein Lipase

Lipoprotein lipase, a key enzyme in the clearance of triglyceride from plasma, has become more accessible to study with the development of reliable assays for its measurement in needle biopsy specimens of human adipose tissue and muscle (70,71). In the fed state, the activity of rat adipose tissue enzyme increases as does that of the adipocytes (72), while the muscle enzyme decreases (73); this response, which correlates with plasma insulin: glucagon ratios, suggests that the activity of the enzyme may play a role in the disposition of substrate. Postprandially, triglyceride fatty acids would largely be directed to adipose tissue for storage, while, in the fasted state, muscle would receive more fatty acids. This increased lipoprotein lipase activity of adipose tissue in the fed state appears to be due to induction of increased enzyme protein synthesis (74).

Lipoprotein lipase exists in different molecular forms (75, 76). The b-form, molecular weight 120,000 daltons, appears to be synthesized in adipocytes; the higher molecular weight a-form is transported to and bound at the endothelial surface. In cultured heart cells, lipoprotein lipase activity is greater in mesenchymal cells than in myocytes (77). In a study of heavily trained long-distance runners, lipoprotein lipase activity in muscle (and also in adipose tissue in men) exceeded control values (78). Muscle and adipose tissue enzyme activities showed a modest negative correlation with plasma VLDL levels. On the other hand adipose tissue lipoprotein lipase activity correlated strongly and positively with HDL levels.

Low-Density Lipoprotein (LDL, β-Lipoprotein)

About two thirds of the plasma cholesterol is carried in LDL, the most abundant of the circulating lipoproteins. LDL is a smaller particle than VLDL or IDL; its precursors have a mean particle size of about 22 nm and a protein content of about 22%. The protein moiety is apo B.

In normal subjects, and in most hyperlipidemic subjects too, LDL is derived as a metabolic product of VLDL via IDL. Although VLDL is readily demonstrated in the rat hepatic Golgi apparatus and in rat liver perfusates, LDL has not been detected. However, in one genetic disorder at least, familial hypercholesterolemia, there is evidence of "direct" secretion of LDL into plasma not involving a circulating lipoprotein precursor.

LDL Kinetics

As assessed by LDL-apo B kinetic parameters, the fractional turnover of LDL is far slower than that of its precursors (39,55,56). LDL concentration is negatively correlated with its fractional catabolic rate (FCR). The FCR of LDL appears not to be a function of the pool size of this lipoprotein (7). Hence, this relationship indicates that catabolism of LDL is an important determinant of its plasma concentration (55). Variations in the synthetic rate of LDL account for the elevated levels seen in some patients with familial combined hyperlipidemia (79) and contribute to some extent to the high LDL concentrations seen in familial hypercholesterolemia (79,53). As mentioned previously, the synthetic rate of LDL-apo B correlates moderately ($r = 0.41$) with that of its precursor, VLDL-apo B.

The site of LDL degradation and its mechanisms and their regulation have been considerably classified during the past six years. Previously it was thought that LDL was both secreted from and degraded within the liver, in a rather futile cycle. The demonstration that the FCR of LDL was not decreased but, if anything, enhanced by hepatectomy in swine indicated that its major sites of catabolism were extra-hepatic (80). This finding has been fruitful in stimulating a large volume of consequential research. Recently the possibility has again arisen that the liver plays a measurable role in LDL catabolism (81). Nevertheless, most physiological studies and tissue culture observations conform with the view that an important role of LDL is the centrifugal transport of cholesterol to peripheral, extra-hepatic tissues.

Receptor-Mediated LDL Catabolism

The organs or tissues responsible for LDL catabolism *in vivo* have yet to be identified with certainty. But observations on several lines of cultured or surviving cells have shown that many extrahepatic tissues are capable of taking up and degrading LDL. Studies in many laboratories (82), and in particular the distinguished work of Brown, Goldstein, and their colleagues (83,84) have led to an increasingly clear picture of the mechanisms involved. The key studies were performed on cultured human fibroblasts (83,84), but similar mechanisms have been identified in human peripheral blood mononuclear cells (85), in cultured arterial smooth muscle cells (86), and in endothelial cells (87).

Membrane Transport of LDL

LDL catabolism by these cell lines is initiated by binding of the lipoprotein to a cell-surface receptor. This receptor appears to "recognize" apo B leading to binding of LDL and VLDL; an apo E-containing subclass of HDL also interacts with the receptor probably due to the existence in both apoproteins of homologous sequences containing arginine (88). The receptors have been localized by electron microscopy after they interacted with ferritin-labeled LDL (89); they are chiefly situated in concavities in the plasma membrane of fibroblasts, the "fuzzy pits." This localization appears to depend on the integrity of cytoplasmic microfilaments (90).

Receptor binding of LDL is characterized by high affinity, saturability, and by the specificities listed in the previous paragraph. The affinity is greater at 4°C than at 37°C. Binding is dependent on calcium or other divalent cations.

At 37°C (but not at 4°C), binding is followed by internalization of the lipoprotein by invagination of the fuzzy pit region of the plasma membrane to form an endocytotic vesicle. These fuse with lysosomes, within which hydrolysis of apo B, cholesteryl ester, and phospholipids takes place. Free cholesterol is thus made available to the cell for plasma membrane synthesis and other purposes.

Control Mechanisms

Several homeostatic mechanisms exist that tend to stabilize the cholesterol content of the cell. The number of receptors on the cell surface varies over a 10-fold range in response to the LDL content of the incubation medium (91,83), decreasing as the availability of LDL increases. Cholesterol synthesis by the cell, regulated by the rate-limiting enzyme hydroxymethylglutaryl coenzyme A reductase, is inhibited as the concentration of LDL in the medium is increased; it appears that synthesis of the enzyme protein is regulated by intracellular free cholesterol (92). It has been calculated that at normal plasma LDL concentration cholesterol synthesis will be relatively suppressed in most nonhepatic cells, and their LDL receptor number will be low; such cells would be meeting their cholesterol requirements from plasma LDL uptake rather than from *in situ* synthesis (83).

Several inborn errors in this pathway have been documented (84,93), constituting the molecular bases of familial hypercholesterolemia. Absence of receptors in the homozygous state was the first lesion to be characterized (heterozygotes having approximately half the normal receptor activity). Defective function is a distinct disorder, and impairment of the internalization process has also been described. Reduction of the fractional catabolic rate of LDL, well recognized in *in vivo* studies (94,55), may reasonably be explained by disorders of the LDL receptor mechanism. The hypercholesterolemia is largely due, then, to impaired LDL clearance from plasma. But evidence has also been obtained of increased LDL-apo B secretion in heterozygotes (79,53) and homozygotes (95). This is not due to increased VLDL conversion to LDL but represents an abnormal "direct" pathway of *de novo* LDL synthesis (79,53,95). Antenatal diagnosis of receptor deficiency has been reported in a 15-week fetus (96); the family was known to be affected. Although receptor studies have not yet been found to be sufficiently sensitive to distinguish heterozygotous persons from normal persons, or from patients with polygenic hypercholesterolemia, there is a strong reason to anticipate that this will become possible.

For extrahepatic cells, then, LDL concentration in plasma, or more specifically in tissue fluid, is an important determinant of their rate of cholesterol synthesis. As discussed previously, hepatic cholesterol synthesis appears to be regulated by the cholesterol of chylomicron remnants; remnant uptake by the hepatocyte is also mediated by a specific high-affinity receptor (58,59).

Cholesterol Metabolism

A number of recent papers have dealt with the effects of changes in dietary cholesterol on the plasma cholesterol level and on cholesterol balance and with the homeostatic mechanisms involved. A modest mean effect of dietary cholesterol on plasma cholesterol levels has long been recognized (97,98). Extensive experiments have revealed a very wide individual variation in the plasma lipid and lipoprotein response to dietary cholesterol in man (99). In normal man a short-term twofold increase in dietary cholesterol leads to an increment of plasma cholesterol concentration ranging from zero to about 60 mg/100 ml at 21 days (99). Most of the increment is due to a rise in LDL-cholesterol levels, but IDL and HDL_2 levels also increase.

To some extent heterogeneity of response is explained by wide variation in cholesterol absorption (about 37 to 75% in free-living subjects on a western diet, about 25 to 70% in vegetarians); the group difference was statistically significant (100). In this study there was no conspicuous relationships between absorbed dietary cholesterol and plasma cholesterol concentration. Another, now classic paper (101) on the effects of very-high-cholesterol intakes in humans revealed a varying degree of compensatory reduction in cholesterol synthesis and of enhanced sterol excretion, also without a predictable degree of hypercholesterolemia. Recently, the effects of changes in dietary cholesterol, more typical of normal human diets, have been described (102). The nine subjects who were studied fell into two distinct groups: in six subjects who compensated for increased absorption chiefly by suppression of cholesterol synthesis, plasma cholesterol increased by 3-26 mg/100 ml; in three whose main compensatory mechanism was enhanced excretion the plasma cholesterol increment was 47-142 mg/100 ml.

Hence, regulation of cholesterol synthesis appears to play the major role in homeostasis in the face of varying dietary cholesterol intake. Mistry and his colleagues (99,103) have investigated the mechanism of this regulation in mononuclear cells from 18 ambulatory subjects whose plasma cholesterol homeostasis while on a high-cholesterol diety was subsequently studied. A negative correlation was observed ($r = -0.74$, $P < 0.001$) between high-affinity LDL receptor-mediated suppression of cholesterol synthesis and the rate of increase in plasma cholesterol levels. This implies that, at least in this cell line, more effective homeostasis is associated with greater LDL receptor activity. These individual differences could not be predicted from a knowledge of such baseline characteristics as plasma lipoprotein levels or dietary patterns as assessed by 24-hour recall.

The determinants of homeostasis against dietary variations are therefore still very incompletely defined. One intriguing hypothesis holds that dietary patterns during the neonatal period may influence the lifelong capacity for and setting of homeostatic mechanisms, a concept of "locking in" of metabolic patterns. Evidence for such a process has been elusive in humans, with one suggestive positive report (104) and one negative study (105). But an experiment on neonatal guinea pigs is likely to stimulate considerable interest (106): from week 2 to week 7 one group of animals received cholestyramine, which increased fecal bile acid exretion threefold and fecal sterol excretion slightly. On a subsequent normal diet bile acid secretion remained higher than in control animals, while plasma cholesterol levels were similar. Thereafter a period of high cholesterol intake began; the rise in plasma cholesterol concentration was smaller in the animals previously exposed to cholestyramine. In this species it appears that a long-term alteration in the rate of catabolism of cholesterol can be induced during neonatal life, a conclusion pregnant with implications for future work.

Regulation of Hepatic Cholesterol Synthesis

The existence of efficient and closely regulated mechanisms for acquiring cholesterol by receptor-mediated LDL internalization is now well established. Efficient cholesterol uptake probably explains the low rates of cholesterol synthesis (107) in most organs other than the liver and small intestine. But in these two sites, sterol synthesis continues at a substantial rate despite similar exposure to LDL (107). Hepatic cholesterol synthesis is regulated by the cholesterol content of the hepatocyte or, probably the content of a cholesterol pool which has a regulatory function. When cholesterol is ingested, it is transported in chylomicrons and subsequently in chylomicron remnants which, at least in some species, are avidly taken up by the liver cell via a selective transport system (58,59). As we shall see, HDL cholesterol is also a source of hepatic cholesterol, and it has been suggested that some LDL, or LDL-cholesterol (108), enters the liver cell. The hepatic cholesterol pools are also affected by variables that alter biliary excretion of cholesterol, for example, thyroxine (1, 109) and by the rate of bile acid synthesis (110,111). The rate of cholesterol secretion by the liver into the plasma as VLDL (which is influenced by many variables including free fatty acid flux, alcohol, and sex hormones) is a further variable altering the cholesterol content of the organ and, in turn, its rate of cholesterol synthesis (90).

Bile acid administration and other procedures that modify the size of the enterohepatic bile acid pool also influence hepatic cholesterol synthesis. The mechanism of this effect has been the subject of prolonged uncertainty. Studies of the interaction between manipulations of cholesterol by the amount of cholesterol reaching the liver and of the bile-acid pool have now established that the bile-acid effects are indirect (91). They are mediated by the influence of bile acids on intestinal absorption of cholesterol and on hepatic conversion of cholesterol to bile acids. Hence, the administration of cholic acid will increase the plasma cholesterol level only if promotion of cholesterol absorption exceeds the consequent inhibition of cholesterol synthesis. Conversely cholestyramine treatment, which increases fecal excretion of bile acids, reduces plasma cholesterol levels by enhancing cholesterol conversion to bile acids, but the effect is offset to a varying extent by the increased hepatic cholesterol synthesis resulting from depletion of hepatic cholesterol.

High-Density Lipoproteins (HDL)

Interest in the metabolic role of HDL and HDL-borne cholesterol, and in regulation of HDL levels and its disorders, has burgeoned since the epidemiological relationship between low HDL-cholesterol levels and high ischemic heart disease (IHD) risk was established (30).

One advance has been the recognition of the considerable heterogeneity of plasma HDL. The existence of two subclasses - - a less-dense HDL_2 and a smaller, denser class with lower lipid: protein ratio, HDL_3 is well recognized (1); the former is isolated in the ultracentrifuge in the density range of 1.063 to 1.125 gm/ml, and the latter between 1.125 and 1.21 gm/ml. HDL_2 is the subclass that displays major sex differences in concentration, which is increased by cholesterol feeding and largely accounts for the lower mean HDL levels in case studies of IHD patients (5).

By analytical ultracentrifugation, HDL_2 has been further classified into HDL_{2b} and HDL_{2a}, the former being larger and less dense (114); their concentrations had similar sex differences and age trends in a population study.

A quantitatively minor subclass of HDL, termed HDL-I in human plasma has been described (88,115). It contains approximately 15% of HDL protein. This lipoprotein, as discussed previously, resembles LDL in its ability to interact with cell surface lipoprotein receptors on fibroblasts and smooth muscle cells. Its protein moiety includes apo E, a related polypeptide "pro-apo E", apo A-I and apo A-II, of which the apo E is responsible for recognition by the receptor (88,115). Hence, this lipoprotein is able to compete with LDL and VLDL for binding and degradation by fibroblasts and thus to influence cellular cholesterol content and cholesterol synthesis.

Nascent HDL

HDL recovered from the perfusate of isolated rat liver differs in structure and composition from HDL of the peripheral plasma (22,117, 29). It is a discoid bilamellar structure with apoprotein (50% of the protein content) located at the edge, the surfaces comprising free cholesterol and phospholipid. Apo A-I and apo C peptides are also present (27). Under the influence of the enzyme lecithin: cholesterol acyltransferase (LCAT) the free cholesterol of the nascent particles is esterified; the cholesteryl ester forms a hydrophobic core for the particle, which is thus remodeled to the mature spherical form. This remodeling appears to include the transfer of apo A-I and apo A-II of intestinal origin from chylomicrons (29). Apo A-II has the property of avidly binding phospholipids (116). Evidence of hepatic production of nascent HDL includes the demonstration that swine guinea pig and rat liver synthesize HDL apolipoproteins (5).

Whether a similar sequence of events occurs in man has yet to be established. In human LCAT deficiency (genetic or due to severe hepatitis) discoid HDL are seen in peripheral plasma (118). Human hepatic venous blood, but not arterial blood, contains discoidal particles with the density range of HDL_2 (119), suggesting production of nascent HDL by splanchnic viscera.

Much recent evidence suggests that the components of "mature" HDL are assembled from a nascent HDL particle, from apo A peptides transferred to it from chylomicrons, and from VLDL surface materials released during the action of lipoprotein lipase (120). Free cholesterol, phospholipid and apo C are released from labeled VLDL by lipoprotein lipase, either in a simple incubation system or in the perfused rat heart (63-67,120). If HDL_3 is present, these products of VLDL catabolism combine with it to form a particle closely resembling HDL_2.

In familial lipoprotein lipase deficiency, plasma HDL levels are strikingly low (121); this suggests that when chylomicron catabolism is impaired, transfer of apo A and apo C peptides, free cholesterol, and phospholipids to HDL or nascent HDL may be impeded. However, HDL levels are also somewhat low in hypertriglyceridemic patients with familial combined hyperlipidemia, in whom there is no evidence of impaired catabolism of lipoproteins; the hyperlipidemia appears to be due to overproduction of VLDL, LDL, or both lipoproteins (79).

During the remodeling of HDL, cholesteryl ester is generated by the action of LCAT. If VLDL or LDL are present when nascent HDL is incubated with this enzyme, much of the newly formed cholesteryl ester is located in these lower-density lipoproteins (122). VLDL and LDL are not substrates for LCAT in the absence of HDL, hence, it is probable that cholesteryl ester, synthesized in HDL, is then transferred to VLDL and LDL. This process is possibly dependent on a serum protein, which facilitates cholesteryl ester transfer between lipoproteins (123,124).

HDL and Centripetal Cholesterol Transport

A review by Glomset in 1968 (125) first explicitly suggested that LCAT, and its preferred substrate HDL provided a mechanism for centripetal transport of free cholesterol from peripheral tissues. The need for such a pathway is evident from the fact that such tissues acquire cholesterol from plasma, as LDL, and have the capacity - - normally largely repressed - - to synthesize this sterol. In the steady state, cholesterol must be transported away from the periphery at a similar rate to this uptake and synthesis. The destination of this pathway must be the liver since this organ is unique in its ability to excrete and to catabolize substantial amounts of cholesterol.

Three major sets of findings have lent support to this concept. *(1)* Miller et al (126) have shown a strong negative correlation in man between plasma HDL-cholesterol levels and the mass of cholesterol in both exchangeable pools of tissue cholesterol. When adipose tissue cholesterol, prelabeled with ^{14}C-cholesterol, is mobilized during weight reduction in obese subjects, cholesterol of relatively high specific activity enters the plasma. The rise in plasma cholesterol specific activity is confined to the lipoprotein fraction of density > 1.063 gm/ml containing HDL (127). After a short period of high cholesterol intake, plasma HDL_2-cholesterol levels remain high for several weeks, although other lipoprotein concentrations decrease rapidly (99).

(2) Incubation of HDL *in vitro* with a variety of cell lines leads to a net transfer of cholesterol to the lipoprotein. This is true of red cell ghosts (128), Landschutz ascites tumor cells (129), cultured human fibroblasts (130), and rat arterial smooth muscle cells (131). Interestingly, LCAT activity is not demonstrably necessary for this uptake process, at least under the *in vitro* experimental conditions (132).

(3) Some recent observations favor the view that HDL-cholesterol is taken up and used by the liver. Rat hepatocytes take up and hydrolyze cholesteryl ester when incubated with HDL, the uptake showing saturation kinetics (133). Receptor-mediated uptake of HDL by the liver is therefore a possibility. *In vitro* studies have shown that rat and human hepatocytes catabolize peptide-labeled HDL (134). However, an attempt to estimate the role of the liver in the catabolism of peptide-labeled HDL *in vivo* suggested a relatively minor role for this organ (135); in this study the removal of HDL from plasma in the intact rat was several-fold faster than removal by the isolated perfused liver. Two studies, each of a patient with a bile fistula, have been reported, in which lipoproteins labeled in the cholesterol moiety were injected and the appearance of label in bile was sought (136,137). HDL-borne cholesterol radioactivity appeared in bile cholesterol and chenodeoxycholic acid earlier than LDL-borne label. Clearly, there is a need for similar studies not involving the abnormal sterol metabolism consequent on a biliary fistula.

Hence, considerable evidence has now accumulated favoring the concept that HDL transports cholesterol from peripheral cells to the liver, where it enters a compartment functioning as a precursor of bile acids and biliary cholesterol. However, much of this evidence is indirect, and examination of the role of HDL in centripetal transport is very incomplete at present.

HDL Kinetics in Man

A number of kinetic studies of HDL metabolism have been carried out using apoliprotein labeled with radioiodine, although the assumed model has varied in different laboratories. In normal subjects the specific activity/time curves for apo A-I and apo A-II are congruent indicating that they are metabolized together (15). In Tangier disease apo A-I is catabolized more rapidly than apo A-II, and evidence has been presented that increased catabolism of HDL peptides is the kinetic abnormality even though there is also some impairment of synthesis (138).

The reduced HDL concentration that accompanies hypertriglyceridemia during short-term high-carbohydrate feeding is associated with rapid fractional catabolism of HDL-apo A peptides (15), and slow fractional removal is seen in patients in whom HDL levels are increased by nicotinic acid therapy (15).

Two studies have shown that apo A-I synthetic rate is also an important determinant of HDL apo A-I levels in plasma (139,140) and HDL cholesterol concentration (140) in normal and hyperlipidemic man; positive correlations ($r = 0.7$) were obtained in both studies between the synthetic rate and apolipoprotein or cholesterol levels. In hypertriglyceridemic patients apo A-I and apo A-II fractional catabolism was unequal, but as in earlier studies they were similar in normal subjects (140).

LIPOPROTEIN METHODOLOGY

Clinical biochemistry laboratories have been faced by a changing pattern of demand, as well as with the familiar growth in workload, in the investigation of plasma lipid disorders and of metabolic cardiovascular risk factors. The use of qualitative lipoprotein electrophoresis for "phenotyping" hyperlipidemic plasma samples, which increased a decade ago, decreased during the mid-seventies. At that time it was being recognized that a particular lipoprotein pattern was not necessarily correlated with a particular genetic disorder; for example, at least four patterns of hyperlipoproteinemia could occur in the affected members of kinships with the probable autosomal dominant disorder familial combined hyperlipidemia (141). Nor did the "phenotype" necessarily indicate an optimal dietary or drug regime in managing the disorder (1,142). In turn, the current resurgence of interest in lipoprotein analyses has arisen largely because of the inverse relationship between HDL-cholesterol concentration and IHD risk (30); the diagnosis of remnant hyperlipoproteinemia (type III; broad beta disease) is also best based on quantitative analysis of VLDL composition (143); and familial elevation of HDL levels (hyperalphalipoproteinemia) has been recognized as a non-infrequent cause of mild hypercholesterolemia (144), not requiring therapy.

Hence, there has been a spate of papers on quantitative lipoprotein fractionation. Several methods for isolation of HDL are available. Most widely used are precipitation procedures using heparin with Mn^{2+}, dextran sulphate with Ca^{+} or Mg^{2+}, or sodium phosphotungstate-Mg^{2+}, and are based on the studies of Burstein and his colleagues (145). Also available are preparative and analytical ultracentrifugation, and the introduction of an air-driven micro-ultracentrifuge (Beckman Instruments Airfuge) has reduced the time and expense of preparative ultrancentrifugation of lipoproteins. Dextran sulphate precipitation, using a high- (146) or low-molecular weight reagent (147), has given satisfactory results in some hands, as has phosphotungstate (148) precipitation of HDL. However, an increasing proportion of laboratories has adopted the heparin-Mn^{2+} method for this purpose. A definitive comparison of three variants of this method has been published (149). The optimal method uses a 2M $MnCl_2$ reagent, double the usual concentration, to obtain almost complete precipitation of VLDL and LDL. It is suitable for EDTA-Plasma in contrast with some other procedures, which have been validated only for serum. This method has been coupled with a simple ultrafiltration procedure for the analysis of heavily lipemic samples (150); even in samples with a mean triglyceride level greater than 5,000 mg/100 ml, contamination of HDL with apo B-containing lipoproteins was minor.

Although denaturation of lower-density lipoproteins occurs on freezing, it appears that reliable HDL-cholesterol analyses may be made on once-frozen plasma samples at one year (151) or six weeks (152).

Measurement of VLDL lipids is usually performed by ultracentrifugation. Precipitation methods have been described, using sodium dodecyl sulphate (145,153,154) or heparin with a low concentration of manganese ions (155). Although SDS precipitation of VLDL is quantitative at plasma triglyceride concentrations up to 350 mg/100 ml, recovery is incomplete above this level (156).

Published experience with the air-driven microultracentrifuge has been satisfactory (157,158). The simplicity of the equipment minimizes breakdown frequency and maintenance costs, items of considerable importance in conventional ultracentrifugation; and the high gravitational force and small sample size permit far shorter separation times than with macromethods. Equipment is available for isolation of fractions by aspiration, although several laboratories have developed satisfactory tube-slicing apparatus. The method is of potential value particularly for VLDL analyses and for HDL subfractionation in clinical biochemistry routine.

Several immunological methods for apolipoprotein measurement are now available, for apo B (159-164), for apo A-I (165), and for other peptides. Although their research usefulness is evident, their place in the routine laboratory has yet to be demonstrated; a probable exception is the value of plasma apo E measurement in the diagnosis of remnant (type III) hyperlipoproteinemia (166).

Advances in lipid measurements have continued, as recorded in a monumental review by Zak (167). The use of cholesterol oxidase in cholesterol measurements has greatly enhanced their specificity in continuous-flow methods (168-170) and in the centrifugal analyzer (171).

REGULATION OF PLASMA LIPOPROTEIN LEVELS

The past two decades have seen a formidable volume of studies on the determinants of plasma lipoprotein concentrations (1), an interest fueled in large part by its relevance to the role of lipoproteins in disease states.

The genetic contribution to regulation of plasma cholesterol concentration has been the subject of divergent reports. Significant modest correlations exist between relatives (173) pointing to a genetic influence or common environmental factors. Formal heritability estimates in Finland based on the concordance between pairs of monozygotic and dizygotic twins, indicated a substantial genetic contribution, and earlier Danish and American studies drew the same conclusion (173). More recent reports from the U.S. National Heart and Lung Institute twin study showed less extensive genetic effects (174,175). The free cholesterol level in plasma and in its LDL and HDL fractions was under genetic influence: but after allowing for bias due to an apparently greater environmental effect on dizygotic twins, no genetically determined variance could be detected in the cholesteryl ester level in plasma or in any lipoprotein fraction. It is a little difficult to envisage a genetically determined control mechanism uniquely affecting the metabolism of unesterified cholesterol. The same study has demonstrated considerable genetic regulation of plasma triglyceride concentration. In view of the negative correlation between triglyceride levels (which are in part genetically determined) and HDL-cholesterol concentration, and knowing the metabolic links between VLDL and HDL metabolism, it is a little surprising that HDL- LDL- and total-cholesterol levels appear to be devoid of genetic control. Blood group studies have revealed significantly higher total serum cholesterol levels in subjects of group A (176,177). A familial form of plasma HDL elevation has also been described (144).

Several recent papers have dealt with age and sex trends affecting plasma lipid levels during childhood (178). Some have reported a pubertal drop in cholesterol levels in boys (179,180), while others (181) noted no age trend between 6 and 18 years and no significant sex difference. A study of cord blood lipoproteins by ultracentrifugation and electrophoresis has considerably supplemented our knowledge of lipid distribution in the lipoprotein classes in neonates (182); on average 50% of the serum cholesterol was transported in HDL. VLDL was consistently present in cord blood serum, albeit with a remarkably wide concentration range. In other studies a relationship has been suggested between prenatal stress and plasma triglyceride levels.

Although interpopulation differences in plasma cholesterol concentration have received attention during the past 25 years, it is only in the past few years that quantitative comparisons of plasma lipoprotein levels have been reported (183,184). In comparisons between carefully screened populations in European cities (184), VLDL-triglyceride and LDL-cholesterol levels showed congruent trends in the four communities, as did total levels of triglyceride and cholesterol in plasma; these trends paralleled IHD mortality rates. HDL-cholesterol concentrations showed no significant interpopulation differences in either sex.

Variables that influence HDL concentrations have received particular attention. An effect of regular vigorous exercise was suggested by reports that marathon runners had HDL-cholesterol levels that were considerably higher than in control subjects (185). When relatively sedentary persons undertook a fairly strenuous exercise program, HDL-cholesterol concentrations increased (186,187). Clearly such a period of training involves changes in body composition and in either food intake or weight or both; the operative factors are not readily identified. Interestingly, HDL-cholesterol levels showed a positive correlation (in a combined group of active and sedentary people) with lipoprotein lipase activity in adipose tissue but not with that in muscle (78). Still, there is an impressive "dose-response" relationship between fitness, assessed as maximum aerobic capacity, and the HDL-cholesterol level (188); the ratio of HDL-cholesterol to apo A-I levels, reflecting the HDL_2:HDL_3 ratio, correlated even more strongly with aerobic capacity (r = 0.88). A direct effect of fitness seems probable in view of the strength of this relationship.

Epidemiological evidence indicates a clear positive correlation between stated alcohol intake and HDL-cholesterol levels (189) and a weaker, but still significant, positive relationship with VLDL-triglyceride concentration (12). This has been confirmed in clinical (190) and experimental studies (191). Other microsomal enzyme-inducing agents had been reported to increase HDL-cholesterol concentration including phenytoin (192) and halogenated hydrocarbon insecticides (193).

Obesity is a common, although far from invariable, finding in hypertriglyceridemic patients (1); and in population studies VLDL-triglyceride levels correlate positively with relative body weight (1,12), although the relationship is seldom strong. The mechanisms of these relationships have been explored. First, most studies of the kinetic basis of endogenous hypertriglyceridemia have revealed overproduction of total- or VLDL-triglyceride (44) and of VLDL-apo B (56), with positive correlations between turnover and concentration. When hypertriglyceridemic patients have been classified according to their genetic basis, those with familial combined hyperlipidemia consistently show overproduction of VLDL-apo B and most of those with familial hypertriglyceridemia have normal VLDL-apo B synthetic rates and very slow FCR (79). Such findings do not necessarily predict the nature of abnormalities of triglyceride kinetics in these disorders. Second, it is known that the elevated mean VLDL-triglyceride synthetic rates in hypertriglyceridemic patients is decreased by weight reduction to a new steady state (194). Hence, it is intriguing to note that relative body weight is not correlated with VLDL-triglyceride synthesis (195). This does not invalidate the therapeutic use of weight reduction in managing hypertriglyceridemic patients but suggests that obesity is not an intrinsic part of the endogenous hypertriglyceridemias as a group.

A link between dietary fiber intake and IHD was first suggested on the basis of the geographical epidemiology of the disease (196) and has initiated extensive nutritional studies. Some of these have been analyzed in a valuable recent review (197). Gel-forming fibers such as pectin and guar reduce plasma cholesterol and LDL levels, possibly by enhancing fecal bile acid and neutral sterol excretion. Plasma triglyceride levels are not affected. Cellulose is negligibly effective, and wheat bran is ineffective in altering plasma cholesterol concentration.

There has been a recurrence of interest in dietary protein as a variable that influences plasma lipid levels. A substantial reduction in elevated plasma cholesterol levels was noted in one study, in which a soybean protein preparation was substituted for all sources of animal protein (198). Some (199), but not all (200), findings had also favored the view that replacement of dietary proteins of animal origin by those of vegetable origin leads to a reduction in plasma cholesterol concentration. The most carefully controlled study to date, using a soybean protein and rigorously excluding other dietary variables, has confirmed a cholesterol-lowering effect (201); however, the reduction was relatively small, about 5% in subjects with normal baseline levels. Dietary protein effects on plasma lipid levels have been critically reviewed (202).

Comprehensive lessening of cardiovascular risk appears to require the reduction, *inter alia,* of elevated blood pressure and elevated plasma LDL (and probably VLDL) levels, and increased HDL-cholesterol concentration (203). Hence, it has been disquieting to note that the most widely used antihypertensive drugs, the thiazide diuretics, tend to increase plasma triglyceride levels (204). A preliminary report has suggested that a thiazide together with propranolol may decrease HDL-cholesterol concentration (205).

An extensive study of the influence of exogenous sex hormones (especially in contraceptive formulations) on HDL-cholesterol levels has been reported (206). Levels were increased by estrogens and decreased by progestogens; oral contraceptive preparations had widely ranging effects, but the majority increased HDL-cholesterol concentration. No data were available about the effects of very-low estrogen formulations.

DISORDERS OF LIPOPROTEIN METABOLISM

A review by the author has discussed the hyperlipidemias in clinical practice and in the context of clinical biochemistry (1). In this section some areas of topical medical interest will be highlighted. Generally, the hyperlipidemias are of interest to the physician in the context of risk factors for IHD and peripheral vascular disease. However, other manifestations of these metabolic disorders are well known, such as cutaneous and tendinous xanthomas and acute relapsing pancreatitis; and with increasing attention to lipid and lipoprotein measurements, further clinical manifestations are becoming apparent.

Some Clinical Manifestations

Neurological Disorders

Abetalipoproteinemia is consistently accompanied, during adolescence or early adult life, by ataxia, motor and sensory deficits, visual changes, and kyphoscoliosis; patients with hypobetalipoproteinemia sometimes develop similar but milder lesions (14). A sensory and motor neuropathy occurs in some patients with α-lipoprotein deficiency (14).

A number of single-case reports, and a recent series of five well-documented cases, have defined the entity of hyperlipidemic peripheral neuropathy (207); long-tract lesions producing ataxia and spasticity have also been recorded. The most consistent finding has been of sensory symptoms - - paresthesia in the extremities, with clinical and electrophysiological evidence of a mild-to-moderate sensorimotor neuropathy. A severe organic dementia has also been described in two patients with hyperlipidemia and without evidence of other causation. The characteristic features of hyperlipidemic neuropathy and dementia are hypertriglyceridemia with relatively normal cholesterol levels. The increase in the triglyceride level is usually but not always gross and is due to raised levels of VLDL, with or without chylomicronemia. In patients with lipemic dementia the lipid abnormality has always been pronounced. The causal nature of the association is attested by conspicuous clinical improvement when the metabolic abnormality is corrected by diet, sometimes with mediation; the neurological problems appear to resolve within about three months.

The mechanism of the neurological abnormality has been the subject of much speculation but is essentially unknown. Clearly the association between nervous system abnormalities and lipoprotein disorders can also be the result of common conditions such as diabetes mellitus, alcoholism, systemic lupus erythematosus, and, of course, cerebral ischemia, which must be excluded before hyperlipidemic neuropathy is diagnosed.

Rheumatological Syndromes

The occurrence of a migratory large-joint polyarthritis in severe familial hypercholesterolemia was well described by Khachadurian (208). Commonly affecting the knee and ankles, and occurring especially in adolescence and early adult life, the syndrome is often misdiagnosed as acute rheumatic fever. The similarity is compounded by the presence of xanthomas, which may be mistaken for rheumatic nodules, and sometimes by an elevated erythrocyte sedimentation rate. Synovial fluid has not been reported to contain crystals, and the basis of this complication of heterozygous and homozygous familial hypercholesterolemia is unknown. The subject has been reviewed (1, 209).

Several other rheumatological syndromes have been described in association with hypertriglyceridemia and hypercholesterolemia (210,211).

Dyspnea

The author's attention was drawn to the possibility that dyspnea may complicate severe hypertriglyceridemia by two patients in whom this symptom was otherwise inexplicable. Both had pronounced elevation of VLDL levels with marked chylomicronemia; and both showed symptomatic improvement within a week of instituting treatment that eliminated the chylomicronemia.

Carbon monoxide diffusing capacity has been measured in clinical and experimental hypertriglyceridemia. In one study no abnormality was observed (212); in another, reduction of diffusing capacity was noted (213). A modest reduction occurred in subjects rendered temporarily hypertriglyceridemic by the infusion of Intralipid (214).

Recurrent venous thromboses have been noted in patients with pronounced hypertriglyceridemia (1) Elevated factor XIII activity in plasma has been reported in 18 patients with a mean plasma triglyceride concentration of 534 mg/100 ml (215); the co-existence of antithrombin III deficiency with moderate hypertriglyceridemia has been reported (237).

Atherosclerosis and Ischemic Heart Disease (IHD)

The preeminence of IHD as a cause of death in men in the United States and most other westernized countries has led not only to one of the most massive research endeavours in medical history but also to increasingly active intervention against the known risk factors for IHD. Hence, the well-documented decrease in IHD mortality in the United States since 1963 (27%) (216,217), and the suspicion that this is also occurring in the United Kingdom (218) and perhaps other countries, is of extreme interest. The relative contributions to this trend of altered dietary habits leading to the reduction of plasma lipid levels and obesity, the improved detection and treatment of hypertension, the lower tobacco consumption in men in certain social classes, and increased leisure-time physical activity remain to be assessed. Since cigarette consumption by women (who show a decrease in mortality equal to that in men) is still increasing, it is relatively unlikely that altered smoking habits are the major reason for the overall trend (217).

In two recent reviews the authors have dealt with the etiology of IHD and have given reasons for believing that the behavioral risk factors play a causal role, as opposed to a merely predictive one (219,203). Plasma cholesterol levels were positively, and HDL-cholesterol levels were inversely, related to the angiographically determined severity of coronary atherosclerosis (220); their independent roles were confirmed by multivariate analysis.

Controlled trials of lipid-reducing agents (diet and drugs) in decreasing IHD incidence have often been marred by the inadequate degree of risk-factor reduction and/or by sample sizes, which, based on present knowledge, have been statistically inadequate. Opinions about their interpretation have varied (291,203,221-226).

Some larger studies, and also those in which effective drugs or drug combinations have produced substantial plasma lipid reduction, have now been completed. One controlled trial on myocardial infarction survivors employed clofibrate with nicotinic acid (227); the relapse rate was approximately halved, but no significant reduction in mortality has yet emerged. Another study of a large number of hypercholesterolemic subjects used the bile-acid sequestrant drug, colestipol (228); reduced fatal and nonfatal events were noted in men but not in women. This was a well-randomized, placebo-controlled study. The very large WHO-supported study of the effect of clofibrate (229) on men in whom the plasma cholesterol level was in the upper tertile involved the intervention group, a set of upper-tertile controls, and a set of lower-tertile controls; attack rate, but not mortality was significantly lessened despite the fairly modest reduction in cholesterol levels. A "dose-response" relationship was evident between resultant cholesterol concentration and risk. Although this study has been interpreted as revealing substantial side effects of clofibrate, the only consistent problem has been an increased risk of gallstone disease (230). Because of the great interest in this problem, there have been reports of a comparison of risk-related variables in Edinburgh and Stockholm (231); the IHD mortality in Edinburgh is threefold higher than in Stockholm. This difference is paralleled by differences in plasma triglyceride levels, obesity, blood pressure, and insulin responsiveness; HDL-cholesterol levels were lower in Edinburgh. The linoleic acid percentage content of plasma triglyceride, and of adipose tissue, was much lower in men in Edinburgh, a finding almost certainly reflecting dietary patterns of fat intake.

Sex hormone abnormalities have been noted in case-control studies of IHD survivors and included altered estrogen:androgen ratios (232), low plasma testosterone levels (233), and high estradiol and estrone levels (234). Whether these variations are of causal significance remains to be determined.

The ongoing debate about the role of plasma triglyceride concentration as a determinant of IHD risk has yet to be resolved. Two recent contributions have favored the existence of a role for triglyceride. Triglyceride levels exceeding 150 mg/100 ml appear to increase the risk of IHD, independently of any variation in plasma cholesterol levels, related body weight, and smoking (235); HDL-cholesterol levels were not measured in this study, however. Independent risk prediction was also indicated in a case-control study in which the joint probability density function was estimated (236).

CONCLUSION

The year 1978 has been remarkable for the number of substantial advances made in our understanding of plasma lipoprotein metabolism and its disorders. Noteworthy has been the unravelling of the relationships between HDL metabolism and that of other lipoproteins. No less substantial has been the recognition that progress is being made in overcoming the problem of ischemic heart disease; this is reassurance indeed for devotees of human lipoprotein metabolism even if they cannot claim all the credit for this trend.

REFERENCES

1. Lewis B: *The Hyperlipidaemias: Clinical and Laboratory Practice*. Oxford, Blackwell Scientific, 1976.

2. Rifkind BM, Levy RI: *Hyperlipidemia Diagnosis and Therapy*. New York, Grune and Stratton, 1977.

3. Levy RI, et al: *Nutrition, Lipids and Coronary Heart Disease: A Global View*. New York, Raven Press, 1979.

4. Schettler G, et al: *Atherosclerosis IV*. New York, Springer-Verlag, 1977.

5. Nicoll A, Miller NE, Lewis B: In *Advances in Lipid Research*.

D Kritchevsky, Paoletti R (eds.): (in press).

6. Jackson RL, Morriset JD, Gotto AM Jr: *Physiol Rev* 56:259, 1976.

7. Eisenberg S, Levy RI: *Adv Lipid Res* 13:1, 1976.

8. Osborne JC Jr, Brewer HB Jr: *Adv Protein Chem* 31:253, 1977.

9. Smith LC, Pownall HJ, Gotto AM Jr: *Ann Rev Biochem* 47:751, 1978.

10. Lewis B, et al: *Atherosclerosis* 17:455, 1973.

11. Fredrickson DG, Levy RI: In *Metabolic Basis of Inherited Disease* ed 3. JB Stanbury, JB Wyngaarden, DS Fredrickson (eds.): New York, McGraw-Hill, 1972, p 545.

12. Lewis B, et al: *Lancet* 1:141, 1974.

13. Carlson LA, Ericsson M: *Atherosclerosis* 21:435, 1975.

14. Fredrickson DS, Gotto AM Jr, Levy RI: In *Metabolic Basis of Inherited Disease* JB Stanbury, JB Wyngaarden, DS Fredrickson (eds.): New York, McGraw-Hill, 1972, chap 26.

15. Blum CB, et al: *J Clin Invest* 60:795, 1977.

16. Wilson DE, Lees RS: *J Clin Invest* 51:1051, 1972.

17. Bradley DD, et al: *N Engl J Med* 299:17, 1978.

18. Belfrage P, et al: *Eur J Clin Invest* 7:127, 1977.

19. Gofman JW, et al: *Plasma* 2:413, 1954.

20. Krauss RM, et al: *Clin Chim Acta* 80:465, 1977.

21. Windmueller HG, Herbert PN, Levy RI: *J Lipid Res* 14:215, 1973.

22. Hamilton RL: *Adv Exp Med Biol* 26:7, 1972.

23. Havel RJ, Kane JP, Kashyap ML: *J Clin Invest* 52:32, 1973.

24. Kashyap ML, et al: *J Clin Invest* 60:171, 1977.

25. La Rosa JC, et al: *Biochem Biophys Res Comm* 41:57, 1970.

26. Felker TE, et al: *J Lipid Res* 18:465, 1977.

27. Hamilton RL, et al: *J Clin Invest* 58:667, 1976.

28. Green PHR, Tall ARJ, Glickman RM: *J Clin Invest* 61:528, 1978.

29. Havel RJ: In *High Density Lipoproteins and Atherosclerosis* AM Gotto Jr, NE Miller, MF Oliver (eds.): Amsterdam, Elsevier-North Holland, 1978, p 21.

30. Miller GJ, Miller NE: In *High Denisty Lipoproteins and Atherosclerosis* AM Gotto Jr, NE Miller, MF Oliver (eds.): Amsterdam, Elsevier-North Holland, 1978, p 85.

31. Kostner GH, Holasek A: *Biochemistry* 11:3419, 1972.

32. Imaizumi K, et al: *J Lipid Res* 19:1038, 1978.

33. Glickman RN, et al: *N Engl J Med* 299:1424, 1978.

34. Schonfeld G, Bell E, Alpers DH: *J Clin Invest* 61:1539, 1978.

35. Schafer EJ, Jenkins LL, Brewer HB: *Biochem Biophys Res Comm* 80:405, 1978.

36. Rao S, et al: (in preparation).

37. Gitlin D, et al: *J Clin Invest* 37:172, 1958.

38. Walton KW, et al: *Clin Sci* 29:217, 1965.

39. Sigurdsson G, Nicoll A, Lewis B: *J Clin Invest* 56:1481, 1975.

40. Sunous LA, et al: *Atherosclerosis* 21:283, 1975.

41. Berman M, et al: *J Lipid Res* 19:38, 1978.

42. Reardon MF, Fidge NH, Nestel PJ: *J Clin Invest* 61:890, 1978.

43. Phair RD, et al: *Fed Proceedings* 34:2263, 1975.

44. Reaven GM, et al: *J Clin Invest* 44:1826, 1965.

45. Olefsky J, Reaven GM, Farquhar JW: *J Clin Invest* 53:65, 1974.

46. Havel RJ: *Proceedings of the 1968 Devel Conference on Lipids* Washington US Government Printing Office, 1968, p 117.

47. Zech LA, Grundy SM, Berman M: *Circulation* 53:II-5, 1976.

48. Higgins JM, Felding CJ: *Biochemistry* 14:2288, 1975.

49. Sreja D, Kallai MA, Steiner G: *Metabolism* 26:1333, 1977.

50. Eisenberg S, et al: *Biochim Biophys Acta* 260:329, 1972.

51. Eisenberg S, Bilheimer DW, Levy RI: *Biochim Biophys Acta* 326:361, 1973.

52. Lewis B: *Biochem Soc Trans* 5:589, 1977.

53. Janus E, et al: *Circulation* 55:III-21, 1977.

54. Janus E, et al: (In preparation).

55. Sigurdsson G, Nicoll A, Lewis B: *Eur J Clin Invest* 6:151, 1976.

56. Sigurdsson G, Nicoll A, Lewis B: *Eur J Clin Invest* 6:167, 1976.

57. Redgrave TG: *J Clin Invest* 49:465, 1970.

58. Cooper AD: *Biochim Biophys Acta* 488:464, 1977.

59. Sherrill BC, Dietschy JM: *J Biol Chem* 253:1859, 1978.

60. Kane JP, et al: *J Clin Invest* 56:1481, 1975.

61. Turner P, et al: *Eur J Clin Invest* (in press).

62. Chait A, et al: *Metabolism* 27:1055, 1978.

63. Eisenberg S: In *High Density Lipoproteins and Atherosclerosis* AM Gotto Jr, NE Miller, MF Oliver (eds.): Amsterdam, Elsevier-North Holland, 1978, p 67.

64. Eisenberg S: *J Lipid Res* 19:229, 1978.

65. Glangeaud MC, Eisenberg S, Olivecrona T: *Biochim Biophys Acta* 486:23, 1977.

66. Deckelbaum R, et al: *Circulation* 54:II-26, 1977.

67. Chajek T, Eisenberg S: *J Clin Invest* 61:1654, 1978.

68. Nicoll A, et al: *Circulation* 55:III-23, 1977.

69. Mjøs O, et al: *J Clin Invest* 56:603, 1975.

70. Pykalisto OJ, et al: *J Clin Invest* 56:1108, 1975.

71. Lithell M, Boberg J: *Scand J Lab Clin Invest* 37:551, 1977.

72. Spencer IM, Hutchinson A, Robinson DS: *Biochim Biophys Acta* 530:375, 1978.

73. Lithell M, et al: *Atherosclerosis* 30:89, 1978.

74. Jansen H, et al: *Biochim Biophys Acta* 531:109, 1978.

75. Robinson DS, Cryer A, Davies P: *Proceedings of the Nutrition Society* 34:211, 1975.

76. Nilsson-Ehle P, Garfinkel AS, Schotz MC: *Biochim Biophys Acta* 431:147, 1976.

77. Chazek R, Stein O, Stein Y: *Biochim Biophys Acta* 528:466, 1978.

78. Nikkilä EA, et al: *Metabolism* 27:1661, 1978.

79. Janus E, Nicoll A, Lewis B: *Protides of the Biological Fluids* 25:171, 1978.

80. Sniderman AD, et al: *Science* 183:526, 1974.

81. Davis RA, et al: *Circulation* 57:II-169, 1978.

82. Williams CD, Avigan J: *Biochim Biophys Acta* 260:413, 1972.

83. Brown MS, Goldstein JL: *Science* 191:150, 1976.

84. Goldstein JL, Brown MS: *Ann Rev Biochem* 46:897, 1977.

85. Kayden HJ, Hatam L, Beratis NG: *Biochemistry* 15:521, 1976.

86. Albers JJ, Bierman EL: *Biochim Biophys Acta* 424:422, 1976.

87. Reckless JPD, Weinstein DB, Steinberg D: *Biochim Biophys Acta* 529:475, 1978.

88. Innerarity TL, et al: *J Biol Chem* 253:6289, 1978.

89. Anderson RGW, Goldstein JL, Brown MS: *Proc Nat Acad Sci USA* 71:788, 1976.

90. Miller NE, Yin JA: *Biochim Biophys Acta* 530:145, 1978.

91. Brown MS, Goldstein JL: *Cell* 6:307, 1975.

92. Brown MS, Dana SE, Goldstein JL: *J Biol Chem* 249:789, 1974.

93. Brown MS, Goldstein JL: *N Engl J Med* 294:1386, 1976.

94. Langer T, Strober W, Levy RI: *J Clin Invest* 51:1528, 1972.

95. Soutar AK, Myant NB, Thompson GR: *Atherosclerosis* 28:247, 1977.

96. Brown MS, et al: *Lancet* 1:526, 1978.

97. Keys A, Anderson J, Grande F: *Metabolism* 14:747, 1965.

98. Hegsted DM, et al: *Amer J Clin Nutr* 17:281, 1968.

99. Mistry P, et al: *Protides of the Biological Fluids* 25:349, 1978.

100. Simons LA, et al: *Amer J Clin Nutr* 31:1334, 1978.

101. Quintao E, Grundy SM, Ahrens EH: *J Lipid Res* 12:233, 1971.

102. Nestel PJ, Poyser A: *Metabolism* 25:1591, 1976.

103. Mistry P, et al: (in preparation).

104. Hodgson PA, et al: *Metabolism* 25:739, 1976.

105. Glueck CJ, et al: *Metabolism* 21:1181, 1972.

106. Li JR, Bale LK, Subbiah MTR: *Atherosclerosis* 32:93, 1979.

107. Dietschy JM, Wilson JD: *N Engl J Med* 282:1128, 1179, 1241, 1970.

108. Sniderman A, et al: *J Clin Invest* 61:867, 1978.

109. Miettinen TA: *J Lab Clin Med* 71:537, 1968.

110. Myant NB, Eder HA: *J Lipid Res* 2:363, 1961.

111. Weis HJ, Dietschy JM: *Biochim Biophys Acta* 398:315, 1975.

112. Goh EH, Heimberg M: *J Biol Chem* 252:2822, 1977.

113. Nervi FO, Dietschy JM: *J Clin Invest* 61:895, 1978.

114. Anderson DW, et al: *Atherosclerosis* 29:161, 1978.

115. Mahley RW, et al: In *High Density Lipoprotein and Atherosclerosis*. A.M. Gotto Jr, NE Miller, MF Oliver (eds.): Amsterdam: Elsevier-North Holland, 1978, p 149.

116. Pownall HJ, Sparrow JT, Gotto AM Jr: In *High Density Lipoprotein and Atherosclerosis*. AM Gotto Jr, NE Miller, MF Oliver (eds.): Amsterdam, Elsevier-North Holland, 1978, p 5.

117. Havel RJ, Hamilton RL: In *Atherosclerosis IV*. G. Schettler, et al (eds.): Berlin, Springer, 1977, p 192.

118. Ragland JB, Bertram PD, Sabesin SM: *Biochem Biophys Res Comm* 80:81, 1978.

119. Turner P, et al: *Lancet* 1:645, 1979.

120. Patsch JR, et al: *Proc Nat Acad Sci USA* 75:4519, 1978.

121. Fredrickson DS, Goldstein JL, Brown MS: In *Metabolic Basis of Inherited Disease ed 5.* JB Stanbury, JB Wyngaarden, DS Fredrickson (eds): New York, McGraw-Hill, 1978, p 608.

122. Glomset JA: In *High Density Lipoprotein and Atherosclerosis* AM Gotto Jr, NE Miller, MF Oliver (eds.): Amsterdam, Elsevier-North Holland, 1978, p 57.

123. Zilversmit DB, Hughes LB, Balmer J: *Biochim Biophys Acta* 409:393, 1975.

124. Barter PJ, Lally JI: *Biochim Biophys Acta* 531:233, 1978.

125. Glomset JA: *J Lipid Res* 9:155, 1968.

126. Miller NE, Nestel PJ, Clifton-Bligh P: *Atherosclerosis* 23:535, 1976.

127. Nester PJ, Miller NE: In *High Density Lipoprotein and Atherosclerosis* AM Gotto Jr, NE Miller, MF Oliver (eds.): Amsterdam, Elsevier-North Holland, 1978, p 51.

128. Glomset JA: *Amer J Clin Nutr* 23:1129, 1970.

129. Stein O, Stein Y: *Biochim Biophys Acta* 326:232, 1973.

130. Stein O, et al: *Biochim Biophys Acta* 450:367, 1976.

131. Stein O, et al: *Biochim Biophys Acta* 431:363, 1976.

132. Stein O, Goren R, Stein Y: *Biochim Biophys Acta* 529:309, 1978.

133. Drevon CA, Berg T, Norum KR: *Biochim Biophys Acta* 487:122, 1977.

134. van Berkel TJC, Koster JF, Hulsmann WC: *Biochim Biophys Acta* 486:586, 1977.

135. Sigurdsson G, Noel S-P, Havel RJ: *Circulation* 56:III-4, 1977.

136. Schwartz CC, et al: *Science* 200:62, 1978.

137. Halloran LG, et al: *Surgery* 84:1, 1978.

138. Schaefer EJ, et al: *N Engl J Med* 299:905, 1978.

139. Fidge N, et al: *Circulation* 57:II-40, 1978.

140. Magill P, et al: (in preparation).

141. Hazzard WR, et al: *J Clin Invest* 52:1569, 1973.

142. Sommariva D, Scotti L, Fasoli A: *Atherosclerosis* 29:43, 1978.

143. Hazzard WR, Porte D, Bierman EL: *Metabolism* 21:1009, 1972.

144. Glueck CJ, et al: *J Lab Clin Med* 88:941, 1976.

145. Burstein M, Scholnick HR: *Adv Lipid Res* 2:67, 1973.

146. Scrinivasan SR, et al: *Clin Chem* 24:157, 1978.

147. Kostner GM *Clin Chem* 22:695, 1976.

148. Lopes-Virella MF,et al: *Clin Chem* 23:882, 1977.

149. Albers JJ, et al: *Clin Chem* 24:853, 1978.

150. Warnick GR, Albers JJ: *Clin Chem* 24:900, 1978.

151. Miller NE: *Lancet* 2:134, 1977.

152. Reckless JPD, et al: *Lancet* 2:350, 1977.

153. Wilson DE, Spiger MJ: *J Lab Clin Med* 82:473, 1973.

154. Ononogbu IC, Lewis B: *Clin Chim Acta* 71:397, 1976.

155. Lampleigh SM, Muirhead RA, Deegan T: *Clin Chim Acta* 86:31, 1978.

156. Hammett F, Lewis B: Unpublished.

157. Bronzert TJ, Brewer HB: *Clin Chem* 23:2089, 1977.

158. Wieland H, Seidel D: *Ärtzl Lab* 23:96, 1977.

159. Schonfeld G, et al: *J Clin Invest* 59:1458, 1974.

160. Bantovich GJ, et al:*Atherosclerosis* 21:217, 1975.

161. Albers JJ, Cabana VG, Harrard WR: *Metabolism* 24:1339, 1975.

162. Thompson GR, et al: *Atherosclerosis* 24:107, 1976.

163. Durrington PM, Bolton CH, Hartog M: *Clin Chim Acta* 82:151, 1978.

164. Karlin JB, et al: *Eur J Clin Invest* 8:19, 1978.

165. Albers JJ, et al: *Metabolism* 25:613, 1976.

166. Kushwaha RS, et al: *Ann Intern Med* 87:509, 1977.

167. Zak B: *Clin Chem* 23:1201, 1977.

168. Richmond W: *Clin Chem* 22:1579, 1976.

169. van Gent CM, et al: *Clin Chim Acta* 75:243, 1977.

170. Lie RF, et al: *Clin Chem* 22:1627, 1976.

171. Wentz PW, Cross RE, Savory J: *Clin Chem* 22:188, 1976.

172. Lewis B: *J Roy Soc Med* 71:809, 1978.

173. Slack J: *Postgrad Med J* 51 (suppl 8):27, 1975.

174. Christian JC, et al: *Amer J Human Genetics* 28:174, 1976.

175. Feinleib M, et al: *Amer J Epidemiol* 106:284, 1977.

176. Oliver MF, et al: *Lancet* 2:605, 1969.

177. Langman MJS, et al: *Lancet* 2:607, 1969.

178. Lewis B: *Postgrad Med J* 54:181, 1978.

179. Ellefson RD, et al: *Mayo Clinic Proc* 53:307, 1978.

180. Frerichs RR, et al: *Circulation* 54:302, 1976.

181. Laver RM, et al: *J Ped* 87:697, 1975.

182. Hardell LI, Carlson LA: *Clin Chim Acta* 90:285, 1978.

183. Castelli WP, et al: *J Chron Dis* 30:147, 1977.

184. Lewis B, et al: *Eur J Clin Invest* 8:165, 1978.

185. Wood PD, et al: *Ann NY Acad Sci* 301:748, 1977.

186. Lopez-s R, et al: *Atherosclerosis* 20:1, 1974.

187. Erkelens DW, et al: *J Amer Med Assoc* (in press).

188. Miller NE, et al: *Lancet* 1:111, 1979.

189. Yano K, Rhodes GG, Kagan A: *N Engl J Med* 297:405, 1977.

190. Johansson BG, Medhus A: *Acta Med Scand* 195:273, 1974.

191. Belfrage P, et al: *Eur J Clin Invest* 7:127, 1977.

192. Nikkila EA, et al: *Brit Med J* 2:99, 1978.

193. Carlson LA, Kolmodin-Hedman B: *Acta Med Scand* 192:29, 1972.

194. Olefsky J, Reaven GM, Farquhar JW: *J Clin Invest* 53:64, 1974.

195. Reaven GM, Bernstein RM: *Metabolism* 27:1047, 1978.

196. Trowell H: *Amer J Clin Nutr* 25:926, 1972.

197. Kay RM, Strasberg SM: *Clin Invest Med* 1:9, 1978.

198. Sirtori CR, et al: *Lancet* 1:275, 1977.

199. Olson RE, et al: *Amer J Clin Nutr* 6:310, 1958.

200. Anderson JT, Grande E, Keys A: *Amer J Clin Nutr* 24:524, 1971.

201. Carroll KK, et al: *Amer J Clin Nutr* 31:1312, 1978.

202. Carroll KK: *Lipids* 13:360, 1977.

203. Lewis B: *J Roy Soc Med* 71:809, 1978.

204. Ames RP, Hill P: *Amer J Med* 61:748, 1976.

205. Helgeland A, et al: *Lancet* 2:403, 1978.

206. Bradley DD, et al: *N Engl J Med* 299:17, 1978.

207. Mathew NT, et al: *Eur Neurology* 14:370, 1976.

208. Khachadurian AK: *Arthr Rheumat* 11:385, 1968.

209. Rooney PJ, et al: *Quart J Med* 47:249, 1978.

210. Buckingham RB, Bole GG, Bassett DR: *Arch Intern Med* 135:286, 1975.

211. Mielants H, et al: *J Rheumat* 2:430, 1975.

212. Newbald HH, et al: *Amer Rev Resp Dis* 112:83, 1975.

213. Enzi G, Bevilacqua M, Crepaldi G: *Bull Eur Physiopathol Resp* 12:433, 1976.

214. Green HL, Hazlett D, Demaree R: *Amer J Clin Nutr* 29:127, 1976.

215. Cucuianu MP, et al: *Thrombosis and Haemostasis* 36:542, 1976.

216. Walker WJ: *N Engl J Med* 297:163, 1977.

217. Gordon T, Thom T: *Prevent Med* 4:115, 1975.

218. Florey C duV, Melia RJW, Darby SC: *Brit Med J* 1:635, 1978.

219. Stamler J: *Circulation* 58:3, 1978.

220. Jenkins PJ, Harper RW, Nestel PJ *Brit Med J* 2:388, 1978.

221. Hegsted DM: *Amer J Clin Nutr* 31:1504, 1978.

222. Glueck CJ, Mattson F, Bierman EL: *N Engl J Med* 298:1471, 1978.

223. Rifkind BM, Levy RI: *Arch Surg* 113:80, 1978.

224. Stone NJ: *J Chron Dis* 31:1, 1978.

225. Mann GV: *N Engl J Med* 297:644, 1977.

226. McMichael J: *Brit Med J* 1:173, 1979.

227. Carlson LA, et al: *Atherosclerosis* 28:81, 1977.

228. Dorr AE, et al: *J Chron Dis* 31:5, 1978.

229. Committee of Principal Investigation: *Brit Heart J* 40:1069, 1978.

230. Lewis B, Miller NE, Brunzell JD: *Lancet* 2:1302, 1978.

231. Logan RL, et al: *Lancet* 1:949, 1978.

232. Phillips GB: *Proc Nat Acad Sci USA* 74:1729, 1977.

233. Poggi UL, et al: *J Steroid Biochem* 7:229, 1976.

234. Entrican JM, et al: *Lancet* 2:487, 1978.

235. Pelkonen R, et al: *Brit Med J* 2:1185, 1977.

236. Scott DW, et al: *J Chron Dis* 31:337, 1978.

237. Gyde OHB, et al: *Brit Med J* 1:621, 1978.

7
Selected Topics in Diabetes Mellitus

George Steiner
Bernard Zinman

INTRODUCTION

We have elected to review four areas of diabetes in which there have been significant advances during 1978. The topics to be discussed are: 1) diabetes, high density lipoproteins and atherosclerosis; 2) insulin resistance, insulin receptors and glucose intolerance; 3) insulin infusion systems for glycemic control; and 4) glycosylated hemoglobins. The most prevalent complication of diabetes is atherosclerosis. In the last year, there has been a reawakening of interest in high-density lipoproteins, a class of lipoproteins that appears to be associated with a decreased risk of atherosclerosis. Evidence is accumulating that, in diabetics, alterations in this class of lipoproteins may be linked to control and may account for much of the prevalence of atherosclerosis. Control is greatly influenced by the body's responsiveness to insulin. A whole new facet in the area of responsiveness and resistance to insulin has been opened up by the ability to study cellular insulin receptors and by the recognition that cellular resistance may be at, or beyond, the level of the receptor. This knowledge about insulin resistance has led to a reexamination of the etiology of diabetes. From this knowledge has emerged the concept that in many patients the initial event underlying diabetes may be insulin resistance. To achieve control, the diabetic must receive amounts of insulin that are appropriate in both timing and quantity. The recognition of this necessity has led to the development of devices designed to deliver insulin continuously. Continuous blood glucose monitoring has demonstrated that randomly obtained blood glucose measurements indicate what is occurring at only a single instant in time and therefore can be extremely misleading guides to control. However, on a population-wide basis, it is impossible to monitor glucose levels continuously. Therefore, one must seek an indicator of the general pattern of glycemia over longer periods of time, one that is not subject to the vagaries of short-term swings, as is the case with blood glucose concentration. Such an indicator may now be available in the assay of glycosylated hemoglobin, the last of the topics to be covered in this review.

DIABETES, HIGH-DENSITY LIPOPROTEIN, AND ATHEROSCLEROSIS

In seeking to understand the cause of the accelerated rate of ischemic heart disease in diabetic patients, attention has recently been focused on high-density lipoproteins (HDL). This family of lipoproteins was first identified as being negatively correlated to the incidence of ischemic heart disease over one-quarter of a century ago (1,2). However, only recently has the interest in HDL been reawakened (3). It has been shown in both retrospective (4) and prospective (5) studies of large kindreds or of communities that high levels of HDL-cholesterol have a protective effect against ischemic heart disease. In this connection it is particularly noteworthy that uncontrolled diabetics have reduced levels of HDL (6-11). In fact the Framingham study suggests that HDL may be the most significant predictor of ischemic heart disease in diabetics (12). There is even some suggestion that the ratio of HDL-cholesterol to low-density lipoprotein (LDL)-cholesterol may be more important that the actual level of HDL in determining the risk of ischemic heart disease. In one study of the effect of HDL on the incidence of ischemic heart disease in a large kindred of hypercholesterolemic subjects, ischemic heart disease was found only in those who had a ratio of HDL-cholesterol/LDL-cholesterol less than 0.2 (4).

It has been suggested that HDL could exert its protective effect in one or both of two ways. HDL has been shown to compete with LDL for receptors on cell surfaces (13). Hence, HDL could interfere with cellular uptake of cholesterol. HDL has also been shown to promote the removal of cholesterol from such diverse sites as fibroblasts and aortic smooth muscle cells *in vitro* (14), liposomes (15), and adipose tissue *in vivo* (16). Thus, HDL could both block cholesterol entry into tissues and stimulate cholesterol removal from tissues. This reduction in cellular cholesterol could result in a decrease in atherogenesis. It is necessary in interpreting data relating to HDL levels in diabetics to be certain that there is no other covariable by which they can be influenced. Of major importance in this respect is the level of plasma triglyceride. In population studies both are seen to be inversely related to each other (17). However, in any person a change in triglyceride concentration need not always be accompanied by an opposite change in HDL concentration (18). It is this inverse relationship between HDL and plasma triglyceride levels that also makes it difficult to interpret the atherosclerosis risk effect of hypertriglyceridemia.

INSULIN RESISTANCE, INSULIN RECEPTORS, AND GLUCOSE INTOLERANCE

Between the pancreatic β-cell and the end-organ response there are a number of levels at which an inadequate insulin-dependent metabolic effect may arise. These levels are depicted schematically in Figure 1. They range from failure of the pancreas to secrete enough, or appropriately timed, insulin, through problems in transport that prevent adequate access to the tissue, to cellular resistance to insulin because of defects at the level of or beyond the insulin receptors.

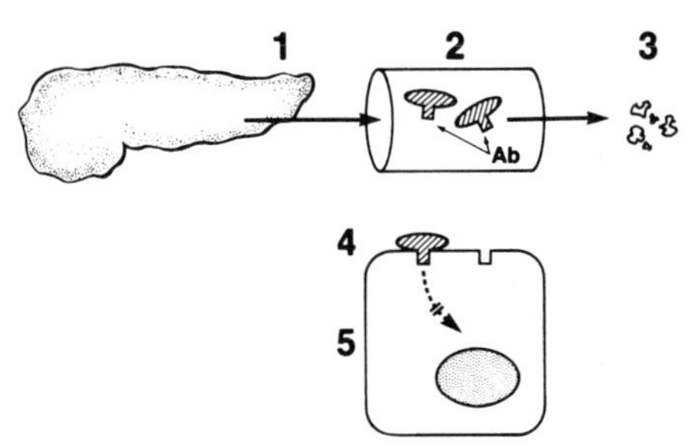

Figure 1. Schematic of sites influencing insulin's effectiveness.

It is difficult to summarize each of these levels. The one that has caused the most novel rethinking about the etiology of diabetes is that relating to the receptor and postreceptor mediation of insulin responses. It has given rise to the notion that the cause underlying the maturity-onset, but not the insulin-dependent, type of diabetes may be resistance to insulin rather than failure of the pancreatic β-cell. Of the variety of definitions of insulin-resistance, the most reasonable was given by Berson and Yalow (19): "a state in which greater than normal amounts of insulin are required to elicit a quantitatively normal response." Reaven and Olefsky have recently summarized the evidence suggesting that insulin resistance may be the cause of that type of diabetes that is not insulin dependent (20). Many of these patients are obese, a state known to be associated with insulin resistance. However, Reaven and Olefsky point out that insulin resistance is also found in such diabetics even in the absence of obesity (20).

Insulin resistance need not result in carbohydrate intolerance if the person secretes sufficient insulin. In fact many insulin -resistant people are hyperinsulinemic. However, the causal role of insulin resistance in increasing insulin secretion is uncertain. Reaven and Olefsky point out the difficulties in trying to decide what is the appropriate amount of insulin for any given plasma glucose concentration (20). Furthermore, it has now been shown that among at least one group of insulin-resistant people, the obese, there are some members who have β-cells that are hyperresponsive to secretagogues (21). This could be the basis of their hyperinsulinemia. The situation is further complicated by the observation that hyperinsulinemia itself can produce insulin resistance.

Although there is now some evidence that insulin can enter the cell, its first metabolic effect appears to depend on an interaction of the hormone with a specific and saturable cell-surface receptor (20).

Much recent work has examined the characteristics of the insulin receptors and their behavior in a variety of tissues. In most of these studies, cells have been examined immediately after their removal from the body. These cells included adipocytes (22,23), muscle cells (24), hepatocytes (24), monocytes (25,26), and erythrocytes (27,28). However, two cell lines, the IM-9 lymphocyte (29) and adipocytes (30), were examined in culture. The former have been sustained through many generations; the latter for 17 hours. Such *in vitro* preparations have been used to test factors that may influence insulin receptors in a direct and isolated manner.

In obesity, insulin binding to fat cells (25) and to monocytes (25) is decreased. This decrease is reversed by fasting (22, 26,31). Alterations in dietary composition, as well as in caloric content, can alter insulin binding. Thus, increasing the proportion of dietary carbohydrate without changing caloric content will decrease insulin binding (32,33). Insulin binding is also decreased by growth hormone and glucocorticoids *in vivo* (34). Most fascinating are those rare cases of very severe, extremely insulin-resistant diabetes that are due to a defect in the binding of insulin to the body's cells. There are two types of such syndromes. In one, associated with acanthosis nigricans, the body's cells lack insulin receptors (35). In the other, there are autoantibodies that block the insulin receptor (36-38). It is interesting that *in vitro* these antibodies can stimulate glucose oxidation and lipolysis in both rat (39) and human (36) adipocytes. Why this can happen even though these antibodies have no apparent insulinlike effect *in vivo* is puzzling.

Two factors influence insulin binding to cells: the number of receptors and their affinity for insulin. For instance, in fasting rats there is an increase in the insulin binding to adipocytes, due to an increase in cellular affinity for insulin (22). One mathematical tool used to examine these two factors is the Scatchard plot. In the case of the insulin-receptor interaction, this plot is typically concave (39). This type of curve can result either from negative cooperativity or from two classes of receptors, one high in affinity and low in capacity and the other low in affinity and high in capacity. Negative cooperativity is seen when there is only one class of receptor that, with a high degree of occupancy, has a low affinity for the ligand and with a low degree of occupancy has a high affinity. There is much support (39,40) for this notion. Adipocytes have been shown to have two classes of receptors, the high-affinity group being sensitive to Ca^{++}, the other not (41). Furthermore, two distinct species of insulin receptors have been isolated from fat-cell plasma membranes (42). However, a considerable body of conflicting data also exists. One possible resolution to this conflict is that both suggestions may be correct. Fat cells may have both a high-affinity, low-capacity receptor that is not subject to negative cooperativity and a low-affinity, high-capacity receptor that is subject to negative cooperativity (43). The former, but not the latter, appears to be involved in the degradation of insulin. Perhaps the most important area of receptor research today is that concerned with examining the physiological implications of changes in insulin binding. In this regard the recent observation that insulin receptors are present on erythrocytes (27) may be the most important. These, like monocytes, are cells that are easily sampled from the body. However, to obtain sufficient erythrocytes for an assay, much less blood is needed. Although the role of insulin receptors in erythrocyte function is uncertain, they may still mirror other receptors in the body.

Hyperinsulinemia is frequently seen in association with insulin resistance and with a decrease in the number of insulin receptors on cells (21,25,26). High levels of insulin added *in vitro* to cultures of IM-9 lymphocytes lead to a decrease in their insulin-binding capacity (29). This raises the possibility that hyperinsulinemia can both produce and be produced by insulin resistance. The relevance of such receptor changes is strengthened by the demonstration of a similar phenomenon *in vitro* in an extremely insulin-sensitive cell, the adipocyte (30). Furthermore, these receptor changes are accompanied by the appropriate shift to the right of the insulin dose/response curve with respect to glucose transport. However, a note of caution is suggested by the demonstration that alloxan-induced insulin deficiency can also produce insulin resistance in dogs (44). Thus hyperinsulinemia is frequently, but not necessarily, associated with insulin resistance and may even aggravate the insulin resistance.

Many investigators have tried to explain the insulin resistance that occurs at the cellular level as a consequence of a receptor defect. It is now apparent that this is an oversimplification. Kahn has pointed out that to analyze insulin resistance properly one must examine an entire insulin dose/response curve (45). One only needs to stimulate a few of the many insulin receptors on a cell to produce a maximum metabolic response. Hence, a receptor defect can be overcome by a sufficient increase in the dose of insulin. Thus, receptor-mediated insulin resistance should show a shift of the dose/response curve to the right, with the magnitude of the maximum response remaining unchanged. On the other hand, the cause of the resistance could be beyond the insulin receptor (i.e., somewhere between the transmission of the receptor's message and the final biochemical effect). In cases of such postreceptor defects, no matter how many receptors are stimulated, the maximal response will be subnormal. Furthermore, the sensitivity of all tissues and of all biochemical processes need not be the same. Thus, it is not enough to draw conclusions about the mechanism of insulin resistance from a single response to only one dose of insulin.

The importance of these considerations should be apparent from the following few experiments. Monocyte insulin binding is directly related to glucose tolerance in humans (46). Incubating adipocytes in the presence of high levels of insulin both decreases the numbers of receptors and shifts the insulin versus glucose transport curve to the right but does not change the maximum rate of transport (30). However, chronic hyperinsulinemia *in vivo* appears to increase the capacity of adipocytes to transport glucose (47). This, then, is a situation in which the cellular sensitivity to insulin and the insulin-binding capacity change in opposite directions. Fasting is another circumstance in which insulin binding and sensitivity to insulin change in opposite directions. Fasting is known to be associated with insulin resistance. However, insulin binding to adipocytes increases (22). It is also interesting that insulin binding and action are parallel to each other in the muscle but not in adipocytes or hepatocytes of obese-hyperglycemic (ob/ob) mice (24). One of the most striking examples of the dissociation between receptors and insulin resistance is seen in the recent description of one young insulin-resistant diabetic with acanthosis nigricans who had normal insulin receptors (48).

To understand the cellular basis of insulin resistance, it is necessary to examine both insulin receptors and insulin effectiveness. The former requires an evaluation of binding, receptor number, and affinity; the latter requires a full dose/response curve carried out to the point of maximal response. Sometimes resistance is a consequence of a receptor defect, and at other times it is a consequence of a postreceptor defect. Whatever the cause, insulin resistance is often accompanied by hyperinsulinemia. This can cause or aggravate the defect in insulin binding.

INSULIN INFUSION SYSTEMS FOR GLYCEMIC CONTROL

The resolution of the controversy over the relationship between diabetic control and the disabling complications of diabetes (49-52) requires the development of methods for attaining metabolic normalization. The administration of insulin to diabetics by subcutaneous injections, although usually effective in preventing major metabolic derangements (e.g., ketoacidosis), fails to achieve complete metabolic normalization (53-57). This is not unexpected when one considers that the pattern of insulin absorption from the subcutaneous depot bears little resembalnce to the secretion of this hormone from the normal pancreas (58). Pancreatic or islet-cell transplantation and the development of mechanical insulin infusion devices have been two methods used in an attempt to restore physiological insulin delivery.

In this section the current studies using insulin-infusion systems for glycemic normalization in diabetes will be reviewed. Mechanical devices can be divided into two types: closed-loop, or feedback-controlled, systems (also referred to as the artificial endocrine pancreas) and open-loop systems, which deliver insulin in a preprogrammed manner. Their current status and development have recently been reviewed (59).

Closed-Loop System

The closed-loop insulin-infusion system consists of three components: glucose analyzer, computer, and infusion pumps (Fig. 2). Blood is continuously withdrawn from a peripheral vein via a double-lumen heparin-infused catheter and is analyzed for glucose concentration. The computer is programmed with one of various possible algorithms (60-62) defining the relationship between glycemia and hormone infusion: insulin if the glycemia is too high and glucagon or glucose if it is too low. The hormone-infusion pumps are connected to the patient through a peripheral vein completing the closed-loop. For glycemic normalization in response to meals insulin must be delivered as a function of both the glucose concentration and its rate of change. Under these circumstances the early portion of rapid insulin release characteristic of the healthy β-cell is reproduced. With careful selection of the insulin-infusion parameters hypoglycemia rarely occurs, and the requirement for glucagon or glucose infusion is all but eliminated. This feedback-controlled insulin-infusion system has previously been shown to normalize glycemia in response to meals (57,60,63), exercise (64), glucose tolerance testing (65), surgery, and obstetrical labor (66).

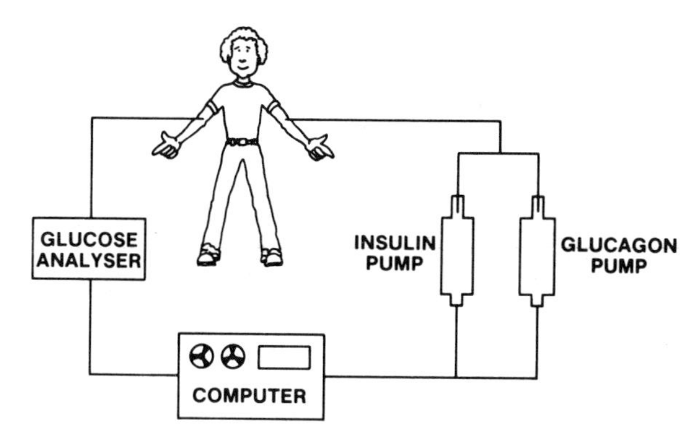

Figure 2. Closed-loop insulin-infusion system illustrating glucose analyzer, computer, and hormone-infusion pumps.

Since abnormalities in the concentrations of circulating intermediates of protein and fat metabolism may be as important as glycemia, as an overall index of metabolic dysfunction in diabetes and, perhaps, in the genesis of diabetic complications, the effect of glycemic normalization on these metabolites was recently examined (57). Complete normalization of glycemia during breakfast, lunch, snack, and supper (Fig. 3) was readily achieved with the closed-loop insulin-infusion system. As a consequence free fatty acid, lactate and pyruvate, and β-hydroxybutyrate profiles measured over the same period were restored toward normal. However, alanine concentrations did not change significantly, as compared with those observed with subcutaneous insulin therapy. Thus, a short period of glycemic normalization during meals can normalize some, but not all, of the other metabolic intermediates influenced by insulin. Attainment of complete metabolic control may require a more prolonged period of normoglycemia. The insulin profiles obtained with feedback-controlled insulin delivery have also been examined (67). Since the subjects under study had insulin antibodies as a consequence of previous insulin therapy, free insulin was measured in polyethylene glycol extracts of plasma. Although the free-insulin patterns were similar in configuration to those seen in normal subjects, the insulin concentrations were threefold to fourfold higher both before and during meals. It was postulated that the hyperinsulinemia may be related to the peripheral intravenous route of administration, as opposed to the normal intraportal release.

By virtue of its size and the need for continuous blood sampling, the closed-loop insulin-infusion system is applicable to short-term studies not exceeding 48 hours. The principal technical barrier to miniaturization and prolonged study is undoubtedly the glucose sensor. As a result the open-loop system was derived by eliminating the glucose sensor and delivering insulin in a preprogrammed fashion in an attempt to mimic normal insulin profiles.

Open-Loop System

The open-loop insulin-infusion systems consist primarily of an insulin pump and reservoir and a flow-rate controller (Fig. 4). Clearly the greatest weakness of this system is the lack of minute-to-

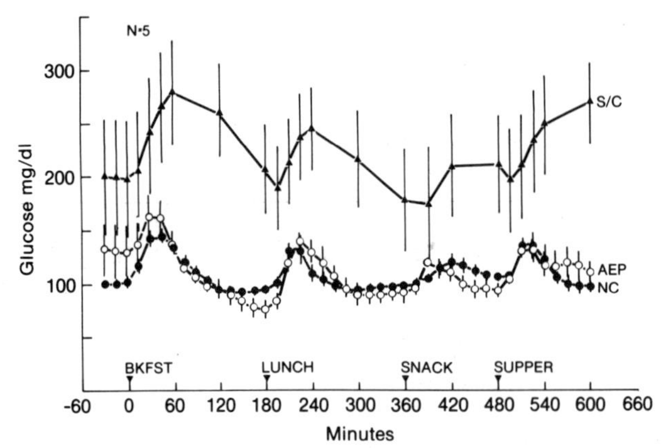

Figure 3. Changes in blood glucose concentrations during breakfast, lunch, snack, and supper for normal control subjects (N/C) and diabetic subjects treated with subcutaneous insulin (S/C) and during artificial pancreas (AEP) control. N = 5; vertical bars represent 1 SEM. (Reproduced with permission from Zinman et al, ref. 57.)

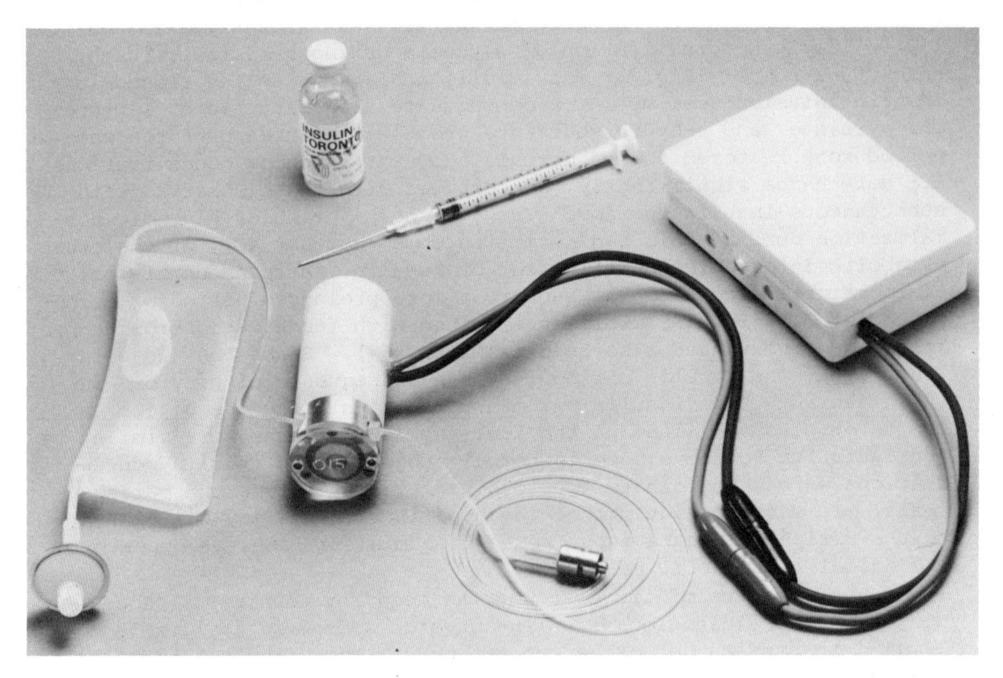

Figure 4. Open-loop insulin-infusion system. The insulin reservoir (left) is refilled with diluted insulin solution through the bacterial filter. Peristaltic pump (center) meters flow of insulin according to a waveform preprogrammed into the battery pack and flow-rate controller (right). Delivery of insulin during a meal is initiated by a magnetic switch in the controller. (Reproduced with permission from Goriya et al, ref. 71.)

minute glycemic feedback control. However, its smaller size and portability make it suitable for long-term studies. The critical question with regard to open-loop systems relates to the effectiveness of preprogrammed insulin infusion in normalizing the metabolic abnormalities of diabetes. Several investigators (68-74) have attempted to answer this question using various modifications of the open-loop system. Using a peripheral intravenous route for insulin delivery, different infusion patterns have been employed. One group (69) used the 24-hour insulin profiles in normal subjects, as well as calculations of insulin distribution and clearance, to derive a rather complicated double-exponential function of insulin delivery to mimic the normal insulin profile. The plasma-insulin excursions with meals in the diabetics receiving the pulsed insulin infusion closely matched the insulin concentrations in the normal controls. Similarly, the glycemic response to meals was markedly improved. However, it must be appreciated that the subjects in this study were obese, noninsulin-dependent diabetics who undoubtedly had significant residual β-cell function. Other investigators have used a simplified approach in which basal insulin is delivered continuously and augmented insulin is initiated for meals. A variety of basal and meal insulin patterns have been used for glycemic control as follows: *(1)* a short pulse of 15-minutes duration with breakfast and of 30-minutes duration with lunch and dinner, thereby raising the basal rate of 26 mU/min to 373 mU/min/during feeding (68); *(2)* a series of pulses varying from a minimal basal rate of 16 mU/min in the early morning period to a peak infusion rate of 67 mU/min for 90 minutes after breakfast (with other meals intermediate amounts of insulin were used with times and durations selected to optimize glycemia [74]); *(3)* a basal rate of 10 mU/min during the day and 20 mU/min overnight, with 100 mU/min for 45 minutes with all meals, beginning 15 minutes after the start of each meal (72); and *(4)* a constant basal insulin rate of 18 ± 2 mU/min with manually augmeted insulin delivered with meals as a constant rate square wave for 45 minutes or as an initial high-rate pulse for 12 minutes followed by a lower sustaining rate for 30 to 75 minutes (70,73). The total amount of insulin required for a particular subject had been derived from his previous subcutaneous insulin requirements or from studies performed with the closed-loop infusion system. In one such study, 12 insulin-dependent diabetics were maintained on preprogrammed continuous intravenous insulin delivery for periods ranging from 3 to 60 days (73). The basal insulin infusion rate was 18 ± 2 mU/min. Total insulin delivered with meals was 12 ± 2, 9 ± 2, and 11 ± 2 units for breakfast, lunch, and dinner, respectively. Fasting and postprandial glycemia were within normal limits, although the normal rates and magnitude of rise and decline in glycemia with meals were not entirely restored. The applicability of this approach for long-term control of glycemia has been demonstrated recently in depancreatized beagle dogs successfully treated for a period of nine months (71).

One of the limitations of prolonged insulin infusion in man relates to the requirements for long-term intravenous access. In an attempt to resolve this problem the effectiveness of continuous subcutaneous insulin infusion was examined (75). Using a battery-driven syringe pump and a fine subcutaneous cannula, insulin was continuously infused at a constant basal rate, and an eightfold increase in the infusion for 17 minutes was given 30 minutes before meals. Of the 14 subjects studied, the mean plasma glucose concentration improved in 6, was unchanged in 6, and deteriorated in 2. The long-term effectiveness of continuous subcutaneous insulin infusion remains to be demonstrated, and it appears unlikely that the absorption characteristics of insulin administered in this manner will result in normal plasma insulin profiles. However, the use of an indwelling subcutaneous catheter may have the advantage of increasing patient compliance for multiple depot insulin administration since multiple injections would not be required. In this regard improvement in glycemic control would be anticipated.

In summary, the available data on open-loop studies indicate that insulin-infusion patterns that simulate normal insulin profiles and result in improved glycemic regulation can be derived. However, the method cannot be considered to be optimal as yet. The variables that require further definition and development include the pump system itself, the insulin-delivery patterns used, the innate variability in the response of the patient to caloric challenges, the route of insulin delivery, and the effect of chronic cannulation. Despite these problems the results obtained with open-loop insulin infusion justify ongoing efforts to improve this system, with the aim of providing an important tool to define further the relationship between metabolic control and the degenerative complications of diabetes mellitus.

GLYCOSYLATED HEMOGLOBINS

As discussed above, the relationship between metabolic control and the occurrence of long-term diabetic complications continues to be controversial, since a method of replacing insulin physiologically has not been perfected. However, of equal importance is the lack of a satisfactory method for the assessment of diabetic control. Such a method would be essential for conducting long-term studies that examine the relationship of diabetic control and complications, as well as being useful for monitoring patients in the clinical setting. The use of random blood sugar estimations, fractional urine collections, 24-hour urinary glucose excretion, or continuous 24-hour blood glucose concentration monitoring are imprecise, unreliable, or impractical. In this context the development of the glycosylated hemoglobin test as an integrated assessment of glycemic control may prove to be a valuable tool in diabetes research. This subject has recently been well reviewed (76,77).

Several chromatographically separable minor hemoglobins are found in human red cells; the common ones are termed AI_a, AI_b, and AI_c. The term HbA_I denotes the sum of all these minor hemoglobins, which normally make up approximately 10% of the total hemoglobin. HbA_{Ic} is the largest component (4 to 6%), while HbA_{Ia} and HbA_{Ib} total 1 to 2%, with the balance being made up by other forms, the structures of which are not clearly defined at the present time. HbA_{Ic} contains a hexose moiety (1-amino, 1-deoxy-fructose) attached to the amino terminal of the B chains by virtue of a Schiff base and Amadori rearrangement (78-80). The precise structure of HbA_{Ia} has not been completely defined. It has been proposed that HbA_{Ia} or HbA_{Ib} are precursors of HbA_{Ic} (81,82), but recent evidence suggests that this is not so (83).

Methodology

The glycosylated hemoglobins can be separated from nonglycosylated HbA by column chromatography with cation exchange resins, since at a slightly acidic pH (6.7), the minor hemoglobins are more negatively charged. The minor hemoglobins, because of differing charge characteristics, can also be separated from each other. However, since there is probably no advantage in the clinical setting in measuring HbA_{Ic} alone, most laboratories measure the sum of the minor hemoglobins (HbA_I). Initially, large columns approximately 30 cm long were used to separate the minor hemoglobins in red blood cell hemolysates (84). Recently the method has been modified (85-87), allowing for shorter columns to be used and thus reducing the assay time. Another approach to the measurement of glycosylated hemoglobins has been the use of high-pressure liquid chromatography (88,89). The time required for the assay of a single sample is greatly reduced, and the results appear encouraging.

Clinical Studies

The presence of increased HbA_{Ic} levels in diabetics has been known for several years (90,91). However, it was only recently demonstrated that the increase in glycosylated hemoglobin in diabetes is the result of a posttranslational modification of HbA. This nonenzymatic process appears to occur slowly throughout the life span of the red cell and is accelerated by an elevated plasma glucose concentration. These observations stimulated many efforts at defining the relationship between glycosylated hemoglobin and diabetes, and some of the recent studies are summarized below.

The HbA_I concentration in diabetics has been shown to correlate significantly with the fasting plasma glucose concentration, mean daily plasma glucose value (calculated from frequent glucose determinations over a 24-hour period), and the highest plasma glucose concentration attained during 12 hours of continuous monitoring (92). A significant correlation was also found with 24-hour urinary glucose excretion measured at monthly intervals in 128 diabetic patients, and, interestingly, the best correlation was noted between the level of HbA_I and the urinary glucose excretion measured two months previously (93). In a prospective study in 10 newly diagnosed diabetics, serial glycosylated hemoglobin measurements were made before and after optimized diet and insulin therapy (94). The average concentration of HbA_{Ic} was 11.4% in the untreated diabetics as compared with 4.3% in healthy controls. With prolonged optimal blood glucose regulation the concentration of HbA_{Ic} decreased to 5.5%. However, the return of the HbA_{Ic} concentration to normal or near-normal values took 25 to 80 days, depending on the initial concentration and on the degree of blood glucose control. This finding supports the hypothesis that the concentration of glycosylated hemoglobin is an expression of the average blood glucose value in the preceding one to two months. In maturity-onset diabetes glycosylated hemoglobin was closely associated with the fasting plasma glucose level ($r = 0.85$, $P < 0.001$) (95). This report suggests that the convenient, rapid, and inexpensive determination of the fasting plasma glucose level should provide an accurate index of past and present glucose homeostasis for stable, noninsulin-dependent diabetics. On the other hand, in insulin-treated diabetics, levels of fasting and mean plasma glucose are subject to day-to-day variability, and in these patients glycosylated hemoglobin measurements may be more informative.

With respect to the relationship between glycosylated hemoglobin concentrations and diabetic complications early cross-sectional studies failed to show a correlation (84). More recently, 20 untreated maturity-onset diabetic patients underwent nerve conduction studies, the results of which were compared with those of normal controls (96). Levels of glycosylated hemoglobin, as an index of long-term glycemic control, correlated well with slowing of peroneal motor conduction velocity. This finding suggests a relationship between the metabolic abnormality of diabetes and motor nerve conduction abnormalities. However, a prospective longitudinal study of diabetic complications using new methods of optimizing glycemia and, therefore, HbA_I has yet to be done.

The glycosylation of hemoglobin also has an effect on hemoglobin function. The oxygen affinity of hemoglobin is reduced and, therefore, tissue availability is increased when 2,3-diphosphoglycerate (2,3-DPG) reacts with the NH_2-terminal amino groups (77). Since this site is blocked by a hexose moiety in glycosylated hemoglobin, the reactivity with 2,3-DPG is decreased, and red cell oxygen affinity is increased. The changes in the oxygen saturation curve tend to be small, and it remains to be demonstrated whether significant tissue hypoxia results.

The demonstration of increased glycosylation of hemoglobin in diabetes has again focused on the hypothesis that a similar phenomenon may be occurring in various organs, and that these glycosylated proteins could play an important role in the genesis of microvascular complications.

In summary, glycosylated hemoglobin assay as a measure of metabolic control in diabetes has certain advantages: *(1)* it is a quantitative test not dependent on the patient's cooperation; *(2)* it is independent of time of day; *(3)* it may simplify follow-up and may provide physicians with a definitive end point toward which therapy should be directed; and *(4)* it will provide a useful test for examining the relationship between metabolic control and diabetic complications. The limitations of glycosylated hemoglobin assay are as follows: *(1)* it is not helpful in deciding on acute insulin changes; *(2)* it does not reflect the occurrence of hypoglycemia; *(3)* in hematological disorders, where red blood cell life span is abnormal, glycosylated hemoglobin levels are significantly altered; and *(4)* in infants and pregnant women HbF may interfere to some degree with glycosylated hemoglobin measurements.

Although the measurement of glycosylated hemoglobins promises to be a valuable tool in diabetes research as well as in the clinical management of diabetes, continued studies will be required to define its impact clearly.

REFERENCES

1. Barr DP, Russ EM, Eder HA: *Am J Med* 11:480, 1951.

2. Nikkila E: *Scand J Clin Lab Invest*, (suppl 5):158, 1953.

3. Miller GJ, Miller NE: *Lancet* 1:16, 1975.

4. Streja D, Steiner G, Keviterovich PO Jr: *Ann Intern Med* 89:871, 1978.

5. Gordon T, et al: *Am J. Med* 62:707, 1977.

6. Nikkila EA: In *High Density Lipoprotein and Atherosclerosis* Gotto AM, Miller NG, Oliver MF (eds): Amsterdam, Elsevier-North Holland, p 177, 1978.

7. Durrington P: *Lancet* 2:206, 1976.

8. Nikkila EA, Hormila P: *Diabetes* 27:1078, 1978.

9. Calvert GD, et al: *Lancet* 2:66, 1978.

10. Elkeles RS, Wu J, Hambley J: *Lancet* 2:547, 1978.

11. Kennedy AL, et al: *Br Med J* 2:1191, 1978.

12. Gordon S, et al: *Ann Intern Med* 87:393, 1977.

13. Miller NE, et al: *J Clin Invest* 60:78, 1977.

14. Stein O, Vanderhoek J, Stein Y: *Biochim Biophys Acta* 431:347, 1976.

15. Tall AR, et al: *Biochemistry* 17:322, 1978.

16. Nestel PJ, Miller NE: In *High Density Lipoprotein and Atherosclerosis* Gotto AM, Miller NG, Oliver MF (eds): Amsterdam, Elsevier-North Holland, p 229, 1978.

17. Schaeffer EJ, et al: *Lancet* 2:391, 1978.

18. Glueck CJ, et al: *Clin Res* 26:680, 1978.

19. Berson SA, Yalow RS: In *Diabetes Mellitus: Theory and Practice* Ellenberg M, Rifkin H (eds): New York, McGraw Hill, p 388, 1970.

20. Reaven GM, Olefsky JM: In *Advances in Modern Nutrition* Katzen HM, Mahler RJ (eds): New York, John Wiley, p 229, 1978.

21. Baxter D, et al: *Europ J Clin Invest* 8:361, 1978.

22. Olefsky JM, Kobayashi M: *J Clin Invest* 61:329, 1978.

23. Olefsky JM, Jen P, Reaven GM: *Diabetes* 23:565, 1974.

24. LeMarchand Y.L., et al: *Diabetes* 26:582, 1977.

25. Olefsky JM: *J Clin Invest* 57:1165, 1976.

26. Bar RS, et al: *J Clin Invest* 58:1123, 1976.

27. Gambhir KK, Archer JA, Bradley CJ: *Diabetes* 27:701, 1978.

28. Thomopoulos P, Bertheillier M, Laudal MH: *Biochem Biophys Res Comm* 85:1460, 1978.

29. Gavin JR III, et al: *Proc Nat Acad Sci USA* 71:84, 1974.

30. Livingstone JN, Purvis BJ, Lockwood DH: *Metabolism* 27:2009, 1978.

31. Archer JA, Gordon P, Roth J: *J Clin Invest* 55:166, 1975.

32. Olefsky JM, Saekow M: *Endocrinology* 103:2252, 1978.

33. Kolterman OG, Reaven GM, Olefsky J: *Clin Res* 26:159, 1978.

34. Kahn CR, et al: *Endocrinology* 103:1054, 1978.

35. Kahn CR, et al: *N Engl J Med* 294:739, 1976.

36. Kosuga M, et al: *Diabetes* 27:938, 1978.

37. Kosuga M, et al: *J Clin Endocrinol Metab* 47:66, 1978.

38. Kahn CR, et al: *J Clin Invest* 60:1094, 1977.

39. DeMeyts P: *J Supramol Struct* 4:241, 1976.

40. DeMeyts P, Bianco AR, Roth J: *J Biol Chem* 251:1877, 1976.

41. Desai KS, et al: *Canadian J Biochem* 56:843, 1978.

42. Krupp MN, Livingstone JN: *Proc Nat Acad Sci USA* 75:2593, 1978.

43. Olefsky JM, Chang H: *Diabetes* 27:946, 1978.

44. Reaven GM, Sageman WS, Swenson RS: *Diabetologia* 13:459, 1978.

45. Kahn RC: *Metabolism* 27:1893, 1978.

46. Beck-Nielsen H, Pederson O: *Diabetologia* 14:159, 1978.

47. Olefsky JM, Kobayashi M: *Metabolism* 27:1917, 1978.

48. Bar RS, et al: *J Clin Endocrinol Metab* 47:620, 1978.

49. Cahill GF Jr, Etzwiler DD, Freinkel N: *N Engl J Med* 294:1004, 1976.

50. Siperstein MD, et al: *N Engl J Med* 296:1060, 1977.

51. Ingelfinger FJ: *N Engl J Med* 296:1228, 1977.

52. Raskin P: *Metabolism* 27:235, 1978.

53. Molnar GD, Taylor WF, Ho MM: *Diabetologia* 8:342, 1972.

54. Alberti KGMM, Dornhurst A, Rowe AS: *Israel J Med Sci* 11:571, 1975.

55. Hansen AaP, Johnasen K: *Diabetologia* 6:27, 1970.

56. Zinman B, et al: *Diabetes* 27(suppl 2):431, 1978.

57. Zinman B, et al: *Metabolism* 28:511, 1979.

58. Molnar GB, Taylor WF, Langworthy AL: *Mayo Clin Proc* 47:709, 1972.

59. Santiago JV, et al: *Diabetes* 28:71, 1979.

60. Albisser AM, et al: *Diabetes* 23:389, 1974.

61. Clemens AH, Chang PH, Myers RW: *Journ Annu Diabetol Hotel Dieu*, 269, 1976.

62. Botz CK: *Biomedical Engineering* 23:252, 1976.

63. Marliss EB, et al: *Diabetes* 26:663, 1977.

64. Zinman B, et al: *J Clin Endocrinol Metab* 45:641, 1977.

65. Albisser AM et al: *Arch Intern Med* 137, 1977.

66. Pfeiffer EF, et al: *Journ Annu Diabetol Hotel Dieu* 279, 1976.

67. Hanna AK, et al: *Clin Res* 26:863, 1978.

68. Slama G, et al: *Diabetes* 23:732, 1974.

69. Genuth S, Martin P: *Diabetes* 26:571, 1977.

70. Albisser AM, et al: *Med Prog Technol* 5:187, 1978.

71. Goriya YB, et al: *Diabetes* 28:558, 1979.

72. Service FJ: *J Lab Clin Med* 91:480, 1978.

73. Hanna AK, et al: *Clin Res* 26:630, 1978.

74. Decker T, Lorup B: *Diabetologia* 12:573, 1976.

75. Pickup JC, et al: *Br Med J* 1:204, 1978.

76. Gonen B, Rubenstein AH: *Diabetologia* 15:1, 1978.

77. Bunn HF, Gabbay KH, Gallop PM: *Science* 200:21, 1978.

78. Bunn HF, et al: *Biochem Biophys Res Commun* 67:103, 1975.

79. Koenig RJ, Blobstein SH, Cerami A: *J Biol Chem* 252:2992, 1977.

80. Stevens VS, et al: *J Biol Chem* 252:2998, 1977.

81. Haney DN, Bunn HF: *Proc Natl Acad Sci USA* 73:3534, 1976.

82. Stevens VJ, et al: *J Biol Chem* 252:2998, 1977.

83. McDonald MJ, et al:*J Biol Chem* 253:2327, 1978.

84. Trivell LA, Ranvey HM, Lai HT: *N Engl J Med* 284:353, 1971.

85. Gabbay KH, et al: *J Clin Endocrinol Metab* 44:859, 1977.

86. Abraham EC, et al: *Diabetes* 27:931, 1978.

87. Chou J, Robinson A, Siegel AL: *Clin Chem* 24:1708, 1978.

88. Davis JE, McDonald JM, Jarret L: *Diabetes* 27:102, 1978.

89. Cole RA, Bunn HF, Soeldner JS: *Diabetes* 26:(suppl 1) 392, 1977.

90. Husman TH, Dozy AM: *J Lab Clin Med* 60:302, 1962.

91. Rahbar S: *Clin Chim Acta* 22:296, 1968.

92. Gonen B, et al: *Lancet* 2:734, 1977.

93. Gabbay KH, et al: *J Clin Endocrinol Metab* 44:859, 1977.

94. Ditzel J, Kjaergaard J: *Brit Med J* 1:741, 1978.

95. Graf RJ, Halter JB, Porte D: *Diabetes* 27:834, 1978.

96. Graf RJ, et al: *Ann Intern Med* 90:298, 1979.

8
Biochemical Tests in Diagnosis and Monitoring of Cancer

Donald J.R. Laurence
Alexander M. Neville

With the increasing recognition that human tumors manufacture and release a wide variety of products, a new subspeciality - - that of tumor markers - - has evolved in the study and practice of oncology.

There can be little doubt that the presently recognized tumor products represent only a fraction of the range of products that further research will uncover. Indeed, there are many aspects of endocrine and metabolic disease in cancer patients for which no known hormone can account (1).

The purpose of this chapter is to outline the tumor products (new and old) reported in 1978 and to attempt to place them in context with other known products that have found an established place in the diagnosis and management of neoplastic disease or in explaining the initial symptoms or resulting complications.

ANTIGENS

Although the term *antigen* may not be strictly correct, it serves to embrace a group of macromolecules detected by immunological techniques and without a known biological function at present. Some, however, on occasion may be antigenic in the tumor-bearing patient.

Oncofetoplacental Antigens

The term *oncofetoplacental antigen* is retained, although it is now increasingly recognized that normal adult tissues may continue to express such materials, thereby negating their tumor, placental, or fetal specificity, as many earlier publications implied.

Alphafetoprotein (AFP)

During the past year, further studies have reported on the purification of AFP (2,3) and on its analysis by immunoassay, both in its enzyme-linked form (4,5) and with radioisotopes (6-10). Additional functional similarities have been reported between AFP and plasma albumin with respect to their abilities to bind copper and fatty acids (11,12).

Liver Disease. In relation to hepatocellular carcinoma and liver cirrhosis, it was found that 29 of 30 patients with carcinoma occurring in a cirrhotic liver gave elevated blood AFP levels, while only 11 out of 20 patients with carcinomas in noncirrhotic liver had raised levels of AFP (13). This suggests that carcinoma accompanied by cirrhosis may be a distinct clinical entity. In cases of cirrhosis in the absence of cancer, only 1 of 100 patients had elevated AFP levels. The test was, therefore, expecially powerful in discriminating those patients with liver cirrhosis who also have cancer. Metastatic liver cancer was seldom associated with raised titers.

The use of AFP levels as a monitor of hepatic tumor activity has received increasing attention recently. In Japan, where liver cancer is more frequent, one study in which patients with chronic liver diseases were followed up, found 18 of 31 cases of cancer (14). Of the group with cancer, 11 of 18 were diagnosed by rising AFP levels alone. The other nine, however, had low AFP levels throughout their illness. Hepatitis B antigen was found to be a poor prognostic sign, and AFP monitoring was considered especially important in these cases. However, the tumor was not resectable in the majority of malignancies detected by AFP. The smallest tumor detected was 4-5 cm in diameter. A further problem relevant to monitoring was that patients who were apparently tumor free could express transitory elevations of AFP levels, although values did not exceed 1,000 ng/ml (14).

In 11 patients with partial hepatectomies no elevations of AFP concentration were found (15). About one third of the patients with acute or chronic active hepatitis had elevated levels, but none was greater than 1,000 ng/ml. With massive necrosis there was a 75% incidence of elevated levels, and, in one sixth of the patients, levels over 1,000 ng/ml were found. From these results it does not seem that elevations in serum AFP levels reflect normal liver regeneration in contrast to some previous reports.

Germ-cell Tumors. AFP and also human chorionic gonadotrophin (HCG) levels are commonly elevated in patients with germ-cell tumors of the ovary (16,17), testis (18,19), and even of the pituitary (20): the markers are associated with different cellular elements of the tumor, and AFP is characteristically produced by yolk-sac (vitelline) elements (16,17).

The yolk sac is known to manufacture transferrin, prealbumin, and α_1-antitrypsin (21). One study evaluated the role of measuring these three indices in plasma by comparing their diagnostic accuracy with that of AFP assays in the same patients. They added no useful clinical information that was not found from the AFP result alone.

The plasma AFP and HCG levels respond to surgery or chemotherapy (see Fig. 1) (16,18,19,23-25). Sometimes if AFP and HCG levels are both raised initially, one marker returns to the normal range while the other remains elevated (18).

Pure seminomas, dysgerminomas, and gonadoblastomas do not produce any alteration in the AFP level. However, all tumors with yolk-sac elements (endodermal sinus tumor) gave AFP-level elevations (16,19). Among teratomas the majority show concordances between elevations or normality of their AFP and HCG levels. In about 18% of the cases one marker is elevated, and the other is normal (19). Surprisingly there is one report of elevated AFP levels in bladder carcinoma (26).

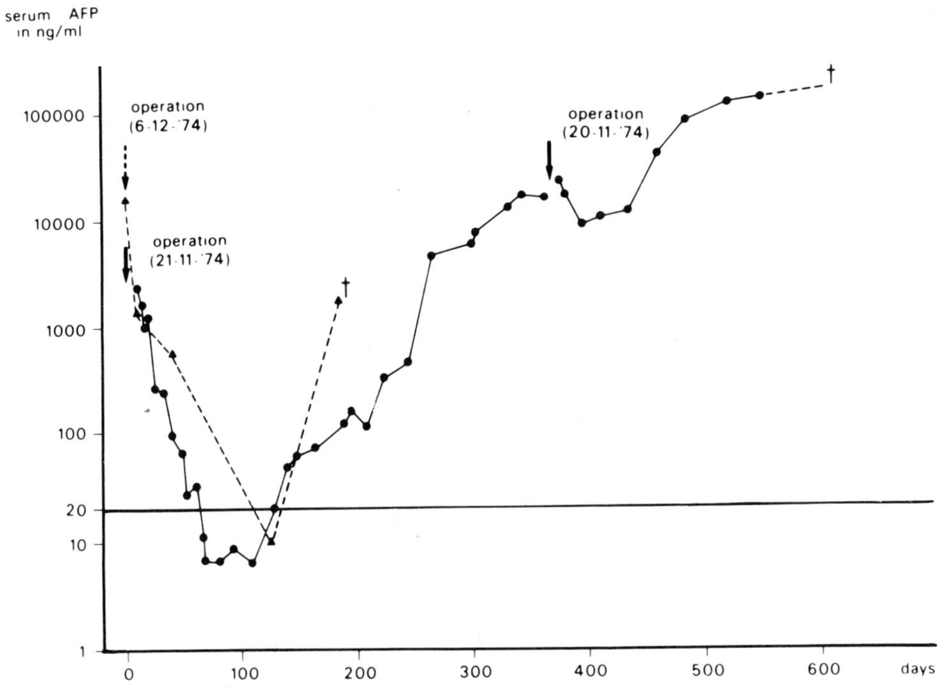

Figure 1. Follow-up results of two patients with mixed germ-cell tumors of the ovary after initial operation. In both cases the AFP level falls below the normal upper limit of 20 ng/ml but subsequently rises over several log increments before death of the patient. (Reproduced with permission from Talerman et al, ref. 16.)

Carcinoembryonic Antigen (CEA)

Work on CEA appears to continue unabated. Investigations have centered on studies to define further its nature both chemically and immunologically and its potential in a variety of clinical cancer situations.

Chemical and Immunological Studies. CEA is a large molecule (molecular mass about 200,000 daltons) with a substantial carbohydrate content. Its usual source for study is hepatic metastases from colorectal carcinomas, from which it is prepared by perchloric acid extraction. The product so obtained has been subjected to trypsin digestion. The partial sequence of seven of the resulting peptides suggests a uniform material with little heterogeneity of the primary peptide sequence (27). None of the peptides had significant immunological activity. This work, together with specific chemical modification of amino acid residues and disturbances of the whole molecule (28), indicates that its immunological activity is dependent on the tertiary structure of the molecule as has been suggested previously. CEA has been treated chemically to modify various amino acid or carbohydrate residues (29). The effects on immunoreactivity demonstrate that some antisera react with carbohydrate and others with peptide sequences in the CEA. The peptide-reacting antisera were the more discriminating in terms of the clinical evaluation of patients' blood CEA levels.

Routine preparations of CEA from a variety of tumors show differences in immunological potency and carbohydrate content but similar molecular mass and amino acid composition (30). Replacement of the standard perchloric acid extraction by a saline method gives some additional active material (31-33). Samples also differ in binding to a lectin column; in two studies a fetally derived CEA did not bind (31,32).

A substance with an N-terminal sequence similar to CEA has been identified in colonic lavages from healthy persons (34).

Evaluation and comparison of CEA assays have been reported. Most of the clinical work has used the zirconyl phosphate (Z-gel) assay (35,36). Solid-phase methods with rapid removal of the major serum proteins have been described (37,38). Double-antibody methods with unextracted plasma tend to give somewhat higher normal values (36,39,40). All methods give similar clinical interpretative results if the normal threshold range for each test is taken into account. The threshold value has also been varied for a single assay depending on the type of discrimination that is required.

Screening and Diagnosis. A series of studies have tended to reaffirm that assays for CEA are not very practicable as screening tests for cancer when applied to randomly selected apparently healthy persons (41). Seven years of screening a large work force (3,024 persons) detected two well-documented cases of unsuspected cancer (42), and one patient underwent surgery for an apparently localized cancer. In an at-risk population in a chemical factory a higher incidence of CEA elevations was found in the exposed group than in a control group (43). There are other isolated reports of the presence of malignancy being positively established or strongly suggested by elevated CEA levels (44-46).

The outcome of CEA testing in a 400 to 500-patient sample in a hospital service is discussed (47,48). Only four patients had an objective change in management as a result of the CEA test, representing an average cost of $5,000 for each of these four patients when the total cost of the survey was taken into account. In a study of 381 patients with clinical problems suggestive of gastrointestinal malignancy, there was a subgroup of 74 patients with signs of liver disorder, and in 38 of these 74 patients the presence of malignancy was determined by CEA testing (49).

Tumor Localization. Using anti-CEA sera, tumors may be localized either in tissue sections (50-52) or in patients *in vivo* (53,54). With immunoperoxidase histochemical procedures, it is possible to detect micrometastases within lymph nodes and to obtain a tumor correlate of the CEA blood levels. *In vivo* injection of ^{131}I-labeled anti-CEA antibodies into patients can facilitate tumor location when the results of physical diagnostic tests are negative (53,54). Anti-CEA antibodies are lytic to CEA-producing cells *in vitro* (55), and a possible role of homing antibodies *in vivo* might be considered.

Analysis of Ascites and Exudates. With a suitable threshold the CEA test can be used to discriminate between benign and malignant lesions in ascites, trasudates, and pleural effusions (46-58) although there is some contrary evidence (59). A more efficient discrimination is possible with pleural effusions if an additional marker, β_2-microglobulin or lysozyme, is included (58,60).

Colorectal Cancer. In 1978 one of the major reported uses of the CEA blood test was in the follow-up of colorectal cancer after surgery. Results give an early indication of the possible benefit to the patient of such a procedure.

Much attention has been given to the interpretation of the rising CEA levels that presage or accompany evident recurrence of colorectal cancers (61,62). A number of false alarms are possible due to transient, idiosyncratic, or moderate elevations in CEA levels. These may occur in the first few months after surgery or in the presence of complicating diseases, such as kidney disease (62-66). Mathematical assessment by nomogram or computer has been used to analyze the time curves (47,67). The lack of a rising titer in recurrent cases is considered more likely if the preoperative CEA level was normal (68,69) or if immunocytochemical studies of the primary tumor show the absence of CEA (50). Even if the preoperative level was elevated, a subsequent recurrence may not be accompanied by a rise in the CEA level (61,68).

Preoperative and postoperative levels of CEA reflect the extent of tumor spread and the presence of residual tumor and so can have prognostic value (70-73). The preoperative level appears to correlate with the success of surgery (69) as the patients with lower CEA values are more likely to have resectable tumors.

In a number of hospital centers, surgical intervention ("second-look" surgery) has been reported as a response to a rising CEA level (see Fig. 2). The presence of a tumor has been confirmed in most reported cases, and some patients have been thought to survive for an extended period because of this second-look operation (69,74-79). More time is needed to establish whether or not the use of these methods has resulted in a cure.

Monitoring of CEA levels during follow-up requires more frequent testing in the first 18 months, the time during which recurrence is most likely, and a set of preoperative levels is also desirable (67,69). The slopes of the rising levels fall into two groups, and the patients with the flatter time curves appear to be those more likely to benefit from further surgery (70,77). Because of the balance of success relative to patient distress, the "second-look" philosophy has not been accepted by all authors (80).

As a monitor of chemotherapy of gastrointestinal cancer the CEA test has been found to be somewhat equivocal (81-84). There are frequent correlations between CEA level and objective response, but there are also substantial discordances. The CEA level showed a fall to the normal range in only 1 of 30 cases studied, and this reflects the need for more efficient chemotherapy in this condition (81).

Radiation therapy as a preliminary to surgery has been found to result in a brief fall in CEA level (85). This has been interpreted as indicating that surgery should not be delayed too long after radiation treatment.

Esophageal, Liver, and Pancreatic Cancer. Increases in the CEA level have been noted in the majority (70%) of the cases of esophageal cancer studied (86), and the value was related to survival time. In contrast, moderate elevations were found in only 4% of the patients with hepatocellular cancer. In patients with pancreatic cancer there was a clear correlation between the CEA levels and the extent of tumor spread. Considerable overlapping of values was noted in cases of pancreatitis and localized disease (88,89). A fall to normal after surgery did not correlate well with survival, and chemotherapy or radiation had little effect on the CEA level (90).

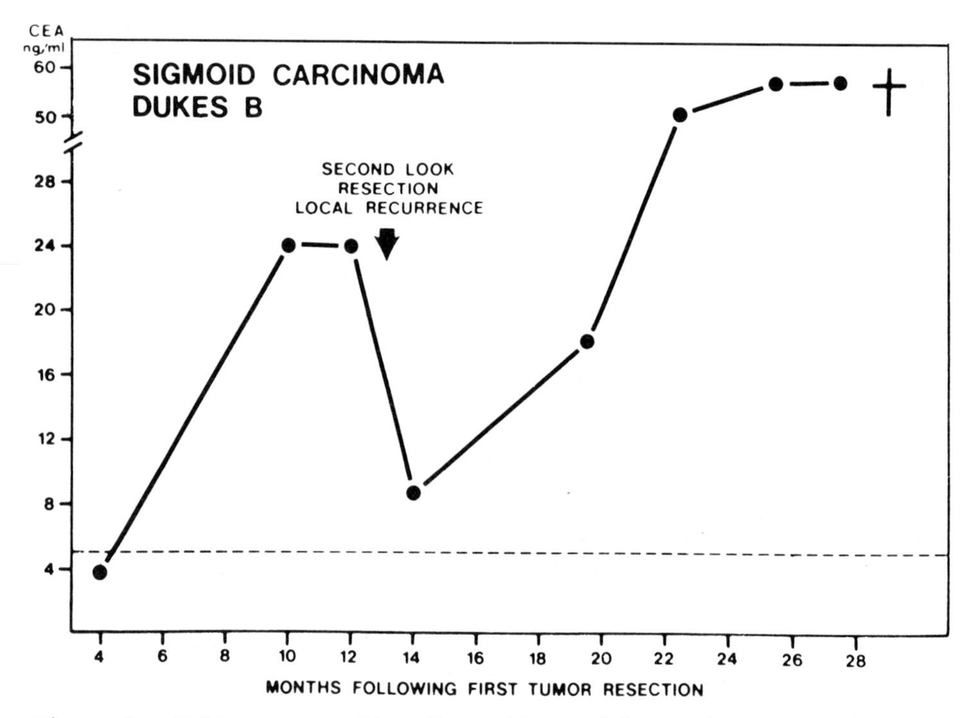

Figure 2. Follow-up results of a patient with complete resection of a Dukes' B rectosigmoid adenocarcinoma. After rise in the CEA level and a positive liver scan, a recurrent local pelvic tumor was removed. The marker level fell but rose again before death of the patient. Liver metastases were found at autopsy. (Reproduced with permission from Mach et al, ref. 62.)

Breast Cancer. Levels of CEA are rarely even moderately elevated in the presence of localized disease (91,92,93), high levels therefore suggest metastases (90,93-95). Because of this, other markers have been added (see Table 1) to increase the likelihood of elevation of at least one marker (93,95-99). This also increases the incidence of false-positive cases, and some sort of discriminant function is desirable to circumvent this (98). Although the use of more markers helps to identify widespread cancer, it does little to detect local disease. The CEA level depends on the site of the secondary tumor deposits (91).

Of the additional markers urinary hydroxyproline levels may be important; 15 to 25% of the cases show elevations in the absence of overt bone metastases (100). Experience has shown that the majority of these patients will later develop overt metastases.

Table 1. Biochemical Markers in Breast Cancer

Oncofetal antigens	CEA
Other antigens	See references 95 and 129-133
Hormones	HCG (96,97) Calcitonin (93)
Enzymes	Alkaline phosphate (93,97) Glycosyl transferases (93) γ-glutamyl transpeptidase (97)
Other substances	Milk proteins (93,97) Serum glycoproteins (93,96) Acute phase proteins (93,97) Polyamines (93,98) Minor nucleosides (93,98) Hydroxyproline (93,99) Ferritin (93)

Lung Cancer. About one third of operable lung cancers led to elevated CEA levels (101,102). A low preoperative value of CEA was considered to be a good prognostic sign (103). The initial CEA level was of doubtful significance in selecting patients for surgery (101-103). Patients with poorly differentiated tumors, whether small-cell or large-cell, were found to have uniformly low blood levels of CEA (101). Two cases of a three-year survival after second-look surgery have been recorded (102).

Gynecological and Bladder Cancer. Ovarian and cervix uteri tumors in patients with elevated plasma CEA levels may or may not show CEA in tissue sections (104), but a tumor that does not cause elevation of the plasma CEA level is likely to be CEA negative by immunocytochemical testing (105). Surgical cure is accompanied by a fall in the CEA level to normal (104,106), and a persistent rise in titer after surgery indicates recurrence (105). Of 204 cases of invasive carcinoma of the cervix followed by CEA testing after initial therapy, 30 had rising CEA levels and 29 of these gave evidence of clinical recurrence (107). The CEA test was predictive of recurrence with a period of 1 to 23 months between rise in CEA level and clinical detection. Bladder tumors rarely show elevations of plasma CEA levels with recurrence. In tissue sections of bladder tumors, not all cells are CEA positive, and there are no obvious histological correlates to distinguish the positive and negative cells (108).

Medullary Carcinoma of the Thyroid (MTC) and Neuroblastoma. MTC is associated with a high incidence of elevated plasma CEA levels comparable with the incidence of elevations of basal calcitonin levels. However, unlike the latter, no stimulation test is available and calcitonin is the preferred marker for distinguishing occult tumors

(109,110). A report on neuroblastoma shows a disappointing (1 of 12) incidence of elevations in this disease compared with previous reports (111).

CEA-Related Substances

Antigens co-isolated with CEA include a tumor-associated antigen "TEX" (112) and the normal cross-reacting antigen (NCA) (113). TEX shares antigenic properties and a N-terminal sequence with CEA (112). The molecular mass is between that of CEA and NCA. A method of developing an antiserum for NCA that does not cross-react with CEA is described (113). No advantage in measuring plasma NCA in addition to CEA could be observed as has been reported previously (61). A further cross-reacting substance has been isolated from the bile, namely, the biliary glycoprotein I (BGPI) (114).

The substance CEA-S is a subfraction of CEA. Radioimmunoassay for this substance is reported to be less subject to positivity in patients with noncolorectal tumors and in the presence of inflammatory conditions, although sensitivity in cases of colorectal cancer was better than for CEA (115).

Pregnancy-Specific β_1*-Glycoprotein (PSBGP)*

An assay for this substance based on electroimmunodiffusion has been described (116). Plasma levels of PSBGP are raised in the presence of hydatidiform mole, invasive mole, choriocarcinoma, and teratoma (117). The concentrations are less dramatically increased than with HCG (117-119), but in some cases the level remains elevated after the HCG level has returned to normal during treatment (117). Lesser elevations have been recorded in patients with carcinomas of the colon, breast, and ovary. Antisera to PSBGP have been found to cross-react with another protein, the α_2-pregnancy-enhanced protein (120), so further studies are needed to clarify the precise materials being measured.

Embryonic Prealbumin

Embryonic prealbumin (EPA) was found in 122 of 205 tumor homogenates, including 32 of 66 cases of breast carcinoma (121). An especially high content of EPA was found in sarcomas, and EPA appeared in the serum of 7 of 20 patients with connective tissue tumors.

New Antigens

Tumor Associated and Serum Markers

The colon mucoprotein antigen (CMA) may be extracted from normal colonic epithelium or colonic tumors by a phenol-water extraction followed by ethanol precipitation. The material gives a single peak on glass bead chromatography with a molecular mass of more than 10^6 daltons. Significant differences in carbohydrate content and composition were found between the normal and the tumor antigen (122). A colon-specific determinant not present in mucus from other tissues was identified in this material (123).

The zinc glycinate marker (ZGM) isolated from colorectal cancers and apparently unrelated to CEA is useful histologically in defining adenocarcinomas of the gastrointestinal tract in tissue sections (124-126); other tumors did not react with the appropriate test reagents.

An alcian-blue-staining intestinal goblet-cell antigen (GOA) (127) was separated from gastric signet-ring-cell carcinoma by perchloric acid extraction, zone electrophoresis, and chromatography. The antibody developed against the material reacts with gastric signet-ring-cell cancers and also with colonic colloidal adenocarcinomas.

Another material derived by papain digestion from cancerous gastric mucosa is the sulphated glycoprotein antigen (SGA) (128). The amino acid composition showed high proportions of threonine and proline, but little serine and no aromatic amino acids were found. Anti-SGA antisera reacted with 50% of the colonic tumors and with 5 of 19 gastric carcinomas, the presence of antigen being correlated with presence of acidic mucous secretion.

A number of substances associated with breast cancer have been recognized. One such product is the mammary-tumor-associated glycoprotein (MTGP), extracted with perchloric acid (129,130). There are two such substances of similar molecular mass (19,500 daltons) and antigenicity but differing in isoelectric point, sedimentation constant, carbohydrate composition, and buoyant density. The antibody to MTGP identified the antigen in ductal cells of 22 of 26 samples of mammary carcinoma but not in normal mammary tissue. Another group of substances was prepared from BOT-2 cells derived from a ductal carcinoma (131). Antibodies to these cell products were present in the serum of more than 45% of the patients with untreated mammary carcinoma. In a second report breast carcinoma was used as a source of antigens (132). The antigens are present in a particle-free supernatant of the tumor and antibodies to the substances are present in the sera of patients with breast cancer. Another antigen was present in SW 527 breast carcinoma plasma membranes (133). Antibodies to "right-side-out" membranes were used to characterize the substance "TA3" extracted from a breast adenocarcinoma, and antibodies to this substance were present in the sera of patients with breast cancer but were absent in the sera of patients with ovarian cancer. The human breast gross cystic disease fluid protein (GC DFP-15) has a molecular mass of 15,000 daltons and a radioimmunoassay is established. The substance was found to be a useful adjunct to CEA in monitoring the progress of metastatic mammary carcinoma (96).

A substance with a molecular mass of 77,000 daltons was found to be present in lung carcinoma but not in normal lung extracts (134). Another substance with a molecular mass of 17,000 daltons was isolated from pleural effusions of patients with squamous-cell lung carcinoma (135).

The pancreatic oncofetal antigen (POA) is a glycoprotein present in fetal pancreas and pancreatic carcinoma with a molecular mass of 8-9 x 10^5 daltons. By quantitative immunoelectrophoresis the substance was detected in the plasma in a number of cancer patients, but the highest incidence of elevated levels (38 of 40 cases) occurred in patients with pancreatic cancer (136,137). Since the levels did not correlate with CEA levels, the determinations of this substance could complement the measurement of CEA in these patients.

Radioimmunoassays were developed for two substances identified in ovarian tumors. An antigen associated with ovarian cystadenocarcinoma (CAA) was found in the sera of patients with advanced ovarian cancer (138) and a correlation was observed between tumor volume and clinical status. A different substance, the tumor-associated antigen fraction "OCA" sharing determinants with CEA was assayed (139) and compared with CEA determinations in the plasma of patients with ovarian cancer. No correlation of OCA with CEA was found, and OCA appeared to be more sensitive than CEA to the presence of ovarian malignancy.

The substance CSA_p (colon-specific antigen-p) was further characterized as being present in both inflammatory and cancerous lesions of the gastrointestinal tract (140).

Antisera to bovine glial fibrillary acid protein (GFA) has been applied to the classification of human tumors of the central nervous system (141,142). It is possible to identify glial elements in astroblastomas and to distinguish these from nonglial tumors with similar morphologic characteristics.

Urinary Markers of Malignancy

The urinary glycoprotein EDC 1 is normally shed in the urine at a rate of less than 1 mg/day. In patients with metastatic cancer the excretion rate is often 100-500 mg/day. An immunological cross reaction has been identified between this substance and the serum inter α-trypsin inhibitor (143,144). A similar, but not identical, substance (HNC 1β) was discovered in the urine of one patient with acute monocytic leukemia (145). In another laboratory, a colored glycoprotein (α_1-microglycoprotein) was recognized (146,147) in the urine of a patient with plasma-cell leukemia. This substance also cross reacts immunologically with the inter α-trypsin inhibitor.

OTHER TUMOR-ASSOCIATED MACROMOLECULES

Although many macromolecules have been reported in the past to be associated with tumors, only a few such reports have been published this year.

Ferritin

Most developments in assay techniques have been to replace the radial immunodiffusion assay by a direct radioimmunoassay (148-150). The earlier method sometimes gave difficulties of interpretation (151). One problem with the new methods is to iodinate ferritin without damaging it (148,149). Even mild purification methods can change the biological half-life of the substance (152). High ferritin levels are found in acute and chronic myeloid leukemia (153) and sometimes in other malignant conditions (154). The ferritin found in malignancy has been reported to be similar to acidic ferritin associated with the normal heart. An assay with specificity for this acidic heart ferritin does not give a positive reaction with acidic tumor-associated ferritin (155). The two forms also differ in their binding by ion exchange resins. Increases in serum ferritin in chronic myelogenous leukemia were found to be correlated with renewal of tissue iron and damage to the erythropoietic system (156).

Milk Proteins

The previously reported finding of α-lactalbumin in the blood of both men and women has been shown to be due to a commonly occurring cross-reacting antibody (157,158). The problems and applications of the casein serum assay have been further defined (159), but, in general, its role in the management of breast cancer would appear to be limited.

Serum Glycoproteins and Proteins

Protein-bound fucose (160,161) and sialic acid (160,162) are increased in concentration in the plasma in patients with widespread malignancy and may have a limited role as an index of recurrent disease.

A fall in the prealbumin level and a rise in the C-reactive protein concentration has been found to occur two to three months before death in patients with colorectal cancer (163). The effect is not due to superimposed infection.

The level of α_1-antitrypsin (α_1AT) is sometimes elevated in the sera of cancer patients. The carbohydrate composition of the tumor-associated product may be abnormal with respect to lectin binding (164). In germ-cell tumors and hepatocellular cancers, α_1AT granules may be found, similar to those seen in the normal liver in α_1AT deficiency in which there is a failure to secrete the ZZ isoprotein (165). The granules in the cancer are not of ZZ type, and failure to secrete in this case must be due to some other cause. The presence of these granules has been suggested as a useful marker for neoplastic and preneoplastic liver lesions.

β_2-microglobulin has been found to be a useful marker in colorectal cancer, having about the same degree of positivity, in relation to the extent of disease, as CEA (166). The serum level is raised in pancreatic cancer but not in pancreatitis, allowing a better discrimination to be made than is possible with CEA (167). The use of this marker in discriminating pleural effusions of malignant and inflammatory origin has already been noted (58). The applications of this marker to cancer medicine have been the subject of a symposium (168).

Polyamines

Polyamines are found in higher concentrations in growing tissues and in certain tumors than in adult nondividing tissues (169,170) and are suspected of being involved in growth control. These substances combine with and may activate transfer RNA during the normal process of protein synthesis (171).

Early reports had suggested that polyamines could be a useful adjunct in the detection of malignant tumors, but this seems to be true only when widespread disease is present (170). From the analytical viewpoint it is easier to measure polyamine ratios rather than absolute concentrations (172). The tissue spermidine/spermine ratio increases progressively from normal kidney through well-differentiated renal carcinomas to the less differentiated tumors (173,174).

Viral Products

Herpes Simplex Virus (HSV-2)

The occurrence of genital herpes is believed to indicate women at risk of developing cervical cancer. The presence of the antigen is correlated with multiple sexual relationships. Antibodies to herpes simplex virus-2 were found in the blood of 10 of 18 prostitutes but in 0 of 18 nuns (175). Of the 1 of 80 cervical smears that showed the presence of HSV-2 about 20% are expected to reveal dyskaryotic cells over a 12-year follow-up period (176).

Epstein-Barr (EB) Capsular Antigen (VCA)

Evidence continues to accumulate for the involvement of EB virus in various human tumors. In a prospective study in Africa it was found that positivity for VCA increased the chance of developing Burkitt lymphoma by 30-fold (177). The presence of IgA antibody to VCA was found in 93% of the patients with nasopharyngeal cancer and was considered to be a useful clinical marker of the condition. Family associates of the patients also showed a high incidence as compared with unassociated normal people (178). Antibodies to VCA have also been found in a group of patients with cancer of the head and neck (179). The levels reflected the stage of disease and recurrence.

Mouse Mammary Tumor Virus (MMTV)

Immunocytochemical studies have tended to suggest that viruses related to MMTV may be involved with human breast cancer (180,181). Antibodies to MMTV glycoproteins reacted with components of breast carcinoma in 38% of the tissue sections from 191 different patients (182). Only 1% of 99 nonbreast tumor sections was positive. Benign breast lesions were nonreactive, but tissues from 1 of 18 normal breasts were reactive.

HORMONES

The inappropriate or ectopic production of hormones by tumors remains one of the most fascinating current biopathological problems in oncology. Their frequency and significance in lung cancer have been the subjects of a recent review (183). The range of hormones and other tumor products is illustrated in Table 2.

Table 2. Biochemical Markers in Bronchogenic Carcinoma[a]

Hormones	ACTH, β-LPH, PTH, HCG, insulin, VIP, calcitonin, ADH, prolactin
Oncofoetal antigens	CEA, AFP
Other antigens	See references 134 and 138
Enzymes	Alkaline phosphatase Placental alkaline phosphatase Salivary amylase Glycosyl transferase γ-glutamyl transpeptidase
Other substances	β_2-microglobulin Acute phase proteins Polyamines Minor nucleotides Hydroxyproline Ferritins

[a]A more detailed list is given in Coombes et al (183).

ACTH and its Congener Peptides

In 1978 evidence accumulated to reaffirm the original contention of Cushing in 1932 that the disease named after him was truly due to pituitary tumors (184). In a small series, pituitary adenomas, frequently microscopic in size, were identified as the source of the increased ACTH level (185). Interestingly, some basophil pituitary tumors may be found without overt endocrine signs or symptoms, a feature previously associated with the chromophobe tumors. In one case the adenoma contained ACTH, but the patient was eucorticoid and responded normally to tests of feedback control (186). The other secreted C-terminal ACTH but little or no bioactive hormone (187).

Cushing's syndrome caused by lung tumor secreting ectopic ACTH can, apart from metabolic changes, lead to a spectacular lowering of immunity (188). Often, in the absence of ACTH determinations, Cushing's syndrome due to a lung tumor is overlooked, but at post-mortem examination Crooke changes in the pituitary and adrenal hyperplasia are detected (189). The ectopic syndrome from a more slowly growing tumor, such as a medullary carcinoma of the thyroid, has more time to produce a frank Cushing syndrome than the more rapidly lethal lung tumors (190). Differential diagnosis between the pituitary-driven and ectopic ACTH syndromes is difficult except in extreme cases Measuring peripheral blood ACTH levels is one recommended method. More information can be derived from selective blood sampling of the right side of the heart, the brachial artery, the inferior petrosal sinus, and a peripheral vein, which may suggest the lung as the likely site of production (191).

Biosynthesis of ACTH involves a precursor molecule that is split by endopeptidases to give a glycopeptide, ACTH, and β-lipotropin (β-LPH) (192). Both ACTH and β-LPH are secreted, and the β-LPH levels in the blood closely reflect ACTH levels for Cushing disease, Nelson syndrome, and the ectopic syndromes (193,194). Since β-LPH is more stable, its assay is more robust than that for ACTH and is considered a useful alternative.

Further cleavage of β-LPH releases β-endorphin, an event that seems to occur in the normal pituitary (195). The level of this substance becomes elevated in the blood in diseases resulting in overproduction of ACTH (196), but β-endorphin has also been detected in the blood of normal persons (197).

Another source of β-endorphin exists outside the pituitary, since it is found in the cerebrospinal fluid when there is a failure of pituitary function leading to an absence of ACTH and β-lipotropin (198,199). Enkephalins (derived from endorphins) can be detected in adrenal medullary tumors, but they are not detectable in the blood (200).

Antidiuretic Hormone (ADH)

ADH has been implicated as the cause of hyponatremia in cancer. In 10 patients with oat-cell carcinoma of the lung and hyponatremia, raised ADH levels were found in tumor and plasma (201). However, raised ADH levels, as has been noted with ACTH, can also be found in tumors in the absence of any signs of hyponatremia. This may be related to the molecular heterogeneity of ADH with higher molecular weight bioinert forms being detected in some lesions, while material of similar size to authentic ADH is found in association with overt signs of hyponatremia. In some instances both ADH and ACTH may be produced by the same lung tumor; in the examples studied, ADH secretion dominated the clinical features (201).

Prolactin

Prolactin is the most reliable indicator of a pituitary chromophobe adenoma and of the efficiency of hypophysectomy, but there are a number of other conditions that need to be taken into account when interpreting the prolactin levels (202-205). If an adenoma exists it may be a consequence of some other disorder such as primary hypothyroidism, where there is an excess of thyrotropin-releasing factor (TRF) in the pituitary portal blood. This excess of TRF can stimulate the prolactin cells and may lead to a pituitary tumor. With adequate replacement therapy for the thyroid deficiency, the adenoma ceases to cause prolactin levels to rise, and peripheral symptoms of hormone excess are completely reversed (206).

Examination of the pituitary cells in and around the tumor by immunocytochemical techniques have revealed that most of the adenoma cells contained prolactin with appropriate secretory granules (207), but that the normal prolactin cells do not become involuted (208). Elevated prolactin levels may also be found with testicular tumors associated with gynecomastia and with renal disease and stress.

As with many other hormones, prolactin is synthesized from a prohormone (209), and an actively secreting tumor may secrete immunoreactive hormone, of varying molecular sizes, into the plasma (210).

Prolactin secretory tumors are often manifested by amenorrhea, impotence, and hypogonadism rather than by galactorrhea and gynecomastia (211,212). Visual field defects and headaches are also initial symptoms (213,214).

A prolactin level of 300 ng/ml (215) or 150 ng/ml (216) has been taken as a threshold value for discrimination of adenoma from a functional defect. Absence of a circadian rhythm is another sign of abnormality (215). Impotence is correlated with low testosterone levels that normalize when the adenoma is suppressed by surgery or drugs (217). Adrenal androgen levels are within normal limits (218). Restoration of the testosterone level alone does not appear to restore potency when the prolactin level is also elevated (219). However, in one case normal testicular function was found in a patient with galactorrhea and a high prolactin level (220).

Surgery is more successful with small tumors (221,222). The fall in prolactin level after operation can be used as an indicator of successful surgery (221,223). An elevated prolactin level may be indicative of a tumor even when the appearance of the sella turcica is normal (224). The prolactin level has a useful monitoring role in surgery performed to remove other types of pituitary tumors (225). If the tumor is not resectable or in the presence of persistently high prolactin levels after surgery, the drug bromocryptine can be used to inhibit prolactin secretion and, hence, the adverse peripheral effects of the hormone (222,226,227). However, if the drug is used, the hormone level can no longer be used as an indication of tumor progression (227). Bromocryptine appears to have long-term beneficial effects on the activity and size of the tumor in some cases (222,228,229).

Since most pituitary tumors associated with multiple endocrine neoplasia type I are functionless with respect to the other pituitary hormones, determination of the prolactin level may be a useful test when this condition is suspected (230).

Calcitonin

Medullary carcinoma of the thyroid (MTC) is a tumor of the C cells of the thyroid whose main secretory product is calcitonin. MTC is not infrequently associated with the familial disease of multiple endocrine neoplasia type II (Sipple syndrome)(231). Calcitonin is a useful plasma marker in this disorder, as is the enzyme histaminase. Histochemical studies have shown that most or all cells of the tumor contain calcitonin but not all contain histaminase (232). Unlike calcitonin, histaminase was absent from cells of benign C-cell hyperplasia. Other substances such as serotonin, somatostatin, ACTH, and CEA may be identified in MTC cells (110,233). Calcitonin has also been found as a component of the intermediary lobe of rat pituitary (234).

In a cell-free system, immunoreactive calcitonin may be synthesized in the presence of MTC messenger RNA. The molecular mass of the synthesized product was 68,000 daltons (235). The hormone of the normal gland and in plasma and urine is heterogeneous with molecular mass ranging from 2,500 to over 30,000 daltons (236) as though it were derived by breakdown from such a large precursor molecule. Some of the hormone may also be aggregated by S-S bonding (237). The existence of heterogeneity complicates the determination of the substance, since the different molecular forms respond differently in radioimmunoassay and radioreceptor assay (236,238).

It is important to recognize the frequently familial character of MTC, which is still greatly underdiagnosed (239). For this purpose, calcitonin is a good marker. In a pedigree of 107 related persons, the first year's screening produced 12 confirmed cases of which 7 had local metastases. Subsequent study of the group has resulted in the discovery of smaller tumors, with no metastases, that are unilateral and that occur in a younger age group (240). There were also a number of C-cell hyperplasias. It is not known whether C-cell hyperplasia is a premalignant lesion. Delay in diagnosis results in an increased incidence of frank malignancy (241). Detection of MTC is facilitated by stimulating the release of calcitonin with pentagastrin or calcium (Fig. 3) (240,242). Combining the two stimulants may be more effective than using either alone (243). Pheochromocytoma with an elevated plasma calcitonin level can accompany C-cell hyperplasia (244). Sometimes the adrenal medullary tumor may be the source of the calcitonin (245). If this is the case, the calcitonin level will fall when the adrenal tumor or tumors are removed.

Residual MTC left in unexplored nodes or as more distant metastases is suggested if calcitonin levels remain high after surgery. It is possible for these to remain clinically dormant despite calcitonin levels which, though elevated, remain stable (246). Selective catheterization has been used to localize and diagnose such tumors (247,248).

A variety of tumors produce calcitonin ectopically, mainly oat-cell carcinomas of the lung (249-251) and pancreatic islet tumors (252, 253). Lung tumors producing calcitonin also cause elevated plasma histaminase and L-dopa carboxylase levels. Since there is heterogeneity in the cellular distribution of these latter substances in the primary tumor, it is not surprising that the liver metastases often contain less than the primary tumor (251). In these cases, circulating levels do not reflect tumor load. The ectopically produced calcitonin reveals a wider degree of structural heterogeneity than the product of MTC; this could prove useful in the future in differential diagnosis (251).

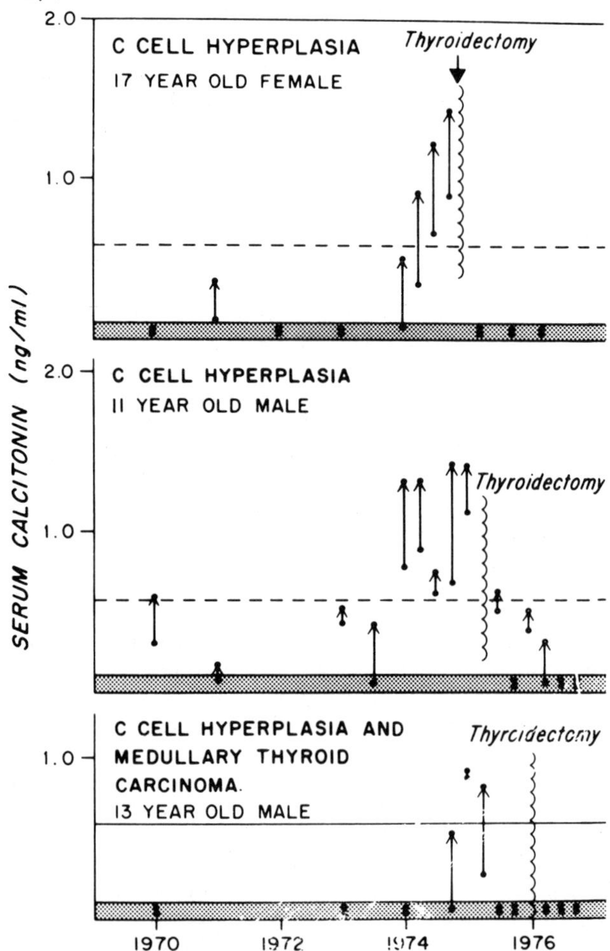

Figure 3. Familial medullary carcinoma of the thyroid monitored by sequential tests of calcitonin secretory reserve. The development of abnormally high reserves was followed by surgical confirmation of pathologic changes in the C-cells. The area shaded includes normal response limits, and one case showed normal limits after an initially abnormal response. (Reprinted with permission, from the *New England Journal of Medicine* 299:980, 1978.)

Gastrin

The normal physiology of gastrin and its relationships and applications to various diseases have been well reviewed (254). In common with many other hormones, gastrin exists in several molecular forms. Gastrin has amino acid sequences in common with cholecystokinin (255).

Hence, it is desirable to develop specific antisera by using that part of the sequence specific to gastrin.

The most important use of gastrin assays is in the detection of gastrin-producing pancreatic tumors in association with the Zollinger-Ellison (Z-E) syndrome. The current use of gastrin assays in its diagnosis has changed the disease profile. When first seen now, patients are younger and have a lower frequency of virulent ulcer, malignant gastrin-producing tumors, and associated endocrinopathies (256). Basal gastrin levels are of less value for this group than are calcium or secretin-stimulated levels. A new basic peptide from nonantral tissue that stimulates gastrin release has been described (257).

Because of the relative absence of multiple symptoms the disease is tending to disappear into the background of duodenal ulcer disease with ulcer or corrosive duodenitis as the only presenting feature (258, 259). According to the more optimistic assessments the diagnosis of Z-E syndrome can now be made before surgery and is not dependent on the findings at laparotomy. The function of the latter is to see if the tumor is resectable or not (256). However, another report suggests that a positive gastrin test is neither sufficient or necessary as a criterion for the syndrome (259). Since the islet cells may be derived from pancreatic ductal cells, it is of interest to find a mucinous ductal-cell carcinoma of the pancreas as a source of the Z-E syndrome (260).

With the advent of H_2-receptor blocking agents such as cimetidine it is no longer as necessary to consider gastrectomy for unresectable tumor (261-263). The gastrin level can be used as a monitor of tumor spread in the long-term follow-up of the patients.

Gastrinoma is frequently accompanied by other endocrine abnormalities in the same patient and in the related family. An intensive study is described using a screen of six hormones (insulin, cortisol, gastrin, PTH, thyroxin, and prolactin) as well as glucose, phosphate, and calcium. In the same family grouping four pituitary tumors, 12 hyperparathyroid states, three additional Z-E syndromes and an insulinoma were found (264). In another case an islet-cell tumor was found to be secreting peripherally active ACTH and melanocyte-stimulating hormone (MSH), together with large amounts of immunoreactive gastrin with little or no biological effect (265). The parietal cells of this patient, however, were capable of responding to pentagastrin.

Pancreatic Polypeptide

This peptide has been considered as an additional marker for the Z-E syndrome (266,267) and may be useful in suitable selected cases (268). There is no evidence that pancreatic polypeptide assays will be used to supplement gastrin assays in cases in which the results of the latter are equivocal (269).

Vasoactive Intestinal Peptide (VIP)

VIP was first recognized in the small intestine but has since been found in the brain, peripheral nerves, and pituitary portal blood (270). *In vitro*, VIP produces a rise in cyclic AMP in colonic carcinoma cells (271). This effect may be related to the mechanism by which VIP causes pancreatic cholera (Verner-Morrison syndrome), a severe watery diarrhea syndrome associated with islet-cell tumors of the pancreas (272). Watery diarrhea and associated hypokalemia may be produced in animals by injection of VIP (273).

Detection of VIP-containing tumors is necessary because about half are malignant, and the tumors may have other lethal peripheral effects (274). Diarrhea with elevated VIP levels can also occur with pheochromocytoma (275) or with pancreatic cysts (276). Diagnosis of the Verner-Morrison syndrome is possible on the basis of blood VIP levels (277), which can also be used to test the efficacy of surgery or chemotherapy (274,278).

Although VIP levels usually correlate well with the patient's symptoms, there are a number of patients with apparently identical symptoms who do not show an elevated VIP level (279,280). In one patient who had a pancreatic islet-cell tumor, the VIP and gastrin levels were normal but the pancreatic polypeptide level was 2,000 times above normal (281).

Insulin and the C-Peptide

Although it has been long recognized that some islet-cell tumors, particularly the β-cell type, can secrete insulin, recent data have revealed that there are errors in its storage and processing (282). Normally, before release the C-peptide is split off from the prohormone and can provide a useful insulin tolerance test of islet-cell function (283-286).

With the C-peptide test it has been possible to suggest strongly the presence of an insulinoma in a patient with a 23-year documented history of hypoglycemia (286). Other features that can be used for monitoring of chemotherapy of insulinomas are an inappropriate insulin level and an abnormally high level of proinsulin. In one case the normalization of these parameters was not accompanied by growth inhibition of the tumor (287).

If small, insulinomas are not detectable by arteriography or by palpation of the pancreas at operation (288). Localization can be made by catheterization via the splenic and portal veins at laparotomy, percutaneously (288), or via the pancreatic vein (Fig. 4) (289). Since there is a possibility of multiple islet-cell tumors, especially in the presence of multiple endocrine neoplasia type I, further blood samples are always essential after the surgery to monitor insulin and C-peptide levels (290).

Parathyroid Hormone (PTH)

Messenger RNA isolated from the parathyroid gland induces the synthesis of preproparathyroid hormone in cell-free systems (291). In the normal gland before secretion, the hydrophobic "presequence" and usually the "prosequence" are removed (292,293). Antisera have been developed against the prosequence but have so far failed to identify the prohormone as a feature of adenoma secretion (294). Earlier claims of tumor prohormone secretion need to be reconsidered in the light of this new assay system.

Much of the immunoreactive PTH in plasma consists of biologically inactive C-terminal fragments (295), which may be more immunoreactive against certain antisera than is PTH itself (296). Breakdown into fragments may be the normal activation step for PTH. The 1-34 amino acid sequence is adsorbed by bone as judged by arteriovenous gradients while the complete PTH molecule is not (297).

PTH used as a marker in studies of parathyroid status usually re-

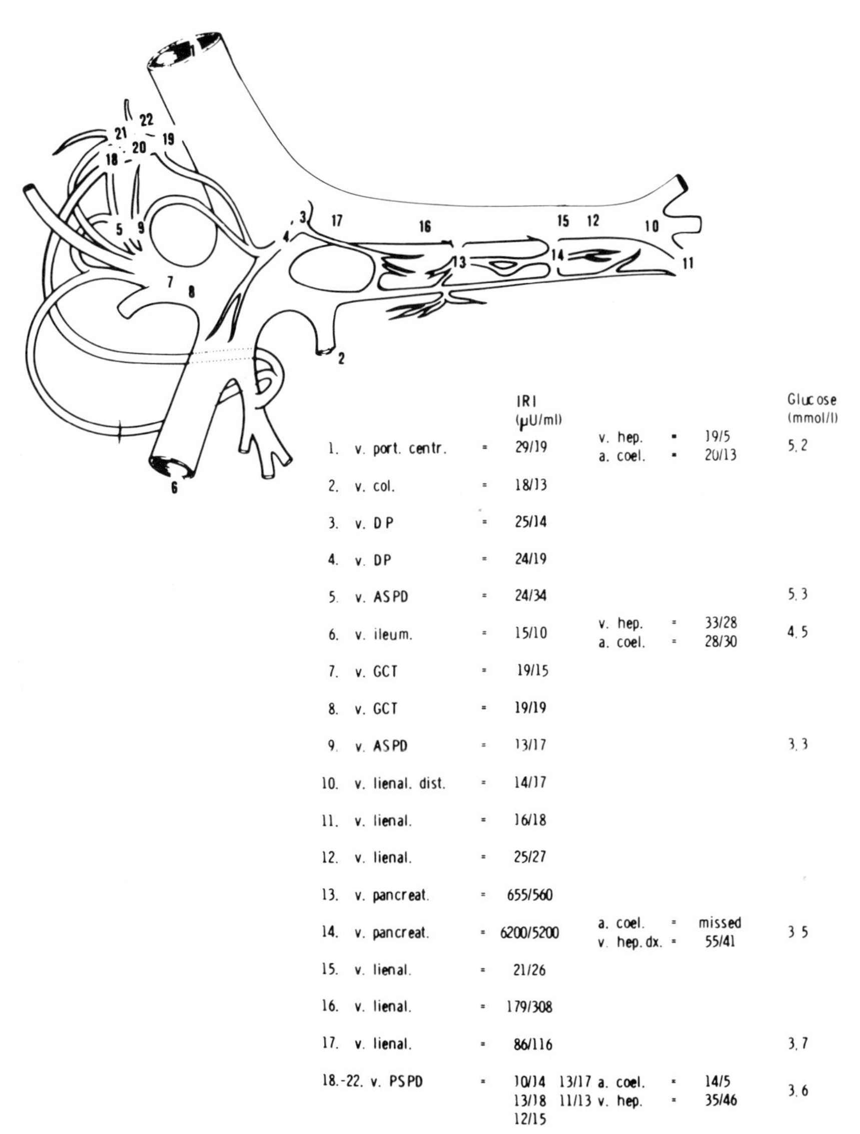

		IRI (μU/ml)			Glucose (mmol/l)
1.	v. port. centr.	= 29/19	v. hep. = 19/5 a. coel. = 20/13		5,2
2.	v. col.	= 18/13			
3.	v. D P	= 25/14			
4.	v. DP	= 24/19			
5.	v. ASPD	= 24/34			5,3
6.	v. ileum.	= 15/10	v. hep. = 33/28 a. coel. = 28/30		4,5
7.	v. GCT	= 19/15			
8.	v. GCT	= 19/19			
9.	v. ASPD	= 13/17			3,3
10.	v. lienal. dist.	= 14/17			
11.	v. lienal.	= 16/18			
12.	v. lienal.	= 25/27			
13.	v. pancreat.	= 655/560			
14.	v. pancreat.	= 6200/5200	a. coel. = missed v. hep. dx. = 55/41		3,5
15.	v. lienal.	= 21/26			
16.	v. lienal.	= 179/308			
17.	v. lienal.	= 86/116			3,7
18.-22.	v. PSPD	= 10/14 13/17 13/18 11/13 12/15	a. coel. = 14/5 v. hep. = 35/46		3,6

Figure 4. Localization of an insulinoma by pancreatic vein catheterization. The immunoreactive insulin (IRI) values by two different assays (results separated by a forward slash) show considerable elevations over a well-defined pancreatic region. (Reproduced from Ingemansson et al, ref. 289. By permission of *Surgery, Gynecology and Obstetrics.)*

quires antisera to the more stable C-terminal region of the molecule (298). To be sure of obtaining such antisera the fragment may be used instead of the PTH as an immunizing source (299). Reports continue to be made of parathyroid adenomas in association with other tumors in both forms of multiple endocrine neoplasia (300) but most commonly with the type 1 (301,302). Occasionally parathyroid carcinomas have been documented (303).

Differential diagnosis of primary versus secondary parathyroidism can be achieved by assay of PTH, calcium, and blood urea levels (304,305). By suitable combinations of routine blood chemical determinations, good evidence for or against primary hyperparathyroidism may be accumulated in the absence of PTH determinations (308). It is also possible to discriminate between parathyroid disease and hypercalcemia caused by tumors by measuring PTH and calcium levels simultaneously (299,306,307).

A second secretory product of the parathyroid glands has been described and more detailed studies are awaited with interest (309).

Hypercalcemia is frequently found in association with malignant disease, but in these cases, PTH levels are rarely elevated (310-312). The peripheral effects of the hypercalcemia are serious if the calcium level is above 13 mg/dl and especially if the rise is acute rather than chronic (305,310). Systemic calcinosis is an occasional complication (313).

Human Chorionic Gonadotrophin (HCG)

While radioimmunoassays of HCG, and more especially of the β-subunit, have become firmly established and of proven value, several new technical approaches to their measurement have been evolved. These include immunoassays using fluorescence intensity (314) and polarization (315). Antisera to HCG suitable for immunoassay have also been developed with the hormone bound to red blood cells as immunogen (316). The β-subunit assay has been applied to estimations of HCG excretion in the urine (317,318). By this sensitive assay an apparently normal HCG has been found in the urine of nonpregnant women (319).

The cost benefits of urinary HCG monitoring in choriocarcinoma have been subjected to detailed analysis with particular respect to the benefit of detecting early disease (320). It was found that drug and hospital care costs could be greatly reduced if the disease was detected earlier in more patients. The addition of an assay for the α-subunit did not appear to aid significantly in the detection of disease (321).

Usually the disease follows pregnancy, with metastases as a later event. In one patient, cerebral metastases accompanied a first pregnancy and were detected by angiography and HCG-level determinations (322). Subsequent surgery and chemotherapy reduced the HCG levels to normal, and these low values have persisted for a year.

Serum β-subunit determinations were preferred to 24-hour HCG excretion determinations for the detection of testicular tumors (16). Elevation of plasma HCG but not AFP levels occurs in about 40% of the patients with seminoma (19). Many patients with teratomas (embryonal carcinomas) may also exhibit raised plasma β-HCG levels. These changes in plasma HCG and β-HCG levels in patients with seminomas and embryonal carcinomas, respectively, have been shown to correlate, using immunoperoxidase cytochemical techniques, with the presence of HCG-positive cells in such tumors (323). Classical trophoblastic tissue was not present. Determination of the HCG level within the first six weeks after surgery has been found to be of prognostic value in cases of advanced testicular disease (324).

Melatonin

Melatonin has been suggested as a marker for tumors of the pineal gland. Since there is a strong circadian rhythm, the daytime values are considered to be more discriminating (325,326). Five cases have been described in which melatonin levels were above normal (327). However, since these tumors exhibit cellular heterogeneity, not all cases may be detected in this manner (328).

STEROID HORMONE RECEPTORS

Since their first recognition, steroid hormone receptors have been studied in detail by many workers. All the recent data point to their value in distinguishing patients with breast cancer who are more likely to show a response to hormone therapy.

Technical Developments

Receptor determinations do not require large amounts of tissue and may be done with biopsy samples of tumors. The storage of the material and its preparation for analysis have been subjects of recent reports (329-332). With biopsy material, it is important to check the sample histologically for the presence of tumor cells (330). Another problem relates to the presence in biopsy material of serum contamination with its accompanying steroid binding globulins (333). Separation of the receptors from the other binding proteins can be achieved by agar gel electrophoresis (334) or by isoelectric focusing (335,336). In the latter case, a brief trypsinization is required to prevent precipitation of the receptor at the isoelectric point. The synthetic estrogen R2858 is useful, since it does not bind to the serum binding proteins (333), and it is also less rapidly metabolized by the tumor cytosol than are the natural estrogens (337,338). A rapid assay using a hydroxyapatite column requiring only 50 mg of tumor tissue has been described (339).

Tumors are often heterogeneous in their cellular populations, and it is important to appreciate that the result from a homogenate is an average taken over the population. Hence, additional immunocytochemical methods have been evolved to determine the variability of receptor content between different cells and to ascertain the cell types involved. This can be done by staining tumor sections with a reagent that binds to the receptor. Anti-estrogen antibodies with immunofluorescence or peroxidase labeling have been used (340,341). There is perhaps some difficulty in accepting that the receptor and the antibody both bind to such a small molecule as an estrogen in a specific manner. Alternatively, a fluorescent tag may be coupled to the estrogen either directly or through a macromolecular bridge (342,343). Another approach to localization of estrogen receptors would be to use a reagent that recognizes this receptor independent of its binding region To develop an antibody against the receptor its isolation is necessary. This has recently been done using dextran coupled to estradiol. The dextran-receptor complex is separated by gel filtration, and the receptor may then be released by adding estradiol (344). The use of these reagents to study pathology of breast and related cancers is awaited with interest.

Breast Cancer

About 50% of breast cancer patients have detectable estrogen receptors (ER) in their tumors; the levels may vary somewhat among different races (345). ER positivity is higher in postmenopausal women, and it was believed that this was the result of the blocking of receptors with endogenous estrogen before menopause. Recent work has found normal premenopausal levels of estrogen (of adrenal origin) in breast cancer patients after the menopause (346). This report attributes the higher receptor level to the absence of progesterone in the blood of the older women. However, support for the blocking hypothesis comes from a comparison of ER and estradiol levels in a group of premenopausal and postmenopausal breast cancer patients (347). Apparently low ER levels were obtained for all patients with serum estradiol values above about 150 picomoles per liter.

As mentioned above, tumors are heterogenous in their estrogen-receptor content as determined by current cytochemical methods (340,343). There are also discrepancies between the ER content of some primary tumors and of their nodal deposits or subsequent metastases, although in general the contents are similar (348-350). There was no indication for a regular progression in discordant cases toward ER negativity.

Levels of ER positivity are a function of the cellularity of the tumor, but were found to be independent of the histological grade (349) although contrary reports have also appeared (351-353).

After surgery, patients with ER-positive tumors have a better prognosis, with fewer recurrences and a longer disease-free interval (351,352). The difference was most noticeable when the adjacent lymph nodes were involved (352). A patient with an ER-negative, poorly differentiated tumor has an especially poor prognosis. ER-negative tumors may have a higher growth rate and mitotic index (354).

About 50 to 70% of the ER-positive patients show a response to endocrine therapy, but only 5 to 10% of the ER-negative patients respond (355). The differential response persists, although the absolute values vary among different reports (330,358). The level of positivity in ER between patients varies over a 400-fold range in terms of steroid-binding capacity (331,358). The level correlates well with response; the more receptor, the better the response (355,359). ER determinations on deposits of metastatic breast carcinoma, if these are accessible to biopsy, have also been found to be of value (358).

Patients may be scored for tumor progesterone receptor (PR) levels, as well as for ER levels. The ER-positive, PR-positive group shows high responsiveness (357), but, even with high levels of both, some patients do not respond (331,357). Absence of estrogen receptor does not preclude a response, although in some of these cases, poor storage of the biopsy material may have caused false-negative result. There is also a small group of ER-negative, PR-positive patients (359). A small pilot study of breast cancer patients treated with the antiestrogen tamoxifen showed a partial correlation of ER positivity and response (360). Of the 38 ER-positive patients, 16 responded to the drug, and 4 of the 17 ER-negative patients responded.

Since the ER-negative tumors seem to behave more aggressively and may have a larger growth rate, they might be expected to respond better to chemotherapy. This is the result obtained by one group (354,361,362). Weakly positive or negative tumors showed objective response to chemotherapy in 34 of 45 cases while ER-positive tumors showed a response in only 2 of 25 cases. The ER-negative group fared even better with 21 of 24 tumors responding to chemotherapy. These results have been both partially confirmed (356) and denied (363-365) by other groups. The conflicting reports may reflect differences in patient selection and in pretreatment of the tumor with other therapies (366) or in methods of assessing responsiveness although the correlation between steroid receptors and response to hormone therapy is firmly established. The discrepancies in correlation with chemotherapy remain to be fully resolved. Together with the suggested cellular heterogeneity of breast tumors with respect to receptor content, this indicates that some combination hormone and chemotherapy might be reconsidered (367-369).

Prostatic Cancer

The effects of anti-androgen therapy on prostatic cancer have been studied in comparison with the presence or absence of 5α-dihydrotestosterone receptors in the tumor. The results with relatively small numbers of patients are not entirely concordant (370-372). Estrogen receptors have also been found in some prostatic tumors.

Renal Carcinoma

Specific steroid receptors have been identified in a number of these tumors, and a response to progesterone therapy was found in the majority of cases (373). In three of the four cases in which no response occurred, estrogen and progesterone receptors were lacking (374).

Hepatic Cancer

In light of the possible association between hepatic neoplasms and the use of contraceptive steroids, it is of interest that a focal nodular liver hyperplasia in one woman on the Pill was found to be richer in estrogen receptors than the normal liver (375).

ENZYMES

A significant interest continues to be shown in the role of enzymes as markers not only of early experimental neoplastic change but also of the type and stage of disease and its responses to therapy.

Alkaline Phosphatase

Alkaline phosphatase is known to exist in several isoenzymic forms. In neoplasia in general, the placental and related forms have received significant attention in the past. Additional methods, including radioimmunoassay and polyacrylamide electrophoresis, have been developed to detect and to measure the placental isoenzyme (376,377). Variants of the isoenzyme have been associated with osteosarcoma, ovarian tumors, and a few hepatic cancers (378-380). However, the placental isoenzyme is still only infrequently associated with general neoplasia.

Elevated plasma levels of bone alkaline phosphatase are encountered in metastatic prostatic carcinoma (381). The initial enzyme level correlates with response to therapy, and if a greater than 25% decrease in level occurs during therapy, there is more subsequent likelihood of response or of the disease becoming stabilized. Associated measurements of plasma γ-glutamyl transpeptidase and urinary hydroxyproline levels are useful in deciding whether an elevated total alkaline phosphatase level is of bone or hepatic origin when isoenzyme patterns cannot be obtained (382,383).

Prostatic Acid Phosphatase

After reports in 1977 of a radioimmunoassay for prostatic acid phosphatase (384-389), technical developments have continued, since it was realized that the immunological method has advantages over the previous enzymatic assay. The enzyme has been purified on an affinity column containing L(+)tartrate, an inhibitor of the enzyme, bound to an agarose matrix (390-392). A solid-phase fluorescence immunoassay has been described that is reported to detect increases in enzyme level above normal in early localized neoplastic disease (393). For discrimination of cancer from benign hyperplasia a less sensitive, but more economical, counterelectrophoresis method may be suitable; this detects a few stage II tumors and the majority of those in the more advanced stages (394,395).

The detection of prostatic acid phosphatase in bone marrow samples has been examined further (396), and the problems in clinical specificity and sensitivity to bone metastases have been explored (397). By immunological methods it is possible to reduce the number of false-positive results (398), but there are still problems with negative findings in the presence of metastases (399). Immunological methods can be used to detect cells of prostatic origin in histological sections of organs such as bone (400).

The ectopic production of prostatic acid phosphatase by a pancreatic islet cell tumor also has been described (401).

Glycosyl Transferases

An isoenzyme of galactosyl transferase, absent from the plasma of normal persons, has been found in 165 of 232 cancer patients (402). Malignant effusions are a rich source of the enzyme. With colon cancer, the incidence of detectable plasma levels rose with the stage of disease, although some widely extended cancers could give negative results.

Two fucosyl transferases have been studied in the serum of patients with Hodgkin disease and acute myeloid leukemia. One appears to reflect tumor burden, while the other may be determined by bone marrow regeneration. They vary independently during therapy (403,404).

Amylase

The isoenzymes associated with pancreatic inflammatory disease have been compared with those found in malignancy (405). A radioimmunoassay specific for the salivary isoenzyme has been established to detect this substance even in the presence of elevated pancreatic isoenzyme levels (406). Tumors of the lung and ovary are sometimes found to be associated with high salivary amylase levels, apparently of tumor origin (407-410).

γ-Glutamyl Transpeptidase (γ-GT)

This enzyme is a cell-membrane component, and its release from the cell requires proteolytic action to remove a fat-soluble peptide sequence (411). Increased levels of γ-GT can reflect various types of hepatic disease or metastatic hepatic tumors. Fractionation procedures have been described to make the determination more liver-specific (412,413). The marker can provide useful evidence of hepatic metastases but does not give an early warning preoperatively or relative to other methods of detection (414,415). The γ-GT level may also be raised by hepatotoxic drugs (416).

A novel isoenzyme of γ-GT is found in the plasma of 35% of the patients with hepatocellular carcinoma but not in a number of other hepatic disorders including metastatic carcinoma, inflammatory disease, and bile duct or pancreatic malignancy (417).

Contrary to an earlier report, γ-GT levels in ascites were not found to be of value in discriminating between hepatoma and non-malignant hepatic disease, especially if a serum value was available (418,419).

The enzyme is also found in the kidney and the γ-GT in renal carcinoma appears to differ from the normal renal or hepatic enzymes in its electrophoretic properties (420).

Other Enzymes

A barely detectable level of hexokinase was found in the plasma of normal healthy persons, but 15 to 25-fold elevations occurred in a variety of malignancies and stages of disease (421). There are also less dramatic rises in inflammatory conditions. Further studies appear to be merited.

Serum levels of lactate dehydrogenase were found to be elevated with primary carcinoma of the ovary but not in benign disease or with a number of other cancers of similar location in the body. It was possible to detect pelvic masses in 18 patients by this procedure (422). After removal of the tumors, the plasma levels returned to normal in one case.

An abnormal isoenzyme of tyrosinase has been found in the serum of patients with melanomas (423).

LYMPHOPROLIFERATIVE NEOPLASIA

Many attempts are being made to define cell-surface characteristics and specific "antigens" of the various cell types of the lymphoid and myeloid series. These studies are aimed at improving our appreciation of the origin and degrees of differentiation of the leukemias and lymphomas and, therefore, may have fundamental and prognostic value (424).

Most of the current tests are directed to the cells, but there is also a spillover of markers into the plasma and urine (e.g., lysozyme, β_2-microglobulin, and the immunoglobulins). The tests on biochemically defined substances supplement more empirical tests such as red cell rosetting, but there is a tendency for undefined markers to be investigated biochemically (425).

The Myeloid and Histiocytic Series

Lysozyme

This enzyme is a good marker of cellular differentiation in the myeloid and histiocytic series (426-428). Observations may be made on serum, urine, or in cell lysates. In the remission phase of chronic myeloid leukemia, the lysozyme level of the cells is high, but during blast crisis, the level falls as the cells become less mature. The level is very high in acute monocytic leukemia and, as expected, low in acute myeloblastic leukemia (429,430). With both chronic lymphocytic and acute lymphoblastic leukemia, a normal or low lysozyme level is found. A radioimmunoassay for urinary lysozyme has been described and applied to monocytic leukemia (431).

Nonspecific Esterase
Nonspecific esterase is found in the cells of histiocytic and monocytic type but not in leukemic blast cells (432,433) and is a good marker for cells of the monocyte lineage (434). Significant levels are not found in Hodgkin disease, although "Hodgkin cells" have been regarded by some to be of histiocytic origin. With changed reaction conditions, the enzyme is reported to be a marker for mature T cells, and a tumor of T cell type was found to react positively (435).

Specific Esterase
Specific esterase is a cytochemical marker for cells of the granulocytic and monocytic series (436). Cells of chronic granulocytic leukemia have been found to contain additional "fast-running" isoenzymes, not found in normal polymorphonuclear leukocytes (437).

Acid Ribonuclease
The acid ribonuclease level is increased in the serum of patients with acute and chronic myeloid leukemia; the level falls after successful treatment (438).

Terminal Deoxynucleotidyl Transferase
Unlike lysozyme, terminal deoxynucleotidyl transferase is present in cell lysates and homogenates in most cases of acute lymphoblastic leukemia (439,440) but not in peripheral T cells (441). Since the enzyme is absent from typical B-cell cancers, for example, B-type chronic lymphocytic leukemia and Burkitt lymphoma (442), it has been regarded as a marker for immature T cells. However, cells of a B-type acute lymphoblastic leukemia have been found to carry the enzyme (443), and it is likely that the presence of the enzyme is a more general sign of cell immaturity. In fact a clinial use has been found for the enzyme in cases of chronic myeloid leukemia. The enzyme appears during blast crisis but disappears during remission (442). The presence of the enzyme during blast crisis correlates in a relatively small number of cases with a good therapeutic response (444).

Acute Lymphoblastic Leukemia

Acid Phosphatase
The presence of acid phosphatase is a useful cytochemical correlate of cells of T lineage, but the relationship is not close enough for the enzyme test to replace the practice of testing for E-rosetting (442,445-558). The marker is a useful additional test together with periodic acid Schiff staining (usually absent in T cell disease) to distinguish the T cell group of diseases from non-B, non-T cell disease, which often has a better response to current therapy. It is also necessary to exclude B-cell disease, which is discussed in a following section. The finding of the tartrate-resistant form of acid phosphatase is a fairly specific cytochemical confirmation of hairy-cell leukemia (449,450).

Glucocorticoid Receptors
High levels of intracellular glucocorticoid receptors in acute lymphoblastic leukemia are indicative of a better prognosis (451) and are found to be of the non-B, non-T type. It is not known whether

the presence of the receptors is a causative factor of therapeutic response in this case since there is no correlation over the whole range of leukemias between drug responsiveness and the level of receptors (452,453).

Hexoseaminidase

An "intermediary" isoenzyme was observed in white cell lysates in 23 of 27 cases of leukemia with the non-B and non-T common antigen (454). In the doubtful cases the presence of an abnormal α-L-fucosidase in the cells was helpful.

B-Cell Diseases

Immunoglobulin (Ig)

Patients with the rare type of acute lymphoblastic leukemia with Burkitt morphology and Sudan staining have a relatively poor prognosis (439). The disease may be distinguished by the presence of surface-membrane immunoglobulin (SmIg) (445,455). One problem is that the SmIg can be lost if transport of the samples to the laboratory is delayed. Since ambient Ig may be bound to the cell by Fc receptors, it is necessary to remove them with proteolytic enzymes and to allow the SmIg to regenerate to be sure that it is endogenous to the cell (455, 456).

The majority of the cases of chronic lymphocytic leukemia, in contrast with the acute form, appear to be of B-cell lineage (457-460). The SmIg is monoclonal in heavy and light chains. This characteristic can distinguish the disease from benign lymphocytosis, in which the B cells are polyclonal and there is an accompanying increase in the T-cell population. Further evidence for the monoclonal nature of the disease comes from studies on glucose-6-phosphate dehydrogenase (461).

Chronic lymphocytic leukemia cells synthesize Ig as a small fraction of their total protein, the SmIg is weakly expressed, and there is imbalanced synthesis with overwhelming expression of light chains (462). The more mature follicular center-cell lymphomas are a contrasting group with a high proportion of Ig in the proteins synthesized, strong SmIg expression, and a more balanced heavy-to-light chain ratio in the cells.

In Hodgkin disease and with non-Hodgkin lymphomas, the B-cell nature of the disease is suggested by accumulation of intracellular Ig. Since this Ig is sometimes polyclonal, there is some doubt of the ability of the cells to synthesize rather than to accumulate the substance (428,463-466).

The more mature B-cell tumors are able to secrete Ig, which may be found in large amounts in the blood plasma and sometimes in the urine. A typical myeloma may start when a patient is 40 years of age and may only become apparent at age 56, when the tumor size has reached 10^9 cells because of a monoclonal spike of Ig (467). Synthesis of Ig is a complex genetic event possibly involving three separate genes to code for a single peptide chain (468).

To assist in the early separation of malignant and benign disorders, it is helpful to consider a number of parameters in bone marrow and blood Ig and total protein as discriminant functions (469). With frank myeloma, the blood urea, uric acid, and albumin levels have been found to be prognostically valuable (470). With current therapies the disease may have a more favorable prognosis than previously (471).

Of 516 cases of paraproteinemia (472), 61% were cases of myeloma and 5%, Waldenstrom's macroglobulinemia. Overall the majority (67%) involved IgG; IgA and IgM were found in 20% and 10% of the cases, respectively, and only one involved IgD (472). In another group of 169 elderly patients with raised serum M (monoclonal immunoglobulin) component (473), 47% had lymphoproliferative neoplasms, and 8% had myelomas. The serum IgM level was greater than 1 gm/dl in 86% of cases of diagnosed Waldenstrom macroglobulinemia. Since the disease often progresses slowly in elderly patients, the treatment suggested was mainly designed to avoid peripheral effects of high plasma viscosity (474).

Heavy-chain disease is a B-cell proliferative disorder with diffuse distribution in the gastrointestinal tract (475,476). The heavy chain often has deletions of functional sequences, but in one case there was a large addition insertion of peptide chain (477).

Light-chain disease accompanied by Bence Jones proteinuria can be studied therapeutically using the urinary protein levels (478, 479). A considerable loss of protein can occur in an exceptional case (480). The light chain may be accompanied by a heavy-chain fragment (481).

Elevations of serum IgD were observed in Hodgkin disease, especially in the lymphocyte-depleted form of this tumor (482).

β_2-Microglobulin

In chronic lymphocytic leukemia the serum level of this marker is raised, but the cell surface level is reduced suggesting increased turnover of the cell-surface components (483,484). A high β_2-microglobulin plasma level is more likely to be associated with the need for therapeutic intervention within three years (485).

It was observed that Hodgkin and non-Hodgkin lymphomas differed in β_2-microglobulin plasma levels, the level being lower in Hodgkin disease (486-488). However, a low level may be a feature of the disease in relapse rather than of the untreated disease (487,488). A high marker level is regarded as a bad prognostic sign (485). The plasma levels of this marker also have been suggested to discriminate between myeloma and benign gammopathies (487).

De Novo and Salvage Enzyme Markers

The success of chemotherapy with antimetabolites can depend on the relative importance of the *de novo* and salvage pathways of nucleic acid synthesis, and these may change in the course of the disease (489). Recently, a cytochemical marker for dihydrofolate reductase has been described (490); this may be a useful growth point in establishing an optimal use of these drugs and in studying relevant cellular heterogeneity.

CONCLUSIONS

It is now evident that the number of known products of human tumors is increasing every year. However, it would be not untimely to draw attention to the need for adequate comparative studies before claims for a "new" marker are made. In a clinical setting it is also important that an adequate number of patients with a variety of nonneoplastic and neoplastic diseases are sampled when the value of a marker is being assessed. A basis of patients at different phases of evolution of the tumor under study is also a prerequisite. Tentatively, many of these caveats are now appreciated by workers in the tumor-marker field. New and exciting research and its outcome in this field should be profitable clinically and in appreciating the fundamental biology of neoplasms.

REFERENCES

1. Odell WD, Wolfsen AR: *Ann Rev Med* 29:379, 1978.

2. Gold P, et al: *Cancer Res* 38:6, 1978.

3. Zizkovsky V, et al: *Neoplasma* 25:559, 1978.

4. Hibi N: *Gann* 69:67, 1978.

5. McDonald DJ, Kelly AM: *Clin Chim Acta* 87:367, 1978.

6. Schiller HS, et al: *Clin Chem* 24:275, 1978.

7. Johansson SGO, et al: *Acta Obstet Gynecol Scand* 69 (Suppl):15, 1978.

8. Blank-Liss WE, et al: *Clin Chim Acta* 86:67, 1978.

9. Khadempour MH, et al: *Ann Clin Biochem* 15:213, 1978.

10. Grenier A, et al: *Clin Chem.* 24:2158, 1978.

11. Aoyagi Y, et al: *Cancer Res* 38:3483, 1978.

12. Parmelee DC, et al: *J Biol Chem* 253:2114, 1978.

13. Johnson PJ, et al: *Brit Med J* 2:661, 1978.

14. Kubo Y, et al: *Gastroenterology* 74:578, 1978.

15. Alpert E, Feller ER: *Gastroenterology* 74:856, 1978.

16. Talerman A, et al: *Cancer* 41:272, 1978.

17. Flamant F, et al: *Eur J Cancer* 14:901, 1978.

18. Javadpur N, et al: *J Urol* 119:759, 1978.

19. Mann K, et al: *Acta Endocrinol* 87 (suppl):130, 1978.

20. Norgaard-Pedersen B, et al: *Cancer* 41:2315, 1978.

21. Gitlin D: *Scand J Immunol* 8 (suppl 8):91, 1978.

22. Tsuchida Y, et al: *J Pediat Surg* 13:25, 1978.

23. Norgaard-Pedersen B, et al: *Lancet* 2:1042, 1978.

24. Schultz H, et al: *Cancer* 42:2182, 1978.

25. Kohn J: *Scand J Immunol* 8(suppl 8):103, 1978.

26. Alsabti EA, Safo M: *Jap J Exp Med* 48:283, 1978.

27. Shiveley JE, et al: *Cancer Res* 38:2301, 1978.

28. Westwood JH, et al: *Brit J Cancer* 37:183, 1978.

29. Ormerod M: *Scand J Immunol* 8(suppl 8):433, 1978.

30. Deyoung NJ, Ashman LK: *Aust J Exp Biol Med Sci* 56:321, 1978.

31. Keep PA, et al: *Brit J Cancer* 37:171, 1978.

32. Rule AH: *Methods Cancer Res* 14:87, 1978.

33. Kimball PM, Brattain MG: *Cancer Res* 38:619, 1978.

34. Shively JE, et al: *Cancer Res* 38:503, 1978.

35. Reynoso G: *Cancer* 42:1406, 1978.

36. Azzolina LS, et al: *Tumori* 64:151, 1978.

37. Kim YD, et al: *J Immunol Methods* 19:309, 1978.

38. Bell-Isles M, et al: *Clin Chem* 24:727, 1978.

39. Kupchik HZ, et al: *Cancer* 42:1589, 1978.

40. Fritsche HA, et al: *Amer J Clin Pathol* 69:140, 1978.

41. Gold P: *Canad J Surg* 21:212, 1978.

42. Chu TM, et al: *NY State J Med* 78:879, 1978.

43. Anderson HA, et al: *Cancer* 42:1560, 1978.

44. Denslow GT, Kielar RA: *Cancer* 42:1504, 1978.

45. Denslow GT, Kielar RA: *Amer J Ophthalm* 85:363, 1978.

46. Baker AL: *JAMA* 240:385, 1978.

47. Meeker WR: *Cancer* 41:854, 1978.

48. Meeker WR: *Cancer* 42:1463, 1978.

49. Hine KR, et al: *Lancet* 2:1337, 1978.

50. Goldenberg DM, et al: *Cancer* 42:1554, 1978.

51. Isaacson P, Judd MA: *Cancer* 42:1554, 1978.

52. Primus FJ, et al: *Cancer* 42:1540, 1978.

53. Goldenberg DM, et al: *N Engl J Med* 298:1384, 1978.

54. Editorial: *Lancet* 2:461, 1978.

55. Carrel ST, et al: *Schweiz Med Wochenschr* 108:955, 1978.

56. Loewenstein MS, et al: *Ann Intern Med* 88:635, 1978.

57. Rittgers RA, et al: *Ann Intern Med* 88:631, 1978.

58. Vladutiu AO: *Lancet* 2:423, 1978.

59. Stanford CF: *Lancet* 2:53, 1978.

60. Klockars M, et al: *Lancet* 2:1057, 1978.

61. Neville AM, et al: *Cancer* 42:1448, 1978.

62. Mach JP, et al: *Cancer* 42:1439, 1978.

63. Rittgers RA, et al: *J Natl Cancer Inst* 61:315, 1978.

64. Gerber MA, Thung SN: *Amer J Pathol* 92:671, 1978.

65. Loewenstein MS, Zamcheck N: *Cancer* 42:1412, 1978.

66. Mavligit GM, et al: *Cancer* 42:1437, 1978.

67. Staab HJ, et al: *Amer J Surg* 136:322, 1978.

68. Reynoso G: *JAMA* 239:870, 1978.

69. Minton JP, Martin EW: *Cancer* 42:1422, 1978.

70. Wanebo HJ, et al: *N Engl J Med* 299:448, 1978.

71. Valdivieso M, Mavligit GM: *Surg Clin N Amer* 58:619, 1978.

72. Jubert AV, et al: *Cancer* 42:635, 1978.

73. Staab HJ, et al: *Scand J Immunol* 8(suppl 8):459, 1978.

74. Evans JT, et al: *Cancer* 42:1419, 1978.

75. Moertel CG, et al: *JAMA* 239:1065, 1978.

76. Nicholson JR, Aust JC: *Dis Colon Rectum* 21:163, 1978.

77. Staab HJ, et al: *J Surg Oncol* 10:273, 1978.

78. Wanebo HJ, et al: *Ann Surg* 188:481, 1978.

79. Welch JP, Donaldson GA: *Amer J Surg* 135:505, 1978.

80. Mavligit GM, et al: *JAMA* 240:1713, 1978.

81. Mayer RJ, et al: *Cancer* 42:1428, 1978.

82. Shani A, et al: *Ann Intern Med* 88:627, 1978.

83. Moertel CG, Connell MJO: *Ann Intern Med* 89:572, 1978.

84. Ellis DJ, et al: *Cancer* 42:623, 1978.

85. Sugarbaker PH, et al: *Cancer* 42:1434, 1978.

86. Alexander JC, et al: *Cancer* 42:1492, 1978.

87. Macnab GM, et al: *Brit J Cancer* 38:51, 1978.

88. Kaiser MH, et al: *Cancer* 42:1468, 1978.

89. Fitzgerald PJ, et al: *Cancer* 42:868, 1978.

90. Barkin JS, et al: *Cancer* 42:1472, 1978.

91. Haagensen DE, et al: *Cancer* 42:1512, 1978.

92. Myers RE, et al: *Cancer* 42:1520, 1978.

93. Coombes RC: *Invest Cell Pathol* 1:347, 1978.

94. Tormey DC, Waalkes TP: *Cancer* 42:1507, 1978.

95. Falkson HC, et al: *Cancer* 42:1308, 1978.

96. Haagensen DE, et al: *Cancer* 42:1646, 1978.

97. Waalkes TP, et al: *Cancer* 41:1871, 1978.

98. Ward AM, et al: *Eur J Cancer* 14:885, 1978.

99. Woo KB, et al: *Cancer* 41:1685, 1978.

100. Cuschieri A, et al: *Brit J Cancer* 37:1002, 1978.

101. Concannon JP, et al: *Cancer* 42:1477, 1978.

102. Dent PB, et al: *Cancer* 42:1484, 1978.

103. Vincent RG, et al: *J Thorac Cardiovasc Surg* 75:734, 1978.

104. Vannagell JR, et al: *Cancer* 42:1527, 1978

105. Vannagell JR, et al: *Cancer* 42:2335, 1978.

106. Rutanen EM, et al: *Cancer* 42:581, 1978.

107. Vannagell JR, et al: *Cancer* 42:2428, 1978.

108. Wahren B: *Cancer* 42:1533, 1978.

109. Wells SA, et al: *Cancer* 42:1498, 1978.

110. DeLellis RA, et al: *Amer J Clin Pathol* 70:587, 1978.

111. Mann JR, et al: *Arch Dis Child* 53:366, 1978.

112. Kessler MJ, et al: *Cancer Res*. 38:1041, 1978.

113. Burtin P, Gendron MC: *Immunochemistry* 15:245, 1978.

114. Svenberg T, et al: *Scand J Immunol* 8(suppl 8):429, 1978.

115. Nakamura RM, et al: *Ann Clin Lab Sci* 8:4, 1978.

116. Bruce D, Klopper A: *Clin Chim Acta* 84:107, 1978.

117. Searle F, et al: *Lancet* 1:579, 1978.

118. Seppala M, et al: *Int J Cancer* 21:265, 1978.

119. Bagshawe KD, et al: *Eur J Cancer* 14:1331, 1978.

120. Teisner B, et al: *Amer J Obstet Gynecol* 131:262, 1978.

121. Tartarinov YS, et al: *Lancet* 2:1122, 1978.

122. Gold DV, Miller F: *Cancer Res* 38:3204, 1978.

123. Gold DV, Miller F: *Tissue Antigens* 11:362, 1978.

124. Doos WG, et al: *J Natl Cancer Inst* 60:1375, 1978.

125. Saravis CA, et al: *J Natl Cancer Inst* 60:1371, 1978.

126. Saravis CA, et al: *Cancer* 42:1621, 1978.

127. Rapp W, Wurster K: *Klin Wochenschr* 23:1185, 1978.

128. Bara J, et al: *Int J Cancer* 21:133, 1978.

129. Leung JP, et al: *Fed Proc* 37:1485, 1978.

130. Leung JP, et al: *J Immunol* 121:1287, 1978.

131. Lerner MP, et al: *J Natl Cancer Inst* 60:39, 1978.

132. Lee CK, et al: *Fed Proc* 37:1485, 1978.

133. Holton OD, et al: *Fed Proc* 37:1485, 1978.

134. Braatz JA, et al: *J Natl Cancer Inst* 60:1035, 1978.

135. Wolf A: *Brit J Cancer* 37:1046, 1978.

136. Gelder FB, et al: *Cancer Res* 38:313, 1978.

137. Gelder F, et al: *Cancer* 42:1635, 1978.

138. Bhattacharya M, Barlow JJ: *Cancer* 42:1616, 1978.

139. Knauf S, Urbach GI: *Amer J Obstet Gynecol* 131:780, 1978.

140. Pant KD, et al: *Cancer* 42:1626, 1978.

141. Eng LF, Rubinstein LJ: *J Histochem Cytochem* 26:513, 1978.

142. Rueger DC, et al: *Anal Biochem* 89:360, 1978.

143. Chalwa RK, et al: *Fed Proc* 37:1282, 1978.

144. Chalwa RK, et al: *Cancer Res* 38:452, 1978.

145. Rudman D, et al: *Cancer Res* 38:602, 1978.

146. Seon BK, Pressman D: *Biochemistry* 17:2815, 1978.

147. Seon BK, Pressman D: *Fed Proc* 37:1282, 1978.

148. Goldie DJ, Thomas MJ: *Amer Clin Biochem* 15:102, 1978.

149. Deppe WM, et al: *J Clin Pathol* 31:872, 1978.

150. Barnett MD, et al: *J Clin Pathol* 31:742, 1978.

151. Heinrich HC: *Blood* 51:764, 1978.

152. Pollock AS, et al: *Proc Exp Biol Med* 157:481, 1978.

153. Munro HN, Linder MC: *Physiol Rev* 58:317, 1978.

154. Giler S, Moroz C: *Biomedicine* 28:203, 1978.

155. Jones BM, Worwood M: *Clin Chim Acta* 85:81, 1978.

156. Eber M, et al: *Nouv Presse Med* 7:3560, 1978.

157. Laurence DJR: *Invest Cell Pathol* 1:5, 1978.

158. Stevens U, et al: *Clin Chim Acta* 87:149, 1978.

159. Zangerle PF, et al: *Antibiot Chemother* 22:141, 1978.

160. Waalkes TP, et al: *J Natl Cancer Inst* 61:703, 1978.

161. Wallack MK, et al: *J Surg Oncol* 10:39, 1978.

162. Silver HKB, et al: *Cancer* 41:1497, 1978.

163. Milano G, et al: *J Natl Cancer Inst* 61:687, 1978.

164. Rostenberg I, et al: *J Natl Cancer Inst* 61:961, 1978.

165. Palmer PE, Wolfe HJ: *J Histochem Cytochem* 26:523, 1978.

166. Plessner T: *Dan Med Bull* 25:91, 1978.

167. Fateh-Moghadam A, et al: *Klin Wochenschr* 56:267, 1978.

168. *Symposium Pathol Biol (Paris)* 26 (pt 6), 1978.

169. Miller TR: *Radiology* 129:221, 1978.

170. Janne J, et al: *Biochim Biophys Acta* 473:241, 1978.

171. Cohen SS: *Nature* 274:209, 1978.

172. Gittings M, Cooke KB: *Biochem Soc Trans* 6:212, 1978.

173. Matsuda M, et al: *Clin Chim Acta* 87:93, 1978.

174. Dunzendorfer V, Russell DH: *Cancer Res* 38:2321, 1978.

175. Patterson WR, et al: *Proc Soc Exp Biol Med* 157:273, 1978.

176. Cornes JC, et al: *Brit Med J* 1:988, 1978.

177. Epstein MA: *Nature* 274:740, 1978.

178. Ho HC: *Lancet* 1:436, 1978.

179. Halili MR, et al: *J Surg Oncol* 10:457, 1978.

180. Yang N-S, et al: *J Natl Cancer Inst* 61:1205, 1978.

181. Hendrick JC, et al: *Cancer Res* 38:1826, 1978.

182. Mesatejada R, et al: *Proc Natl Acad Sci USA* 75:1529, 1978.

183. Coombes RC, et al: *Brit J Dis Chest* 72:263, 1978.

184. Editorial: *JAMA* 240:865, 1978.

185. Tyrell JB, et al: *N Engl J Med* 298:753, 1978.

186. Kovacs K, et al: *Amer J Med* 64:492, 1978.

187. Tramu G, et al: *Ann Endocrinol (Paris)* 39:51, 1978.

188. Cummings RO, et al: *Arch Intern Med* 138:1005, 1978.

189. Singer W, et al: *J Clin Pathol* 31:591, 1978.

190. Rosenberg EM, et al: *J Clin Endocrinol Metab* 47:255, 1978.

191. Kley HK, et al: *Acta Endocrinol (Kbh)* (suppl) 215:130, 1978.

192. Eipper BA, Mains RE: *J Biol Chem* 253:5732, 1978.

193. Orth DN, et al: *J Clin Endocrinol Metab* 46:849, 1978.

194. Jeffcoate WJ, et al: *J Clin Endocrinol Metab* 46:160, 1978.

195. Allen RG, et al: *Proc Natl Acad Sci USA* 75:4972, 1978.

196. Suda T, et al: *Science* 202:221, 1978.

197. Nakao K, et al: *J Clin Invest* 62:1395, 1978.

198. Jeffcoate WJ, et al: *Lancet* 2:122, 1978.

199. Editorial: *Brit Med J* 2:155, 1978.

200. Sullivan SN: *Lancet* 1:986, 1978.

201. Morton JJ, et al: *Clin Endocrinol* 9:357, 1978.

202. Stepanas AV, et al: *Cancer* 41:369, 1978.

203. Cowden EA, et al: *Clin Endocrinol* 9:241, 1978.

204. Koninckx P: *Lancet* 1:273, 1978.

205. Jeffcoate SL: *Lancet* 2:1245, 1978.

206. Tolis G, et al: *Amer J Obstet Gynecol* 131:850, 1978.

207. Osamura RY, et al: *Acta Endocrinol (Kbh)* 88:643, 1978.

208. Kovacs K, et al: *Hormone Metab Res* 10:409, 1978.

209. Maurer RA, McKean DJ: *J Biol Chem* 253:6315, 1978.

210. Fang VS, Refetoff S: *J Clin Endocr Metab* 47:780, 1978.

211. Van Bogaert LJ: *Nouv Presse Med* 7:105, 1978.

212. Sherman BM, et al: *Lancet* 2:1019, 1978.

213. Swanson JA, et al: *Obstet Gynecol* 52:67, 1978.

214. Frantz AG: *N Engl J Med* 298:201, 1978.

215. de Gennes JL, et al: *Nouv Presse Med* 7:1713, 1978.

216. Tindall GT, et al: *J Neurosurg* 48:849, 1978.

217. Franks S, et al: *Clin Endocrinol* 8:277, 1978.

218. Parker LN, et al: *Clin Endocrinol* 8:1, 1978.

219. Carter JN, et al: *N Engl J Med* 299:847, 1978.

220. Hulugalle RS, et al: *JAMA* 240:2565, 1978.

221. Franks S, Nabarro JDN: *Ann Clin Res* 10:157, 1978.

222. von Werder K, et al: *Acta Endocrinol (Kbh)* 215 (suppl):1, 1978.

223. Grisola F, et al: *Nouv Presse Med* 7:1819, 1978.

224. McKenna TJ, et al: *Acta Endocrinol (Kbh)* 87:225, 1978.

225. Schnall AM, et al: *J Clin Endocrinol Metab* 47:410, 1978.

226. Polatti F, et al: *Amer J Obstet Gynecol* 131:792, 1978.

227. Bergh T, et al: *Brit Med J* 1:875, 1978.

228. Corenblum B: *Lancet* 2:786, 1978.

229. Sobrinho LG, et al: *Lancet* 2:257, 1978.

230. Vandeweghe M, et al: *Postgrad Med J* 54:618, 1978.

231. Zeman V, et al: *Neoplasma* 25:249, 1978.

232. Mendelsohn G, et al: *Amer J Path* 92:35, 1978.

233. Capella C, et al: *Virchows Arch (Pathol Anat)* 377:111, 1978.

234. Deftos LJ, et al: *J Endocrinol Metab* 47:457, 1978.

235. Jullienne A, et al: *Hormone Metabol Res* 10:456, 1978.

236. Becker KL, et al: *Hormone Metabol Res* 10:457, 1978.

237. Goltzman D, Tischler AS: *J Clin Invest* 61:449, 1978.

238. Argemi B, et al: *Acta Endocrinol (Kbh)* 88:75, 1978.

239. Starling JR, et al: *Arch Surg* 112:241, 1978.

240. Graze K, et al: *N Engl J Med* 299:980, 1978.

241. Fletcher DR: *J Royal Soc Med* 71:289, 1978.

242. Hillyard C, et al: *Lancet* 1:1009, 1978.

243. Wells SA, et al: *Ann Surg* 188:139, 1978.

244. Ram MD, et al: *JAMA* 239:2155, 1978.

245. Raue F, et al: *Klin Wochenschr* 56:719, 1978.

246. Block MA, et al: *Arch Surg* 113:368, 1978.

247. Raue F, et al: *Acta Endocrinol (Kbh)* 215 (suppl):118, 1978.

248. Wells SA, et al: *Ann Surg* 188:377, 1978.

249. Baylin SB, et al: *N Engl J Med* 299:105, 1978.

250. Wolfe HJ: *N Engl J Med* 299:146, 1978.

251. Becker KL: *Acta Endocrinol (Kbh)* 89:89, 1978.

252. Rambaud JC, et al: *Lancet* 1:220, 1978.

253. Galmiche JP, et al: *N Engl J Med* 299:1252, 1978.

254. Buchanan KD: *Eur J Clin Invest* 8:3, 1978.

255. Rehfeld JF: *Nature* 271:771, 1978.

256. Deveney CW, et al: *Ann Surg* 188:384, 1978.

257. McDonald TJ, et al: *Gut* 19:767, 1978.

258. Regan PT, Malagelada JR: *Mayo Clin Proc* 53:19, 1978.

259. Stage JG, et al: *Scand J Gastroenterol* 13:500, 1978.

260. Mihas AA, et al: *N Engl J Med* 298:144, 1978.

261. Lamers CBH, et al: *Dtsch Med Wochenschr* 103:356, 1978.

262. Straus E, et al: *Lancet* 2:73, 1978.

263. Vitaux J, Paolaggi JA: *Nouv Presse Med* 7:1856, 1978.

264. Lamers CB, et al: *Amer J Med* 64:607, 1978.

265. Joffe SN, et al: *Brit J Surg* 65:277, 1978.

266. Taylor IL, et al: *Lancet* 1:845, 1978.

267. Schwartz TW: *Lancet* 2:43, 1978.

268. Bloom SR, et al: *Lancet* 1:1155, 1978.

269. Lamers CBH, et al: *Lancet* 2:326, 1978.

270. Samson WK, et al: *Lancet* 2:901, 1978.

271. Laburthe M, et al: *Proc Natl Acad Sci USA* 75:2772, 1978.

272. Domscke S, et al: *Gut* 19:1049, 1978.

273. Modlin IM, et al: *Gastroenterology* 75:1051, 1978.

274. Modlin IM, Bloom SR: *S Afr Med J* 54:53, 1978.

275. Cooperman AM, et al: *Amer Surg* 187:325, 1978.

276. Burbige EH, et al: *Amer J Gastroenterol* 70:136, 1978.

277. Bloom SR: *Amer J Digest Dis* 23:373, 1978.

278. Modlin IM, et al: *Brit J Surg* 65:234, 1978.

279. Ebeid AM, et al: *Ann Surg* 187:411, 1978.

280. Gardner JD: *Amer J Digest Dis* 23:370, 1978.

281. Lundqvist G, et al: *Scand J Gastroenterol* 13:715, 1978.

282. Puzyrev AA, et al: *Tsitologiya* 20:17, 1978.

283. Faber OK, et al: *Diabetes* 27:170, 1978.

284. Horwitz DL, et al: *Diabetes* 27:267, 1978.

285. Zick R, et al: *Dtsch Med Wochenschr* 103:1255, 1978.

286. Nelson RL, et al: *JAMA* 240:1891, 1978.

287. Kiang DT, et al: *N Engl J Med* 299:134, 1978.

288. Turner RC, et al: *Lancet* 1:515, 1978.

289. Ingemansson S, et al: *Surg Gynecol Obstet* 146:725, 1978.

290. Farndon JR, et al: *Lancet* 1:723, 1978.

291. Habener JF, et al: *Proc Natl Acad Sci USA* 75:2616, 1978.

292. Habener JF, Potts JT: *N Engl J Med* 299:580, 1978.

293. Habener JF, Kronenberg HM: *Fed Proc* 37:2561, 1978.

294. Habener JF, Potts JT: *N Engl J Med* 299:635, 1978.

295. Yalow RS: *Science* 200:1236, 1978.

296. Rosenblatt M, et al: *Endocrinology* 103:978, 1978.

297. Martin KJ, et al: *J Clin Invest* 62:256, 1978.

298. DiBella FP, et al: *Clin Chem* 24:451, 1978.

299. Gasser AB, Burkhardt P: *Schweiz Med Wochenschr* 108:1031, 1978.

300. Goltzman D: *Canad J Surg* 21:285, 1978.

301. Jung RT, et al: *Postgrad Med J* 54:92, 1978.

302. Palmer FJ, Sawyers TM: *Arch Intern Med* 138:1402, 1978.

303. Jarman WT, et al: *Arch Surg* 113:123, 1978.

304. Parthemore JG, et al: *J Clin Endocrinol Metab* 47:284, 1978.

305. Schweitzer VG, et al: *Arch Surg* 113:378, 1978.

306. Carswell GF, et al: *J Urol* 119:175, 1978.

307. Martin KJ, et al: *Ir J Med Sci* 147:62, 1978.

308. Lo Cascio V, et al: *Clin Endocrinol* 8:349, 1978.

309. Ravazzola M, et al: *Lancet* 2:371, 1978.

310. Banabe JE, Martinezmaldonado M: *Arch Intern Med* 138:777, 1978.

311. McKay HA, et al: *J Urol* 119:689, 1978.

312. Spiegel A, et al: *Amer J Med* 64:691, 1978.

313. Hall SW, et al: *Med Pediat Oncol* 4:49, 1978.

314. Kikutani M, et al: *J Clin Endocrinol Metab* 47:980, 1978.

315. Urios P, et al: *FEBS Lett* 94:54, 1978.

316. Shahani SK, et al: *J Immunol Methods* 23:91, 1978.

317. McCready J, et al: *Clin Chem* 24:1958, 1978.

318. Ayala AR, et al: *J Clin Endocrinol Metab* 47:767, 1978.

319. Robertson DM, et al: *Acta Endocrinol (Kbh)* 89:492, 1978.

320. Bagshawe KD: *Ann R Coll Surg* 60:36, 1978.

321. Rutanen EM: *Int J Cancer* 22:413, 1978.

322. Greene JB, McCue SA: *Amer J Obstet Gynec* 131:253, 1978.

323. Heyderman E: *Scand J Immunol* 8(suppl 8):119, 1978.

324. Eyben FE: *Cancer* 41:648, 1978.

325. Barber SG, et al: *Brit Med J* 2:328, 1978.

326. Barber SG, et al: *Lancet* 2:372, 1978.

327. Barber SG: *Brit Med J* 2:896, 1978.

328. Arendt J: *Brit Med J* 2:635, 1978.

329. Hahnel R, Twaddle E: *Eur J Cancer* 14:125, 1978.

330. King RJB, et al: *Brit J Cancer* 38:428, 1978.

331. Nordenskjold B: *Acta Med Scand* 204:1, 1978.

332. Keffer JH: *Amer J Clin Pathol* 70:719, 1978.

333. Okret S, et al: *Cancer Res* 38:3904, 1978.

334. Murayama Y: *J Clin Endocrinol Metab* 46:998, 1978.

335. Wrange O, et al: *Anal Biochem* 85:461, 1978.

336. Gustaffson J, et al: *Cancer Res* 38:4225, 1978.

337. Brooks SC, et al: *Cancer Res* 38:4238, 1978.

338. Varela RM, Dao TL: *Cancer Res* 38:2429, 1978.

339. Garola E, McGuire WL: *Cancer Res* 38:2216, 1978.

340. Nenci I, et al: *Tumori* 64:161, 1978.

341. Ghosh L, et al: *J Surg Oncol* 10:221, 1978.

342. Dandliker WB, et al: *Cancer Res* 38:4212, 1978.

343. Lee SH: *Amer J Clin Path* 70:197, 1978.

344. Hubert P, et al: *Proc Natl Acad Sci USA* 75:3143, 1978.

345. Yamasaki S: *Jap J Clin Oncol* 8:37, 1978.

346. Martin PM, Chouvet CD: *Cancer Res* 38:3468, 1978.

347. Theve N-O, et al: *Eur J Cancer* 14:1337, 1978.

348. Tilley WD, et al: *Brit J Cancer* 38:544, 1978.

349. Webster DJT, et al: *Amer J Surg* 136:337, 1978.

350. Masters JRW, et al: *Eur J Cancer* 14:303, 1978.

351. Rich MA, et al: *Cancer Res* 38:4296, 1978.

352. Maynard PV, et al: *Cancer Res* 38:4292, 1978.

353. Maynard PV, et al: *Brit J Cancer* 38:745, 1978.

354. Lippman ME, et al: *N Engl J Med* 298:1223, 1978.

355. Lippman ME, Allegra JC: *N Engl J Med* 299:930, 1978.

356. Jonat W, Maass H: *Cancer Res* 38:4305, 1978.

357. McGuire WL, et al: *J Steroid Biochem* 9:461, 1978.

358. Jochimsen PR, et al: *Surg Gynecol Obstet* 147:842, 1978.

359. Appelbaum J: *Brit Med J* 2:1161, 1978.

360. Cheix F, et al: *Nouv Presse Med* 7:5633, 1978.

361. Allegra JC, et al: *Cancer Res* 38:4299, 1978.

362. Allegra JC, et al: *Cancer Treat Rep* 62:1281, 1978.

363. Webster DJT, et al: *N Engl J Med* 299:604, 1978.

364. Greenspan EM: *N Engl J Med* 299:604, 1978.

365. Kiang DT, et al: *N Engl J Med* 299:1330, 1978.

366. Lippman ME, et al: *N Engl J Med* 299:605, 1978.

367. McGuire WL: *Cancer Res* 38:4289, 1978.

368. Baulieu EE: *Klin Wochenschr* 56:683, 1978.

369. Scheiffarth OF: *Dtsch Med Wochenschr* 103:1619, 1978.

370. Mobbs BG, et al: *J Steroid Biochem* 9:289, 1978.

371. Wagner RK, Schulze KH: *Acta Endocrinol (Kbh)* 87:139, 1978.

372. Gustafsson JA, et al: *Cancer Res* 38:4345, 1978.

373. Concolino G, et al: *J Steroid Biochem* 9:399, 1978.

374. Concolino G, et al: *Cancer Res* 38:4340, 1978.

375. Macdonald JS, et al: *Cancer Chemother Pharmacol* 1:135, 1978.

376. Holmgren PA, et al: *Clin Chim Acta* 83:205, 1978

377. Guilleux F, et al: *Clin Chim Acta* 87:383, 1978

378. Benham FJ, et al: *Clin Chim Acta* 86:201, 1978.

379. Crofton PM, Smith PF: *Clin Chim Acta* 86:81, 1978.

380. Singh I, et al: *Cancer Res* 38:193, 1978.

381. Wajsman Z, et al: *J Urol* 119:244, 1978.

382. Burlina A, Bugiardini R: *Clin Chim Acta* 85:49, 1978.

383. Slaunwhite D, et al: *Ann Clin Lab Sci* 8:117, 1978.

384. Foti AG, et al: *Human Pathol* 9:618, 1978.

385. Romas NA, Tannenbaum M: *Human Pathol* 9:620, 1978.

386. Fink DJ, Galen RS: *Human Pathol* 9:621, 1978.

387. Editorial: *Brit Med J* 2:719, 1978.

388. Editorial: *N Engl J Med* 298:912, 1978.

389. Murphy GP: *CA* 28:104, 1978

390. Vihko P, et al: *Clin Chem* 24:466, 1978.

391. Vihko P: *Clin Chem* 24:1783, 1978.

392. Vihko P, et al: *Clin Chem* 24:1915, 1978.

393. Lee Ch-L, et al: *Cancer Res* 38:2871, 1978.

394. Chu TM, et al: *Invest Urol* 15:319, 1978.

395. Foti AG, et al: *Clin Chem* 24:140, 1978.

396. Belville WD, et al: *Cancer* 41:2286, 1978.

397. Boehme WM, et al: *Cancer* 41:1433, 1978.

398. Pontes JE, et al: *J Urol* 119:772, 1978.

399. Cooper JF, et al: *J Urol* 119:388, 1978.

400. Jobsis AC, et al: *Cancer* 41:1788, 1978.

401. Choe BK, et al: *Invest Urol* 15:312, 1978

402. Podolsky DK, et al: *N Engl J Med* 299:703, 1978.

403. Kessel D, et al: *Biochem Soc Trans* 6:187, 1978.

404. Khilanani P, et al: *Cancer* 41:701, 1978.

405. Berk JE: *Amer J Gastroenterol* 69:417, 1978.

406. Boehm-Truitt M, et al: *Anal Biochem* 85:476, 1978.

407. Flood JG, et al: *Clin Chem* 24:1207, 1978.

408. Braganza JM, et al: *Cancer* 41:1522, 1978.

409. Corlette MB,et al: *Gastroenterol* 74:907, 1978.

410. Le Cam-Sagniez M, LeReste JY: *Clin Chim Acta* 90:225, 1978.

411. Huseby N-E: *Biochim Biophys Acta* 522:354, 1978.

412. Beck PR: *Ann Clin Biochem* 15:151, 1978.

413. Burlina A: *Clin Chem* 24:502, 1978.

414. Beck PR, et al: *Clin Chem* 24:839, 1978.

415. Ordronneau J, et al: *Nouv Presse Med* 7:290, 1978.

416. Persijn JP, van der Slik W: *Clin Chem* 24:727, 1978.

417. Sawabu N, et al: *Gann* 69:601, 1978.

418. Daudet J, et al: *Nouv Presse Med* 7:3559, 1978.

419. Phillips PJ, et al: *Brit Med J* 2:1432, 1978.

420. Hada T, et al: *Clin Chim Acta* 85:267, 1978.

421. Jurga L, et al: *Neoplasma* 25:95, 1978.

422. Awais GM: *Surg Gynec Obstet* 146:893, 1978.

423. Nishioka K, et al: *Cancer Biochem Biophys* 2:145, 1978.

424. Kersey JH, et al: *Amer J Path* 90:487, 1978.

425. Chechik BE, et al: *J Natl Cancer Inst* 60:69, 1978.

426. Breenberger JS, et al: *Blood* 51:1073, 1978.

427. Boehm HR, Tischendorf FW: *Klin Wochenschr* 56:61, 1978.

428. Taylor CR: *J Histochem Cytochem* 26:496, 1978.

429. Parker AC, et al: *Scand J Haematol* 20:467, 1978.

430. Yavorkovsky LI, Grant KY: *Problemy Gematol* 23:13, 1978.

431. Peeters TL, et al: *Clin Chem* 24:2155, 1978.

432. Siegal FP, et al: *Amer J Pathol* 90:451, 1978.

433. Curran RC, Jones EL: *J Pathol* 125:39, 1978.

434. Leder LD: *Klin Wochenschr* 56:1091, 1978.

435. Knowles DM, Holck S: *Lab Invest* 39:70, 1978.

436. Kass L, Peters CL: *Amer J Clin Pathol* 69:57, 1978.

437. Kass L, et al: *Amer J Clin Pathol* 69:329, 1978.

438. Akagi K, et al: *Cancer Res* 38:2168, 1978.

439. Belpomme D, et al: *Brit J Haematol* 38:85, 1978.

440. Mohanakumar T, Raney RB: *Clin Haematol* 7:363, 1978.

441. Yanovich S, et al: *Blood* 51:435, 1978.

442. Meyskens FL, Jones SE: *N Engl J Med* 298:845, 1978.

443. Shaw MT, et al: *Blood* 51:181, 1978.

444. Marks SM, et al: *N Engl J Med* 298:812, 1978.

445. Kumar S, et al: *Lancet* 2:164, 1978.

446. Catovsky D, et al: *Lancet* 1:749, 1978.

447. Lisov V, et al: *Scand J Haematol* 21:167, 1978.

448. Nezelof C, et al: *Nouv Presse Med* 7:4027, 1978.

449. Golomb HM, et al: *Ann Intern Med* 89:677, 1978.

450. Turner A, Kjeldsberg CR: *Medicine* 57:477, 1978

451. Lippman ME, et al: *Cancer Res* 38:4251, 1978.

452. Crabtree GR, et al: *Cancer Res* 38:4268, 1978.

453. Duval D, Homo F: *Cancer Res* 38:4263, 1978.

454. Besley GTN, et al: *Lancet* 2:1311, 1978.

455. Eden OH, Innes EM: *Lancet* 2:378, 1978.

456. Chess L, Schlossman SF: *Adv Immunol* 25:213, 1978.

457. Rudders RA, Howard JP: *Blood* 52:25, 1978.

458. Heller P: *Ann Clin Lab Sci* 8:254, 1978.

459. Dhaliwal HS, et al: *Clin Exp Immunol* 31:226, 1978.

460. Metzgar RS, Mohanakumar T: *Semin Hematol* 15:139, 1978.

461. Fialkow PJ, et al: *Lancet* 2:444, 1978.

462. Gordon J, et al: *Immunology* 34:397, 1978.

463. Bërnnan D, et al: *Brit J Haematol* 40:51, 1978.

464. Davey FR, et al: *Human Pathol* 9:285, 1978.

465. Van den Tweel, et al: *Amer J Clin Path* 69:306, 1978.

466. Kadin ME, et al: *N Engl J Med* 22:1208, 1978.

467. Sonntag RW: *Schweiz Med Wochenschr* 108:1247, 1978.

468. Editorial: *Science* 202:298, 1978.

469. Morell A, et al: *Acta Haematol (Basel)* 60:129, 1978.

470. Matzner Y, et al: *Acta Haematol (Basel)* 60:257, 1978.

471. Malpas JS, Parker D: *Brit Med J* 2:563, 1978.

472. Tichy M, et al: *Neoplasma* 25:477, 1978.

473. Peltonen S, et al: *Acta Med Scand* 203:257, 1978.

474. Messmore HL, et al: *Ann Clin Lab Sci* 8:310, 1978.

475. Cohen HJ, et al: *Cancer* 41:1161, 1978.

476. Lemercier Y, et al: *Nouv Presse Med* 7:1028, 1978.

477. Papac RJ, et al: *Arch Intern Med* 138:1151, 1978.

478. Virella G, et al: *Acta Haematol* 60:269, 1978.

479. Solomon A: *J Clin Invest* 61:97, 1978.

480. Hayes JS, et al: *Arch Intern Med* 138:785, 1978.

481. Pruzanski W, et al: *Amer J Med* 65:334, 1978.

482. Corte G, et al: *Blood* 52:905, 1978.

483. Daver A: *Dan Med Bull* 25:91, 1978.

484. Cooper EH, et al: *Biomedicine* 29:154, 1978.

485. Spati H, et al: *Lancet* 2:987, 1978.

486. Cassuto JP, et al: *Lancet* 2:950, 1978.

487. Cassuto JP, et al: *Lancet* 2:108, 1978.

488. Amlot PL, Adinolfi M: *Lancet* 2:476, 1978.

489. Rustum YM, Higby DJ: *Eur J Cancer* 14:5, 1978.

490. Kaufman RJ, et al: *J Biol Chem* 253:5852, 1978.

9
Clinical Enzymology

Arthur R. Henderson

Progress in clinical enzymology appears to accelerate each year. It would be an impossible task to produce a coherent review of all the progress in this area and, accordingly, I have chosen to review a few of the advances in relation to a small number of enzymes, thus giving, I trust, a better insight into what is happening and why it is important.

To do this means being selective. Browning's lines perhaps best express my intention

> *Here, work enough to watch*
> *The Master work, and catch*
> *Hints of the proper craft, tricks of the*
> *tool's true play . . .*

CATALYTIC ACTIVITY AND ASSAY TEMPERATURE

The Expert Panel on Enzymes of the IFCC Committee on Standards has issued a definitive recommendation on the measurement of catalytic concentration of enzymes (1). Several points raised in this recommendation are worthy of comment. The catalytic activity of an enzyme is to be "measured by the catalysed rate of reaction of a specified chemical reaction, produced in a specified assay system." Catalytic activity is defined as moles per second or by the special non-SI name Katal (kat). The term *enzyme unit* (U) is to be abandoned (*Note:* 1U = 1 μmol/min = 16.67 nmole/sec = 16.67 nkat). The catalytic activity concentration (or just *catalytic concentration)* should be referred to the volume of a liter (of the original enzyme containing solution) thus, katal per liter, or mole s^{-1} l^{-1}. These terms will replace the previously proposed *amount of enzyme, catalytic amount, enzymic activity,* and *catalytic ability* - - a welcome recommendation indeed!

In the section on method selection the problems created by isoezyme mixtures in serum, which may vary with the severity of disease and the organ(s) involved, are faced squarely by suggesting the use of either a well-characterized preparation of a single enzyme form or the use of a "mixture of enzyme forms approximating those in the specimens to which the method is intended to be applied."

The present recommendations are unusual in that there appears to be a belated realization that clinical chemistry exists because patient care requires it. Thus, in addition to the statement quoted above, concern is expressed about concentrations of the enzyme in serum due to disease processes. It seems at last that the undue attention paid to analytical problems is now being replaced by a more balanced view of the total analytical *and* clinical picture.

The new definitive recommendations are similar to the 1974 provisional recommendations. This is true also for the suggested assay temperature of 30.00°C, but recent advances in the field of thermometry (reviewed later in this Chapter) have produced a new secondary reference point for the International Practical Temperature Scale due to the liquid-solid equilibrium of pure gallium. Gallium of 99.9999% (six nines--6N) purity melts at 29.770 ± 0.002°C, and this temperature plateau can be maintained for many hours. Gallium cells are commercially available, and the significance of this secondary reference point is that it can be used as a daily working standard in all laboratories.

To use the gallium cell correctly requires some appreciation of the problems of supercooling and the nucleation of molten gallium. Gallium can supercool by as much as 70°C, but if the molten metal is contained by a polypropylene surface, the low interfacial energy between the plastic and metal surfaces allows nucleation to occur reliably at 10°C. Solidification is achieved under these operating conditions within about 30 minutes when the gallium cell is cooled.

A constant-temperature chamber assembly is now available (3) to cycle a 40 gm gallium cell sequentially through the melt plateau and solidification. Heating and cooling are achieved by thermoelectric modules placed around the cell within a constant temperature chamber. The duration of the liquid-solid equilibrium depends on the difference between the gallium melt temperature and the heating chamber temperature. With a temperature differential of 0.13°C the melt plateau can be sustained for many hours. How can an operator know that the melt plateau has not been exceeded? The chamber assembly is equipped to be rapidly heated by 1°C within 30 seconds. If the liquid-solid equilibrium still exists, the calibrating thermocouple placed within the gallium cell will not register any change in temperature, whereas if all the solid gallium has melted, the thermocouple will register a temperature increase (Fig. 1). When the calibration process is complete, the gallium can be solidified by freezing (the thermoelectric module cools the cell when the polarity of the current is reversed), and the cell is ready for another calibration run.

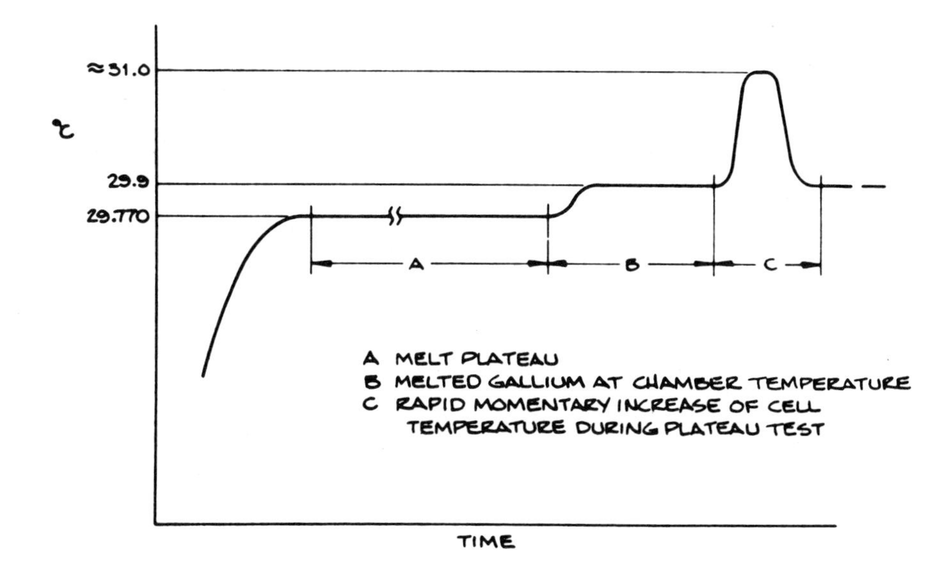

Figure 1. Profile of cell temperature as gallium melts. (Reproduced from Sostman and Manley: *Clinical Chemistry* (1978) 24, 1333, by permission of the authors [H. E. Sostman, Yellow-springs Instrument Co. Inc.] and the editor.)

The advent of the gallium cell and the constant temperature chamber promises to open up an exciting pathway in enzymology, and it has even been suggested that the gallium melting point should be the temperature at which enzyme assays should be made (3). It is hard to refute such a proposal.

ACID PHOSPHATASE

The Purification of Prostatic Acid Phosphatase

It is evident that the technique of affinity chromatography is having a profound influence on the technology of enzyme purification. Depending as this technique does on the unique specificity of biological interaction such as the affinity of an enzyme for its substrate, cofactor, or inhibitor, affinity chromatography has revolutionized the field of enzyme purification.

Two general approaches are used in affinity chromatography, and both will be illustrated with respect to the purification of prostatic acid phosphatase (ACP). The first depends on the specific affinity of the enzyme for an inhibitor or substrate, provided, of course, that these compounds can retain their specific affinities after being covalently bound to the insoluble material (usually agarose). The second, more general technique, is to use group-specific absorbants bound to the insoluble matrix such as Concanavalin A, which absorbs all glycoproteins, or 5'AMP, which can absorb all NAD-dependent dehydrogenases. Although the affinity of such ligands is far from specific for individual enzymes, their resolving power can be enhanced by the use of eluting solvents, which are highly specific substrates, cofactors, or inhibitors of the enzyme being purified thus achieving highly selective separations.

Three examples of the use of Concanavalin A bound to agarose for prostatic ACP (a glycoprotein) have recently been published. The simplest routine (4), a two-step procedure of an ammonium sulphate precipitation "bracket" followed by affinity chromatography with elution of the ACP by means of a mannose gradient, produced a 16-fold purification, with 11% recovery from 50 gm of hypertrophic human prostate. The enzyme was shown to be homogeneous by 9% polyacrylamide gel electrophoresis at pH 8.9.

Choe et al (5) used a more complex four-stage procedure achieving a 77-fold purification with a 76% recovery. Working with either a homogenate of prostate or pooled sperm, the material was dialyzed and concentrated, separated on a CM-Sephadex (cation-exchange) column by a pH gradient and then applied to a Concanavalin A Sepharose 4B column. The ACP was eluted with 0.5 M methyl-α-D-gluco-pyranoside, and the resulting protein was then run on a S-200 Sephacryl (1.8 x 90 cm) column. The purified enzyme was homogeneous by Sephadex G-200 column chromatography and showed a single band (i.e., one subunit) by 1% SDS or 8 M urea polyacrylamide electrophoresis. On isoelectric focusing or DEAE-Sephadex chromatography, using a convex 7.4 → 6.5 pH gradient, charge heterogeneity was evident. The two forms were antigenically similar but gave K_m (p-nitrophenylphosphate) values of 6.2 and 5.6 x 10^{-4} mole/liter respectively. Treatment with neuraminidase abolished the charge heterogeneity without affecting the antigenic specificity. The purified enzyme was stable at 4°C for three months and, when suspended in 20% glycerol, was stable for over one year at -25°C. This particular preparation therefore seems to be well characterized and is not too demanding in terms of equipment or technique.

Another purification by the same group (6) used the same basic manipulations but added another affinity chromatography stage. Cibacron Blue F3Ga, when linked to agarose, has been found to have a high affinity for albumin and many enzymes. This material was used to selectively remove albumin from the preparation of prostatic ACP. The overall purification was 62-fold with a final yield of 21%. The purified prostatic ACP showed a single band by SDS-polyacrylamide gel electrophoresis and was shown, immunochemically, to be uncontaminated with other proteins.

Group-specific affinity chromatography provides good purification of prostatic ACP, but the advantages of using the uniquely specific ligand L(+)-tartrate are obvious. Van Etten and Saini (7) synthesized a long-chain monoamide derivative of L(+)-tartaric acid and attached it to CNBr-activated Sepharose 4B. The prostatic homogenate was clarified by centrifugation, then subjected to dialysis against water and then buffer. This retentate was applied to the affinity column and was washed with buffers at two different pH values to remove protein impurities. The enzyme was eluted with a 0-25 mmole/l sodium phosphate gradient. The resulting purification was 44 times higher, with a 72% recovery. The purified enzyme was homogeneous by electrophoresis on a 9% polyacrlyamide gel at pH 8.9.

It is of interest to contrast this relatively simple procedure (except for the preparation of the affinity chromatography ligand) with a similar affinity chromatography technique (8). L(+)-tartrate was coupled to aminoethyl agarose (AH-Sepharose 4B) by means of a carbodiimide derivative. This preparation has been criticized (7) because the carbodiimide coupling method used can create considerable cross-linking with the tartrate ligand and because the AH-matrix can function as an ion-exchanger. These two factors could reduce the capacity of the column and alter its separative characteristics. Vihko and his colleagues (8), using a five-stage procedure, obtained a 1,300-fold increase in purification with a 19% yield. Before commenting on these results the purification will be described. The prostatic homogenate was clarified and a 30 to 70% ammonium sulphate saturation cut was made. After dialysis the preparation was loaded on to the affinity column, and contaminating protein was eluted with the equilibrating buffer. Then increasing concentrations of L(+)-tartrate (10-40 mmole/l) in buffer were used to elute the ACP activity. These fractions were applied to a Sephadex G-200 column (25 times 325 mm). The most active fractions were pooled and separated on an isoelectric focusing column. Two enzymic fractions were obtained (pI 4.9 and pI about 5.9). The former peak had the major activity, and these fractions were subjected to chromatography on a Sephadex G-50 (15 times 335 mm) column. Fractions of high enzymic activity were concentrated by ultrafiltration and were re-applied to the G-50 column. The ACP isolated from the last purification step was found to be homogeneous by 7.5% polyacrylamide gel electrophoresis (pH 8.4) and by SDS electrophoresis. It was noted that the material at this stage contained only 400 μg of protein and, as pointed out by Van Etten and Saini (7), an error in protein determination at this stage could have profound effects on the calculation of specific activity, recovery, and purification. However, many preparations were made by this procedure, and the results appear to be consistent from one lot to the next. It is also noted that the affinity chromatography step resulted in a purification of 77-fold with a yield of 45%. However, the product contained two well-defined protein zones by polyacrylamide electrophoresis. This finding suggests that the affinity column was much less specific than it should have been, and the fact that several subsequent purification steps were necessary suggests that the purification procedure may need to be altered so the expected exquisite specificity of the L(+)-tartrate ligand on the affinity column could be used to better effect.

It is of some interest to contrast the purifications described above with the accounts of prostatic ACP purification used to develop an antiserum for diagnostic purposes. Thus, the radioimmunoassay (RIA) by Cooper and his colleagues was developed from purified prostatic fluid (9). This fluid was subjected to Sephadex G-100 (2.5 times 100 cm) column chromatography, and the main ACP fractions were used without further purification. This fraction was checked for homogeneity by 7.15% polyacrylamide gel electrophoresis (pH 6.8) and ". . . 90 to 95 per cent of the protein bands were identical with the acid phosphatase enzyme." This observation suggests that these authors' assay may also be detecting protein other than prostatic ACP, which may or may not be unique to prostatic disease.

The solid-phase RIA was set up by this group (10) using a similar purification. After Sephadex G-100 chromatography the protein peak containing up to 95% of the total ACP activity was used for immunization. This peak was shown to possess one band by polyacrylamide electrophoresis (7.15% at pH 6.8), but no further details of the degree of purification of the enzyme were given.

Another preparation used in clinical work is that of Chu et al (11). These workers obtained a 150-fold purification of prostatic ACP, and they used a multistep procedure (homogenate → dialyzed against buffer → concentration by ultrafiltration → affinity chromatography for removal of ribonuclease and other proteins → 75% saturation with ammonium solphate → dialysis → phosphocellulose column for removal of protein → DEAE-Sephadex A-50 ion-exchange column → Sephadex G-200 column chromatography). The final product appeared to have an extremely high protein content (absorbance at 280 nm of > 1.0) but it was homogeneous by polyacrylamide gel electrophoresis (pH 8.3) at an unstated concentration.

This group (12) have also used another purification to develop a rather unusual immunoassay (see the section on prostatic ACP) of prostatic ACP. This preparation was obtained as follows: prostatic tissue homogenate → dialysis → ammonium sulphate → fractionation → dialysis → Concanavalin-A-Sepharose 4B with elution by an α-methyl-D-mannopyranoside gradient → dialysis → DEAE-cellulose column with elution by a linear NaCl gradient → Sephadex G-100 column. This resulted in an 85-fold purification with a 38% recovery. The preparation was shown to be homogeneous by 7.5% polyacrylamide gel electrophoresis.

The Immunoassay of Prostatic Acid Phosphatase

Prostatic ACP is one of the most unstable enzymes. Foti et al (13) showed that even when enzymic activity of prostatic ACP in serum was lost, prostatic ACP antigenicity could still be deted by the RIA technique they used (10). This finding is of great clinical significance. In addition they demonstrated that the immunoassay gave fewer false-positive results (5%) than the enzymatic assay (27%). The RIA used was the solid-phase system (10), which had been shown to be superior to their original double-antibody technique (9).

The solid-phase RIA was further evaluated (14) in a blind study using 113 untreated patients with prostatic cancer, 50 normal men as controls, 83 patients with various nonprostatic cancers, 36 patients with benign prostatic hyperplasia, 20 patients with various gastrointestinal disorders, and 28 patients with a total prostatectomy performed at least one year previously. In the normal controls the RIA level of prostatic ACP in serum did not exceed 6.6 ng/100 μl (which is the mean value + 2 SD), a finding similar to that of earlier studies by these workers on more than 600 normal men. Prostatic ACP in serum was found not to vary with the age of the patient after puberty.

Complete removal of the prostate gland is more likely to be curative if the tumor has not spread beyond the prostatic capsule. Thus prostatectomy on patients in stages 1 and 2 is likely to be more successful than on those in stage 3, where tumor cells have spread beyond the capsule of the gland. The care necessary to establish the clinical staging of the tumor is considerable. A detailed review of carcinoma of the prostate, describing the natural history and staging of the disease, has been prepared by Catalona and Scott (15).

Table 1. Comparison of RIA and Enzymic Assay in the Detection of Elevated Serum Prostatic ACP[a]

Group	No. of Patients	Elevation by RIA				Elev. by Enzymic Assay	
		6.6 ng/ 0.1 ml		8.0 ng/ 0.1 ml		0.2 Sigma unit/1.0 ml	
		No.	*%*	*No.*	*%*	*No.*	*%*
Normal Controls	50	0	0	0	0	0	0
Prostatic Cancer							
Stage 1	24	12	50	8	33	3	12
Stage 2	33	26	79	26	79	5	15
Stage 3	31	25	81	22	71	9	29
Stage 4	25	24	96	23	92	15	60
Benign Prostatic Hyperplasia	36	5	14	2	6	0	0
Total prostatectomy	28	5	18	1	4	0	0
Other Cancers	83	14	17	9	11	7	8
Gastrointestinal Disorders	20	2	10	1	5	0	0

[a]Reproduced by permission from the *New England Journal of Medicine* 297:1357, 1977.

The results of the study by Foti et al (14) are shown in Table 1. At each cancer stage the RIA results correctly classified more patients than the enzymic assay results when the cut-off value was 6.6 ng/100 µl. Between 10 to 20% of the patients with benign prostatic hyperplasia, total prostatectomy, other cancers, and gastrointestinal disorders had RIA values greater than 6.6 ng/ 100 µl. At first sight it may appear remarkable that patients with total prostatectomy should possess detectable prostatic ACP in their serum, but it is pointed out that all these patients had their prostate removed for prostatic malignancy, and it is not known if there was tumor "seeding" and if metastatic disease will yet occur.

If the cut-off value is increased from the mean +2 standard deviations to the mean +4 standard deviations of the normal range, there is a marked reduction in the number of false-positive results in the nonprostatic cancer group (see Table 1). However, in the prostatic cancer groups there was a reduction in the number of those correctly classified in stages 1 and 3, although the performance of the RIA was still better than the enzymatic assay. Careful examination of the data for stages 1 and 2 indicates that the enzymic assay can pick up a few cases that are not detected by the RIA method, and this finding, if confirmed by more extensive studies, suggests that the two assays may, with advantage, be used together. When the cut-off value was increased, the same number of true-positive results in the stage 2 category were obtained; this is a valuable feature of the RIA test because, as explained earlier, tumors in this category are still confined within the capsule of the prostate gland itself, and, therefore, removal of the gland is potentially curative.

Carroll (16) has put these data in terms of the predictive value model, and some of this analysis is shown in Table 2.

Table 2. Sensitivity, Specificity, and Predictive Values of Data Given in Table 1 in Diagnosis of Prostatic Cancer

	All Patients from Table 2 (Prevalence 40%)		Patients in Curable Group of Prostatic Disease - (prevalence 25%)[a]			
Percentage	Enzymic Assay	RIA	Enzymic Assay[b]	RIA[c]	RIA[d]	RIA[e]
Sensitivity	28	69	14	60	32	9
Specificity	96	92	100	95	99.4	100
Predictive Value of Positive Test	82	86	72 (57)	100	94.7 (86)	100 (100)
Predictive Value of Negative Test	66	82	87 (96)	77	81	76

a Data in parentheses show indices when the prevalence is 10%.
b (8 ng/100 μl) *c* (as *b*) *d* (10 ng/100 μl) *e* (20 ng/100 μl)

It is perhaps necessary to remind the reader that test sensitivity indicates the true-positive test rate and test specificity indicates the true-negative test rate. From Table 2 (all patients) the RIA technique is more sensitive (i.e., it detects more cases of disease) than the enzymic assay, but it is marginally less specific (i.e., it detects fewer true-negative values) than the enzymic assay. However, for patients in the potentially curable group of prostatic disease (i.e., stages 1 and 2) alone - - which are the groups that the surgeon hopes to detect - - the sensitivity of the RIA is markedly superior to the enzymic test. However, the enzymic test is a better measure of "health," that is, it detects more true-negative values.

It is possible to increase the specificity of the RIA by increasing the cut-off limit and the results of doing this are shown in Table 2. Naturally, the detection rate drops. The prevalence used in these calculations is between 40 and 25%. However, as Carroll (16) points out, the prevalence of prostatic cancer in men over the age of 60 years is probably about 10%. The data in Table 2 given in parenthesis show the effect of this change in prevalence on the diagnostic performance of the RIA method when used as a screening procedure. At the extreme cut-off limit of 20 ng/100 µl the RIA method has a false-negative rate of 91%, but the predictive value of a high assay is 100%. These values were similar to those for the enzymic assay used by Foti et al (14), which are also shown in Table 2. It is evident that population screening by means of a sensitive RIA technique for early prostatic carcinoma is simply not a realistic endeavor. This general point has been made many times by Galen and Gambino (18). (For editorial reviews of the work of Foti et al, ref. 14, see refs. 18,19.)

In a later paper, Foti and his group (20) present data that are substantially similar - - in terms of the diagnostic power of RIA for prostatic ACP - - to their previous data (14).

Prostatic ACP can be a valuable marker of prostatic cancer metastases. It is known that an elevation in the bone marrow ACP level may be an early indication of metastases even when the blood level of prostatic ACP is normal and the bone survey, the bone scan, and the bone marrow histologic pattern are all normal. However, as a large number of tissue elements contain ACP (erythrocytes, leukocytes, platelets, osteoclasts, and reticuloendothelial cells) false-positive elevations of bone marrow ACP levels are not uncommon. Boehme et al (21) reported higher elevations in bone marrow ACP levels after fast aspirations of bone marrow than when aspiration was gentle and slow. It seems very likely that this effect is due to the extensive cell destruction, and release of ACP, that follows fast aspiration of marrow. In all, however, this source of variability adds another degree of uncertainty to the determination.

The availability of an RIA for prostatic ACP described above (14) suggested to Cooper et al (22) that results by this technique would be less disturbed by blood diseases, by mishandling of blood samples, or by damage to marrow specimens by poor sampling procedures, factors that certainly affect the standard enzymatic assays.

Preliminary analysis of their data suggests that the RIA for prostatic ACP is not perturbed by other acid phosphatases from the bone marrow. Analysis of the results obtained by both RIA and enzymic techniques in 19 cases of benign prostatic hyperplasia and in 25 cases of prostatic cancer (stages 2 to 4) gave the following results (enzymic method in parenthesis):

Sensitivity	80% (56%)
Specificity	100% (89%)
Predictive Value of a positive test	100% (88%)

These results suggest that further evaluation of this technique would be useful.

Another report (23) questions the need for RIA and enzymic determinations of bone marrow ACP when an immunoassay of high sensitivity - - in this case counterimmunoelectrophoresis - - is available (11). Questioning the need for RIA and enzymic assay does not appear to be totally realistic, since the demonstration of prostatic ACP in bone marrow proves metastatic spread, and this marker may precede other indicators of such spread.

Other immunoassays for prostatic ACP have been described. Choe et al used their purification (5) to prepare a double-antibody RIA (24). The rabbit antiserum was used at a 1:1000 dilution, and the normal serum range, determined from values obtained for 162 normal men, aged 23 to 40 years, was 16.0 ± 8 μg/liter. Patients with nonprostatic carcinoma had levels of 18 ± 6 μg/l, and three patients with localized prostatic carcinoma had levels of 18-25 μg/liter, that is, within the normal range. The normal range reported by Choe and coworkers (24) is lower than that reported by Foti et al (14), presumably because of the use of a more purified prostatic ACP preparation. Unfortunately, no effort has been made to ascertain the clinical efficacy of this RIA - - information that is badly needed.

Vihko and his colleagues (25), using their highly purified preparation of prostatic ACP (8), have developed a double-antibody RIA using a 3,000-fold dilution of antiserum (but a 10,000-fold dilution to increase the sensitivity when testing sera obtained from healthy persons). They obtained a normal range of < 1 to 10.0 μg/liter, which is lower than those obtained by other groups (14,24), presumably because of the highly purified prostatic ACP used as the antigen. Disappointingly, no data have been provided for serum levels of prostatic ACP in stage 1 and stage 2 prostatic carcinoma. This information is essential to assess the efficacy of this assay system. One interesting aspect has already emerged however. Vihko et al (25) found that the levels of ACP in sera of patients with metastatic carcinoma were considerably below those reported by the other groups (14,24). It was speculated that this may have been due to the differences in purity of the antigen used.

The RIA approach to measuring prostatic ACP is lengthy, and alternative assay methods have been sought. Because prostatic ACP has a pI in the range 4 to 5, it is negatively charged at alkaline pH values and moves towards the anode. Antibodies move in the opposite direction on electrophoresis, so that the process rapidly leads to the formation of detectable immune complexes. This is the process of counterimmunoelectrophoresis. Foti et al (26) used their antibody preparation (10) but also stained the slide to demonstrate ACP activity. They found that while as little as 0.3 ng of purified prostatic ACP could be detected, the method could not detect serum enzyme levels below 1.5-2.0 ng. Thus, the staining method could not be used to detect patients in the stage 1 and 2 categories, while their RIA method could be used for these patients (14). By contrast, patients in stages 3 and 4 could readily be detected. Other investigators (27) seem to have had similar results, although one group claims a 30% detection rate of stage 2 prostatic cancers (11). It would seem that counterimmunoelectrophoresis is a reasonable technique for detecting prostatic cancer at stages 3 and 4, but more work must be done to assess the value of this technique. Again, since the technique depends on the enzymic activity of ACP, problems of catalytic stability must also be considered.

An extremely interesting solid-phase fluorescent immunoassay has been proposed for prostatic ACP (12). Using a modestly purified enzyme, rabbit antiserum to this protein was obtained. After purifying the IgG by ammonium sulphate precipitation and DEAE-cellulose chromatography, it was linked to CNBr-activated Sepharose 4B, and the resulting IgG-Sepharose beads were used to bind, specifically, prostatic ACP. The enzyme IgG-Sepharose complex was found to be catalytically fully active for 48 hours whereas "free" ACP lost 64% of its activity within the same time period. Optimum conditions were established for binding prostatic ACP from serum (two hours at room temperature, then overnight at 4°C). After washing the Sepharose beads, the amount of enzyme on the beads could be assayed by incubating with α-naphthyl-phosphate for one hour at 37°C. The reaction was stopped by adding NaOH and the α-naphthol detected fluorimetrically. It was found that the IgG-Sepharose could be re-used for three more assays, with a slight diminution in its affinity for the prostatic ACP.

With this immunoassay, normal ranges were similar to those found by Vihko and his colleagues (25), so the assay appears to be comparable to a good radioimmunoassay. The one major disadvantage of this assay is that it depends on the enzymic activity of the prostatic ACP moiety, a dependency not required by the conventional radioimmunoassay techniques. Because IgG-bound enzyme is much more stable (catalytically) than "free" enzyme, it is conceivable that, by careful patient sampling techniques with immediate addition of the serum to the IgG-Sepharose beads, the catalytic lability problem could be overcome.

Is Prostatic Acid Phosphatase a Single Enzyme?

There is considerable evidence to suggest that ACP comprises a series of several enzyme species (5,8,12), but the most fascinating indication is the isoelectric focusing results of Chu et al (28). They examined a homogenous preparation of prostatic ACP (11), using analytical polyacrylamide isoelectric focusing over a pH range of 4 to 6. After separation, the gel was stained for ACP activity. Eight enzymic bands were obtained (Fig. 2) with pI values between 4.4 and 5.3. This was a consistent finding of tests made on several separate preparations of prostatic ACP.

Sera from patients with prostatic carcinoma exhibited a similar pattern (Fig. 3), but as the enzyme activity in serum increased, the pI of the bands became increasingly more acid. Neuramini-

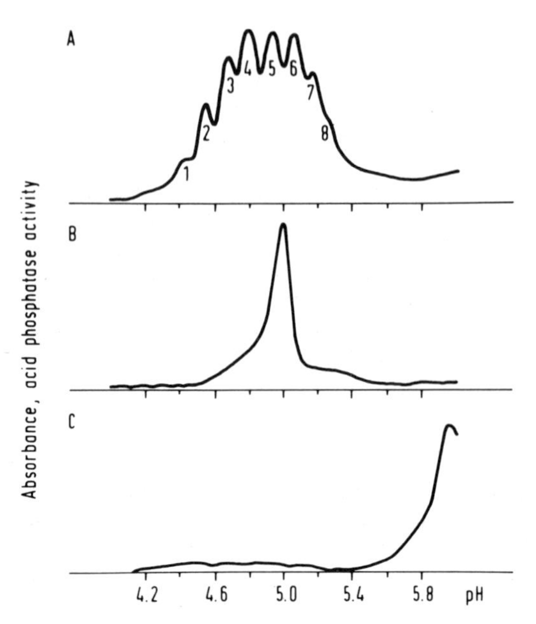

Figure 2. Isoelectric focusing of acid phosphatases. (A) Enzyme purified from human malignant prostate. (B) Serum acid phosphatase from a patient with Gaucher disease. (C) Enzyme purified from human erythrocytes. (Reproduced with permission from Chu et al, (ref. 28).

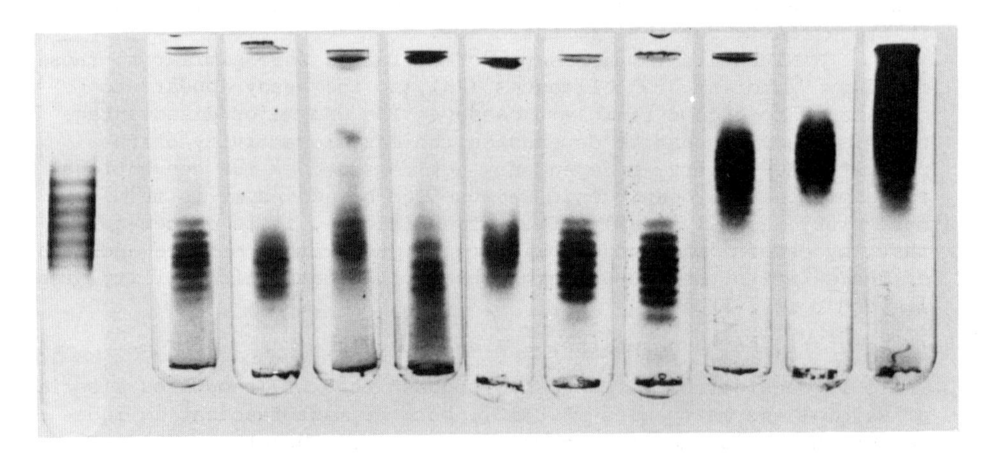

Figure 3. Isoenzymes (isoelectric focusing) of human prostatic acid phosphatase from malignant prostates (far left) and serum acid phosphatase from patients with prostate cancer (remainder). Enzyme activity of the serum specimens increased from 10 to 370 units/l from left to right. The pH ranged from 4.0 (top of gel) to 6.0 (bottom of gel). (Reproduced with permission from Chu et al (ref. 28.)

dase treatment converted the multiple bands into one large and one small zone, focusing at a higher pH. This finding highlights an important area for further investigation.

Tartrate-Labile Acid Phosphatase

Although important advances are being made in the immunological detection of prostatic ACP, the majority of clinical chemists will continue to use tartrate-labile ACP assays for some time. Readers are referred to an excellent case conference by Ladenson and McDonald (29) on acid phosphatase and prostatic carcinoma. This article presents three cases together with a discussion on the problems of choice of substrate and sources of error; there is also a useful review of the use of bone marrow aspirate.

Other Articles of Interest

In 1978, papers appeared dealing with the following themes: some properties of purified prostatic ACP (30); pancreatic islet cell carcinoma secreting ACP immunologically similar to prostatic ACP (31); preservation of ACP activity in medicolegal cases (32); interference with the kinetic determination of ACP (33); multiple forms of ACP in Gaucher's disease (34); tartrate-resistant ACP (35, 36); and a reexamination of the effect of digital examination of the prostate on serum ACP activity (37).

γ-GLUTAMYLTRANSFERASE

The Liver Enzyme

The activity of γ-glutamyltransferase (GGT) in serum arises mostly from the liver. It is therefore essential to characterize the liver enzyme. Huseby (38) has purified the enzyme from normal human liver by an 11-stage procedure with a 7% recovery and a 9400-fold purification. Important steps in this procedure were *(a)* release of the enzyme from membranes by deoxycholate, lubrol W, acetone, and butanol treatment; *(b)* the use of affinity chromatography (the enzyme is a glycoprotein) with Concanavalin-A-Sepharose 4B; and *(c)* treatment with papain to release the enzyme from lipid and protein aggregates. This last treatment, by creating a low-molecular-weight enzyme, which does not form aggregates, allows it to enter the polyacrylamide gel during electrophoresis.

The purified enzyme was shown to be homogeneous regardless of the acrylamide concentration. In dodecyl sulphate gel electrophoresis two bands were found. The relative molecular mass of the purified enzyme was determined to be about 90,000-110,000 daltons and the subunits, 47,000 and 22,000 daltons, respectively. From work on the kidney enzymes of rat and man (39-41), it is thought that the enzyme is a dimer, and this will likely be found to be true for the liver enzyme. Kinetic properties of the purified enzyme were shown to be similar to those of the nonpapain-treated enzyme and the enzyme found in serum.

Shaw et al (42) have shown that the untreated liver enzyme has a relative molecular mass > 200,000 daltons and, after digestion with trypsin, a low-molecular-weight form was found (M_r ~ 120,000 daltons). These molecular mass estimates were by gel filtration, and it has been pointed out that glycoproteins can behave anomalously (38) in molecular sieving procedures.

These authors also purified the liver enzyme (426-fold, 16% yield) by a four-step procedure including a Concanavalin-A-Sepharose affinity chromatography step. Comparisons were made with the human kidney enzyme. For example, while 80% of the liver enzyme binds to immobilized Concanavalin A, only 20% of the kidney enzyme so binds. This and other studies suggested that there were significant structural differences in the carbohydrate moieties of the two enzymes. As might be expected, marked differences were found between the electrophoretic mobilities of the two enzymes (42). Future advances are expected in this area.

Multiple Forms of the Enzyme in Serum

This subject has previously been reviewed (43), and there are some recent papers of note. The techniques of visualizing multiple forms of GGT after electrophoresis have been assessed (44), and suggestions have been made about the optimal concentrations of the reagents. Such recommendations are welcome since there is still much confusion in this field. For example, Burlina (45) identified up to four bands of GGT activity in serum in hepatic disease but only two in serum of healthy people. Kok et al (46) identified two bands of activity in health and a third band (infrequently) in disease. Both groups used cellulose acetate for separation.

It is possible that Huseby's recent work on this subject (47) can provide the answer to the diverse results reported in the literature (43). Serum, bile, and enzyme samples from normal human liver were studied by gel filtration, electrophoresis, and density-gradient centrifugation. After gel filtration the most active "peak" (there are three) elutes in the void volume for all three tissue sources. Treatment with 0.5% deoxycholate and 0.5% lubrol W converts this activity to a species with a Stoke radius of between 55 and 65 Å. Treatment with 1% deoxycholate, with gel filtration in the presence of 0.5% deoxycholate, converts the largest species to a molecule with a relative molecular mass of 175,000 daltons and with a Stoke radius of 48 Å. With this treatment the small molecular form (38) was also formed, in small proportions, in serum and bile. If the original material was treated with papain all the enzyme activity appeared in the small molecular form (i.e., relative molecular mass 80,000 daltons with a Stoke radius of 37 Å). When the void volume peaks of enzyme activity (or the intermediate peak) were treated with Triton X-100 and cetyltrimethylammonium bromide or with Triton X-100 alone, the enzyme was shown to bind detergent by agarose electrophoresis. These experiments showed that two of the three enzyme species contained a hydrophobic domain that can bind detergent, lipids, or amphiphilic proteins in membranes.

Huseby (47), therefore, concluded that there are only two molecular forms of the enzyme in serum - - the hydrophobic form, which can bind lipids and amphiphilic proteins, and a hydrophilic form (the small molecular form). In hepatic disease, for example, there will be elevations of lipid and amphiphilic protein levels, which could result in the production of enzyme heterogeneity because of differential binding by the hydrophobic form of the enzyme. It has already been shown that the existence of chylomicrons in serum can change the electrophoretic pattern of GGT (48). Until the problem of multiple forms of GGT is clarified, it would seem wise to refrain from calling these forms isoenzymes.

ASPARTATE AMINOTRANSFERASE

The Isoenzymes

The prognostic value of the mitochondrial isoenzyme of aspartate aminotransferase (AST-2) in serum has been suggested by Smith et al (49) for postmyocardial infarction patients. The magnitude of elevation of serum AST-2 activities correlated well with the severity of postinfarction heart failure and also with the likelihood of death. This latter association appeared to be stronger than with serum creatine kinase-2 elevations. These observations were made on a small group of patients and need to be validated by further work.

Many methods are available to estimate AST isoenzymes. These have been reviewed by Rej (50). A particularly fine study of one such technique (ion-exchange column chromatographic separation) has been published (51), and the discussion in the report of the systematic approach to separation is of interest. The minicolumn (using DEAE Sephadex A-50 in Pasteur pipettes) was optimized sequentially for bed height, buffer pH value, and NaCl concentration in the eluting buffer. An example of the results obtained by altering the bed height is shown in Figure 4. AST-2 washes through this column, whereas the cytoplasmic AST-1 is adsorbed and requires an increase in the buffer ionic strength to be eluted. From Figure 4 it can be seen that bed heights over 6 cm provide the maximum separation (in terms of fraction number) between AST-1 and AST-2. The contraction of the bed volume after an increase in the ionic strength of the eluting buffer was observed to increase the volume of one fraction by 30%, and this effect was found to be so constant that the enzyme activity in that fraction had to be multiplied by 1.30 for correction. Final optimal operating conditions were found to be, in addition to the bed height of 6 cm, 50 mmole/l Tris buffer (pH 8.50) containing 50 mmole/l NaCl, with a single step to 200 mmole/liter NaCl for eluting AST-1. The high pH value was needed to avoid poor recovery of both isoenzymes from the column. The precision of the method proved to be excellent (Table 3). An interesting finding was that dilution of serum (normally 1 ml was used as sample) with water did not effect the distribution of the isoenzymes. Sampson and his coworkers concluded that dialysis - - the usual preparative step for the sample before loading the ion-exchange column - - was not necessary. The method, as presented, looks robust and up to 15 separations can be completed in 6 hours.

The ability to separate AST isoenzymes in serum begs the question as to their concentrations in heart, liver, and muscle tissue. An elegant study by Rej (50) provides valuable information about the concentration of the enzyme in human liver. Rej first showed that simple homogenization of the liver extracted as much AST as did treatment of the homogenate with 0.2% deoxycholate alone or with sonication in the presence of 0.2% deoxycholate. The iso-

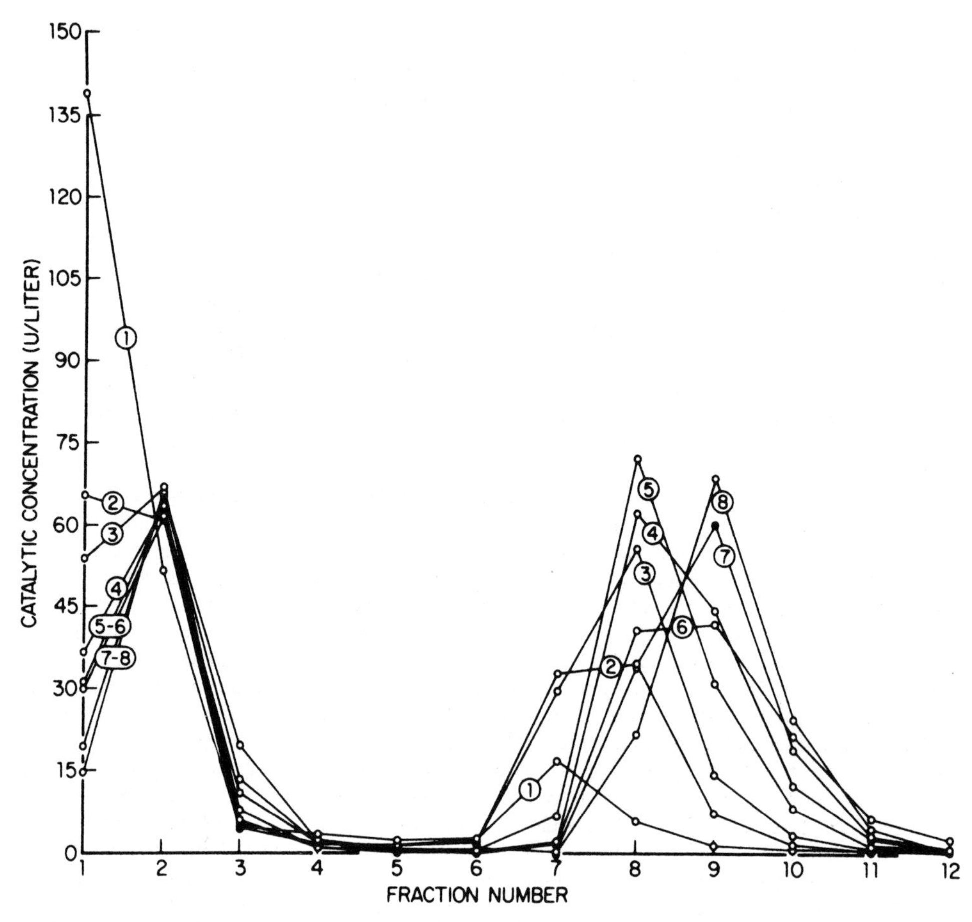

Figure 4. Elution profiles of AST isoenzymes as a function of bed height of DEAE-Sephadex A-50. Numbers in circles correspond to bed height in cm. The second eluting buffer starts at fraction number 7. (Reproduced from Sampson et al: *Clinical Chemistry* 24:1805, 1978, by permission.)

Table 3. Precision of Aspartate Aminotransferase Isoenzyme Measurement by Column Chromatographic Method [a,b]

	AST Activity		
	Mean (units/l)	SD (units/l)	CV (%)
Mitochondrial	111.8	2.92	2.6
Cytoplasmic	120.4	7.12	5.9
Sum of isoenzymes[c]	232.2	8.87	3.8
Direct measurement[d]	228.2	4.16	1.8

a Reproduced from Sampson et al: *Clinical Chemistry* 24:1805, 1978, by permission.

b Ten days, single measurement.

c Sum of mitochondrial and cytoplasmic fractions; mean recovery is 101.7% (SD 3.2%).

d AST activity in specimens added to columns.

---

enzyme distribution was estimated by four techniques, namely:

1. The use of rabbit antiporcine AST-1 precipitating antiserum with enzyme determination of the supernatant (for AST-2 activity)
2. Isoelectric focusing of the homogenate in the range of pH 3 to 11
3. AST-2 estimation by DEAE-Sephadex A-50 exclusion
4. AST-1 estimation by CM-Sephadex C-50 exclusion

The mean results of all these techniques were 81% mitochondrial and 19% cytoplasmic AST for human adult liver tissue. This gives an AST-2:AST-1 ratio of 4.27, and for fetal tissue the ratio was 1.30. Rej noted that for the adult, the ratio was similar for different sections of liver, although it is to be expected that further work will show a change in the AST-2:AST-1 ratio across the acinar zones of the liver.

A short report (52) describes an elevation of AST-2 activity in serum in the early stages of Duchenne muscular dystrophy. As the disease progresses the elevation of AST-2 activity in serum decreased. This serum activity was related to the destruction of mitochondria in the skeletal muscle cell, a change that has been observed by electron microscopy.

Assay Conditions and Pyridoxal Phosphate

Bergmeyer et al (53) have published a detailed account of optimization of the aspartate (and alanine) aminotransferase methods at 30°C. Their findings are embodied in the latest report of the Expert Panel on Enzymes (54).

It is widely recognized that AST is inhibited competitively by 2-oxo-glutarate against L-aspartate. Thus, there is a parabolic interdependence between these substrates. Since 2-oxo-glutarate absorbs significantly at the wavelength of measurement, the concentration should be less than 20 mmole/l; but since hepatic and cardiac enzyme sources also have to be considered, a suitable 2-oxo-glutarate value is 12 mmole/liter, with the concentration of L-aspartate being approximately 240 mmole/liter. Experimentally these concentrations give 96% of the maximum velocity, a result also expected on theoretical grounds. For a more detailed discussion of the rationale for the selection of substrates in this system, the reader is directed to an article by Bergmeyer (55).

For the AST reaction the optimal pH value is between 7.5 and 8.0 in Tris buffer. Since this pH value is close to the pK value of the buffer, an 80 mmole/liter concentration (above which there is a slight inhibition of the reaction) gives good buffering capacity at pH 7.8.

Two anions - - phosphate and chloride - - must be avoided in the reaction mixture, since the former inhibits the recombination of pyridoxal phosphate with the apo-enzyme and the latter inhibits the reaction above a concentration of 125 mmole/liter.

The effect of pyridoxal-5-phosphate on AST creates a number of problems. Bergmeyer et al (53) used 100 μmole/liter with a preincubation period of 10 minutes. Two effects were noted. The cardiac group of sera was activated by a mean increase of 49%, but with a range of stimulation of between 18 and 81%! The hepatic group gave a mean stimulation of 15%, with a range of 4 to 27%. The second effect concerns the blank reaction. Sample blanks increased from less than 2 units/liter (without pyridoxal phosphate) up to 4 units/liter. Further, unpublished, work by Bergmeyer has indicated that in a large group of healthy persons (n = 104) the average sample blank is 2.99 ± 0.55 (units/l) (i.e., mean ± SD), but that in 32 cases of myocardial infarction the sample blank ranges between 3 and 25 units/liter. Similar effects have been reported by many other groups (for references, see ref. 53). In addition the precision of the aspartate aminotransferase assay, with pyridoxal phosphate added, does not improve (see, for example, ref. 56).

That sera from patients with myocardial infarction demonstrate significant stimulation with pyridoxal phosphate, sufficient to improve disease discrimination, is now well documented (57-59), although Jung and Böhm (59) appeared to obtain even greater stimulation in their series than had previously been reported. These authors also showed that the AST isoenzyme content of the sera did not correlate with the magnitude of the stimulation in such cases.

Unfortunately, the addition of pyridoxal phosphate to sera from patients with hepatic disease does not improve diagnostic discrimination (57,60,61). As expected, quality control sera show a pyridoxal stimulation effect (62). Clearly the pyridoxal phosphate effect is creating many new problems, both analytically and diagnostically.

An Interlaboratory Survey of Aspartate Aminotransferase (AST) Assays

One of the most interesting interlaboratory studies to be carried out has been reported (63) for AST using purified cytoplasmic enzyme from human erythrocytes suspended in bovine serum albumin, porcine heart AST-1 also in bovine serum albumin, and a human serum specimen. The stability of the lyophilized materials was studied at various temperatures over 185 days to mimic, at least in part, transport and storage conditions encountered during distribution to laboratories throughout the United States in July. Results are shown in Figure 5. Experiments involving shipment and reconstitution were also carried out to ensure complete stability of the materials. Specimens were distributed to 293 laboratories of which 251 responded (86). Details of the analytical methods used were sought. A wide range of wavelengths were used although the most common were 339, 340, or 340/380 nm (83% of participants). The assay was carried out at 30°C or 37°C by 85% of all participants.

Results from laboratories using NADH-detection methods, when expressed as coefficients of variation, ranged up to 36%.

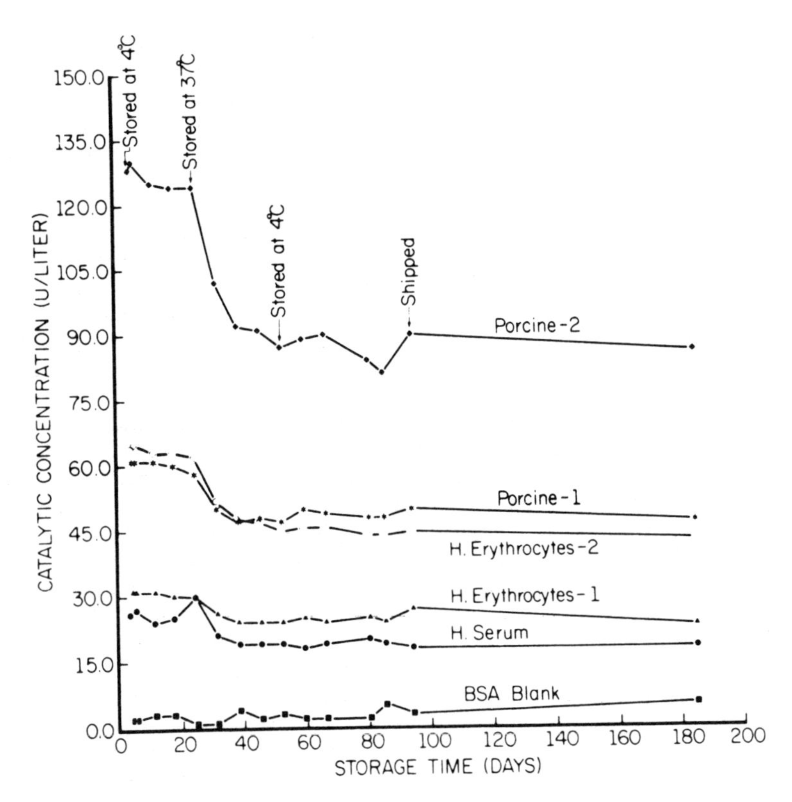

Figure 5. Stability curves for lyophilized specimens of various preparations of AST. (Reproduced from Burtis et al: *Clinical Chemistry* 24:916, 1978, by permission.) Ref. 63.

When seven selected laboratories with special interests in enzymology assayed the material, with pyridoxal phosphate supplementation, coefficients of variation of between 5 to 7% were obtained which should be contrasted with coefficients of variation of between 20 and 27% obtained by the main group. The selected laboratories are all staffed by clinical chemists whose professional interest is in enzymology.

The authors conclude ". . . close agreement can be reached in interlaboratory measurements under controlled reaction conditions. Similar observations have been reported by the Scandinavian Society for Clinical Chemistry and Clinical Physiology; interlaboratory coefficients of variation were as low as 6% when reference methods were used." It is evident that investigators in most countries have a long way to go in achieving high standards of performance in enzyme analysis.

CREATINE KINASE

The Role of Creatine Kinase in the Cell

Enzymes of clinical importance often seem to be in search of a function, and this lack of identity within a physiological framework tends to give a capricious air to the subject, although no one denies the diagnostic usefulness of the enzyme determinations themselves. Therefore, it is helpful, when possible, to integrate the enzymes with a physiological role; and for this reason a recent useful summary (64) of the cellular functions of creatine kinase (CK) is welcome.

The mitochondrial CK bound to the outer side of the inner mitochondrial membrane accounts for 40% of the total enzyme activity of the cardiac cell. Also, creatine kinase-3 (MM-CK) is bound to the myofibrils, sarcoplasmic reticulum, and plasma membranes of cardiac and skeletal muscle cells. In cardiac muscle creatine kinase-2 (MB-CK) is also located on the myofibrils. What are their functions?

It now seems likely that ATP produced in mitochondria during oxidative phosphorylation is used by the mitochondrial CK to produce creatine phosphate. The mitochondrial ATP is translocated across the inner mitochondrial membrane to the site of the mitochondrial CK by an ATP-ADP translocase. This mechanism allows for high efficiency in the production of creatine phosphate by the mitochondria. Since ADP is also a substrate for the translocase, the tendency for the reaction ATP + creatine $\rightarrow$ ADP + creatine phosphate to reverse is markedly decreased. Thus, a high creatine phosphate: creatine ratio can be maintained in the myoplasm, where it is used by the CK *(a)* on the myofibril for the regeneration cycle; *(b)* on the sarcoplasmic reticulum for the regeneration of ATP for the Ca^{2+}-ATPase reaction; and *(c)* on the plasma membrane for the Na^+, K^+-ATPase reaction, which maintains the ion-balance across the surface membrane of the cell.

Thus, it appears that the creatine phosphate pathway is a device for intracellular energy transport. A schema of the pathway is shown in Figure 6. The evidence for this conclusion is discussed at length in an article by Saks et al (64). In addition, a review of the recent advances in the study of the mechanism and subunit behavior of CK has also been published (65).

Inactivation, Assay and Storage

One of the most significant results of the work being done on enzymes is the isolation and identification of a new potent inhibitor of CK from human serum (66). Jacob et al isolated, by anion-exchange chromatography and multiple gel filtration steps (Sephadex G-10, Biogel P-2 100 mesh and then 400 mesh), an active factor from a dialysate of human serum. This factor was positively identified as cystine, and inhibition of CK was shown to be noncompetitive. The inhibition was found to be similar with either human or rabbit CK-3. Compared with urate - - the CK inhibitor identified by Warren (67) - - cystine was at least 10 times more potent, although differing blood levels would influence this inhibitor ratio.

Jacob et al (66) estimated that cystine and urate together accounted for almost 75% of the inhibitory activity present in serum. (The remaining inhibitory activity remains to be identified.) They were able to calculate that, for the upper limit of

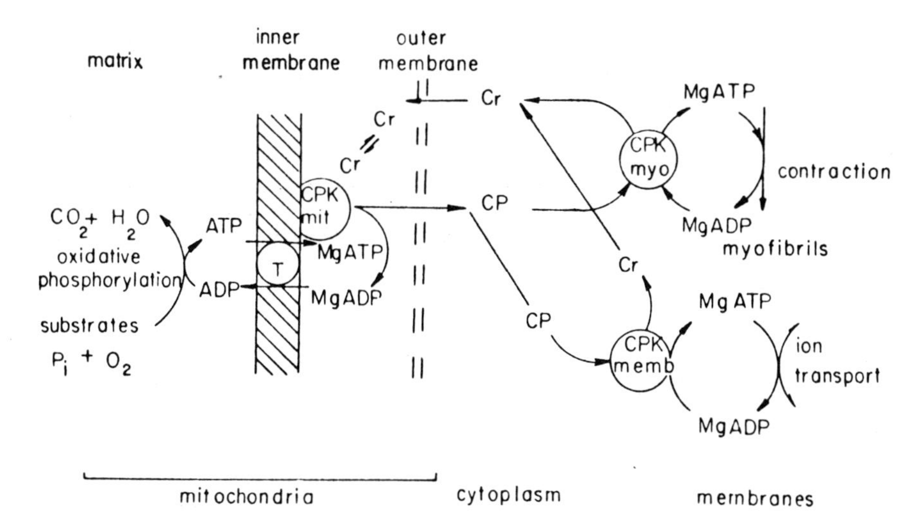

Figure 6. The creatine phosphate pathway for energy transport in the myocardial cell. T: mitochondrial ATP-ADP translocase, CPK mit: mitochondrial creatine kinase, CPK myo: myofibrillar creatine kinase, CPK memb: creatine kinase bound to membrane of sacroplasmic reticulum and to the plasma membrane. (Reproduced by permission of the National Research Council of Canada from the *Canadian Journal of Physiology and Pharmacology*, Volume 57, pp. 697-706.)

the normal values for CK, cystine is present in a 10,000-fold molar excess over the enzyme. They proposed that a disulphide interchange reaction was the cause of the reversible inhibition, and the proposed mechanism is shown in Figure 7. These authors also pointed out that the loss of negative charges would account for subforms of isoenzymes that are more cathodic than their parent forms.

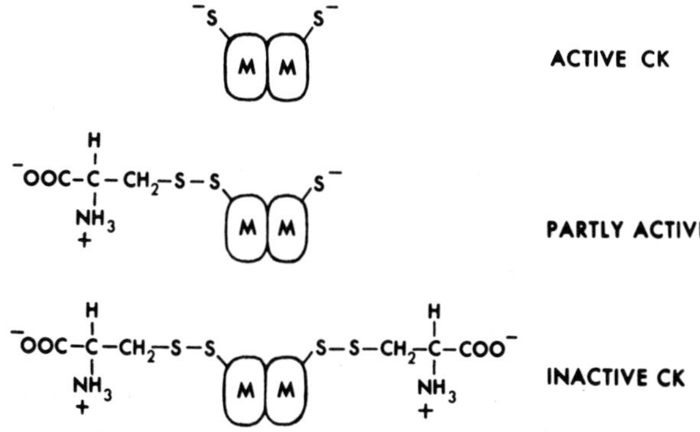

$^{-}$S [M|M] S^{-} ACTIVE CK

$^{-}$OOC–CH(NH_3^+)–CH_2–S–S [M|M] S^{-} PARTLY ACTIVE CK

$^{-}$OOC–CH(NH_3^+)–CH_2–S–S [M|M] S–S–CH_2–CH(NH_3^+)–COO^{-} INACTIVE CK

Figure 7. Proposed mechanism of inactivation of creatine kinase by cystine. (Reproduced with permission from Jacob et al, ref. 66.)

The assay of CK still seems to be presenting problems. Adenylate kinase interference appears to have been solved for the moment by the addition of AMP and diadenosine pentaphosphate to the assay medium (68). Since the three isoenzymes of CK have different kinetic properties, only one isoenzyme can be optimized at a time. If CK-3 is optimized, the other isoenzymes will be inhibited by substrates by as much as 6% (69) if the assay of Szasz et al is used (70). Szasz and Gruber (69) also found that purifying CK-3 altered, by a factor of two, the Michaelis constants (at 30°C) for both creatine phosphate and ADP. They concluded that considerable care must be taken when selecting a preparation of an enzyme form while standardizing an assay.

The buffer used by Szasz et al (70) is 100 mmole/liter imidazole acetate at pH 6.7 (25°C). Morin (71) has suggested that the use of 200 mmole/l bis-tris acetate at pH 6.45 (37°C) gives superior activity. Szasz has countered (72) with the statement that when the assay medium contains 2 mmole/liter EDTA, the activity with the imidazole buffer is comparable with that obtained with the bis-tris buffer (with or without EDTA added), and that this effect - - that is, increased activity with Morin's buffer - - may be due to a weak chelating effect of bis-tris buffer at high concentration.

These assay formulations also differed in the thiol agent used, Morin (71) preferring 20 mmole/liter 1-thioglycerol and Szasz et al (70) using N-acetyl-cysteine, at the same concentration. Recent work by Szasz et al (73) may help to clarify this controversy. They examined the effect of thiol compounds on the storage of the CK isoenzymes. Before their findings can be appreciated it is important to realize that serum rapidly oxidizes thiol. Thus, accumulation of oxidized thiol agents occurs, and Szasz et al found that the oxidation products of 2-mercaptoethanol and 1-thiolglycerol appeared to be more potent inhibitors of CK activity in serum under storage conditions than oxidized N-acetyl-cysteine. This finding may not necessarily apply to the assay, but it does suggest that N-acetyl-cysteine is a superior thiol agent in at least one respect. Another advantage is that N-acetyl-cysteine is less likely to cause serum turbidity than any other thiol.

Recently renewed interest has been shown in the value of using chelating agents for the storage and assay of CK. A stabilizing agent of CK activity, such as EDTA, has been used since 1965 at least, and Gruber (74) has provided an interesting account of the practice of Boehringer Mannheim GmbH of adding 2 mmole/liter EDTA to the glutathione-activated CK kits because the calcium ion was causing interference with certain control preparations. Since the EDTA did not appear to affect the assay method, no announcement of its presence was thought necessary. In passing, readers will be familiar with many test formulations that contain low concentrations of EDTA, added as a precautionary measure against metal ion interference. Gruber pointed out that, in the normal assay medium for CK, calcium is present at a concentration of approximately 100 mmole/liter. Calcium inhibits purified CK preparations at concentrations of 50 mmole/liter in the absence of EDTA and at concentrations of 500 mmole/liter in the presence of 2 mmole/liter EDTA.

However, it is probably the valuable work of Ladenson's group (75,76) that created a resurgence of interest in chelating agents and their effects on CK. They pointed out the excellent stability of CK in frozen sera in the presence of ethylene-glycol-bis-(β-aminoethyl ether)-N,N'-tetracetatic acid (EGTA) found by Roberts et al (77), and they made a systematic study of the stability of endogenous CK in unfrozen sera. In addition to the use of EDTA and EGTA, Rollo et al (76) also used phenylmethylsulfonyl fluoride (PMSF, an inhibitor of cation-dependent serine proteases) and Chelex (a weakly acid cation, chelating, resin).

The addition, one hour before assay, of EDTA or EGTA (10 mmole/l) enhanced CK in sera. If 2-mercapotoethanol (10 mmole/l) was added alone, a similar enhancement occurred. However, when the thiol and EGTA were present together, the effects were additive. The chelator effect was independent of a decrease in lag phase, which was decreased by the addition of thiol. This activation was shown to be due to chelation of the calcium present in the medium because equimolar EGTA and calcium ion abolished this effect.

EGTA 10 mmole/l) has a second effect. It stabilizes CK activity in serum so after incubation for 24 hours at 27°C, 100% activity remains. EDTA-treated sera retained about 95% activity under the same conditions. Treatment of sera with Chelex, thereby removing metal cations, gave similar results. Addition of metal cations (Ca^{2+}, Mn^{2+}, Fe^{3+}, Fe^{2+}, Zn^{2+}, or Cu^{2+}) to Chelex-treated sera did not alter the Chelex-enhanced stability of CK. Experiments with PMSF showed that protease action was not involved in this stabilizing effect. Therefore, Rollo et al (76) concluded that *(a)* 2-mercaptoethanol should be added before assay and *not* before storage (N-acetyl cysteine does not appear to have been tested in this way) and *(b)* EGTA should be added before storage.

Sandifort (78) made substantially similar observations of the effect of EDTA (2 mmole/l) on CK activity but questioned whether or not the Mg^{2+} concentration of 10 mmole/liter in the assay would have to be increased to ensure that the Mg^{2+} concentration, in the presence of EDTA, does not become rate limiting. Sandifort's question has been answered by the work of Gerhardt (79), Gruber et al (80), and Waldenström et al (81). These workers showed that the Mg^{2+} concentration of 10 mmole/l in the CK assay is sufficient, even in the presence of 2 mmole/l EDTA, to obtain maximal activity. They also showed that the instability of N-acetyl-cysteine (because of the formation of auto-oxidation products) was considerably improved, and that the complete assay formulation was stable after the addition of 2 mmole/l EDTA for up to 5 days, as compared with 24 hours without EDTA. They also confirmed that EDTA increased the thermal stability of CK.

The problems associated with the assay conditions discussed previously are echoed in the performance of quality control sera containing CK. This has already been alluded to (74), but several other groups (82-84) have commented on the change in activity obtained by the addition of thiol agents or EDTA to the reconstitution diluent, or due to the temperature of the diluent, there being greater activity when reconsitution occurs at 4°C than at 25°C. The addition of EDTA to the diluent does not abolish this effect, but excess thiol agent (62 mmole/liter) does. These factors are additional sources of variation in quality control aspects of CK assays.

Creatine Kinase Isoenzymes: Technical Aspects of Separation

Isoenzyme separations by electrophoresis are often complicated by the presence of artifacts, which can be misinterpreted. CK isoenzymes are no exception. The most important artifact is that due to fluorescent albumin. In patients with endstage renal diseases, who are on maintenance hemodialysis (an increasing population), albumin fluoresces strongly. This fluorescence disappears or diminishes after successful renal transplantation. While this fluorescent albumin can be seen on lactate dehydrogenase (LDH) isoenzyme separations, the albumin does not usually interfere with LDH-1 estimations. However, in CK isoenzyme work, albumin may comigrate (agarose media) or be very close to (cellulose acetate media) CK-1. This close association can lead to claims that CK-1 is present in certain diseases. From the work to be described it will be seen that such claims should be supported by evidence that "blank" reactions on the electrophoresis plate do not show the band and, preferably, by the demonstration of isoenzyme activity by means of another separative procedure, for example, ion-exchange chromatography.

Coolen et al (85) found that the serum from a patient maintained on renal dialysis had a quite characteristic fluorescence emission spectrum with a peak at 440 nm (excitation 366 nm) (Fig. 8). They were unable to characterize the material (which appears to attach to albumin), and mass spectral analysis only showed increased fatty acids. They concluded that the fluorescence had some relationship to the vitamin supplementation (particularly the B_6 group) that these patients receive.

The "blank" reaction has been advocated in a number of formulations (85-89) as a means of eliminating artifactual interpretation of electrophoretic separations. The commercial availability of an anti-CK-1 serum has allowed the routine use of a confirmatory, immunological, identification of isoenzymes containing the B subunit (86). A patient serum showing any electrophoretic abnormality at or near the CK-1 zone can be incubated with anti-CK-1 serum for 15 minutes at room temperature, and electrophoresis repeated. The presence of anti-CK-1:CK-1 (or CK-2) complexes appears to broaden the CK-3 zone but completely eliminates the CK-1 zone (and 50% of the CK-2 zone). An example of this technique (86) is shown in Figure 9 (86).

An alternative technique, the inhibition of CK activity by 2.5 mmole/liter iodoacetate, has also been advocated (87) for the detection of artifactual "bands". Tonks and his group (87), investigating the occurrence of CK-1 in the sera of patients with chronic renal failure, demonstrated, by this inhibition reaction, that the fluorescent zone appearing near to the CK-1 zone in these patients was not due to CK activity (Fig. 10). They also showed, by the omission of the detecting reagents and by fluorescence intensity measurements before and after application of the detecting reagents, that a non-CK fluorescent entity does exist

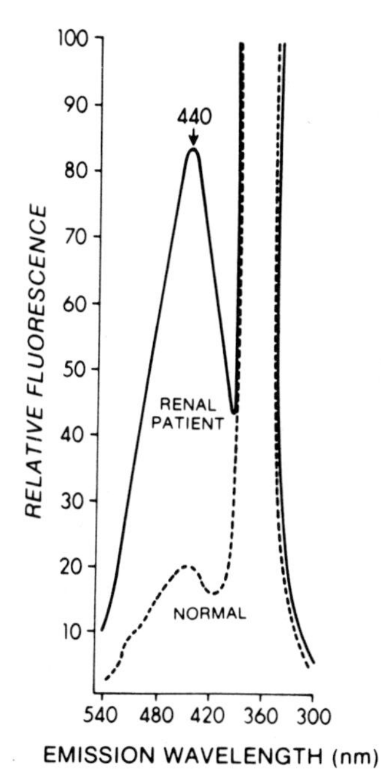

Figure 8. Fluorescence emission spectrum of normal serum (---) and of serum from a patient being maintained by renal dialysis (——). Sera diluted 30-fold with water; excitation wavelength, 366 nm. (Reproduced from Coolen et al: *Clinical Chemistry* 24:1636, 1978, by permission.)

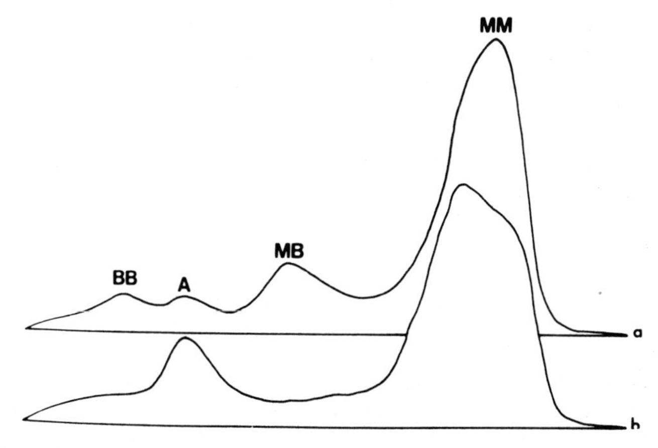

Figure 9. Electrophoretic pattern on cellulose acetate of a patient's sample containing the CK isoenzymes MM, MB, and BB and a fluorescent artifact (A) with (b) and without (a) pre-incubation with anti-BB activity. (Reproduced from Van Lente and Galen: *Clinica Chimica Acta* 87:211, 1978, by permission.)

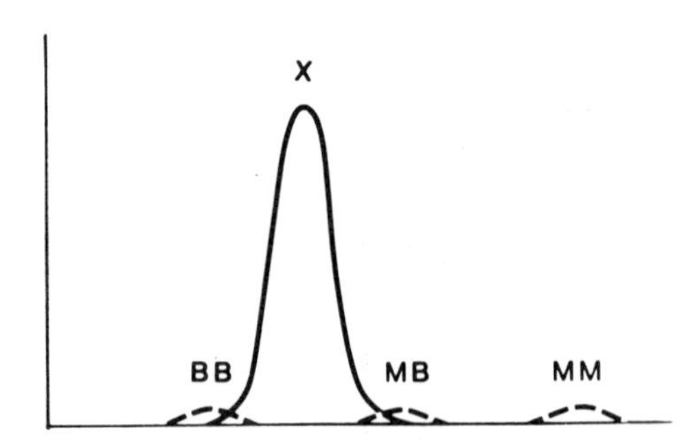

Figure 10. Effect of iodoacetate (2.5 mmole/l) on the BB band in the electrophoresis of sera from hemodialyzed patients. Note the almost complete disappearance of MM, MB, and BB bands in the control marker (---), as well as total inhibition of MM isoenzyme in the patients' sera. The peak X of renal failure remained unaffected (—). (Reproduced from Aleyassine et al: *Clinical Chemistry* 24:492, 1978, by permission.)

in the sera of patients with chronic renal failure (88). This group further demonstrated by using rabbit antihuman-albumin precipitation, that the fluorescent entity is associated with albumin, thus confirming the work of Coolen et al (85). Others, however, while agreeing that there is a non-CK fluorescent entity in the sera of patients with chronic renal failure, have identified CK-1 activity in these patients (86,89) and it seems likely that the disparity results from each group's use of different detection limits.

Radioimmunoassay has been applied, by many workers, to the determination of CK isoenzymes (see for example, refs. 90-95). Recent reports (96,97) have characterized assays for CK-3 and CK-1. The CK-3 radioimmunoassay (double-antibody system) was usable over a range of 80 to 2560 µg/l. The concentration of CK-3 in the sera of healthy (white) men was 102-1668 µg/l (n = 101), and in the sera of white women, it was 36-487 µg/l (n = 101). Both distributions were positively skewed. Interestingly, correlations (96) between the RIA result and enzymatic activity were better in health (r = 0.98) than in disease (r = 0.95) due to a 3-fold larger scatter of the data from patients than from the healthy population. Cross reactivity between this assay and CK-1 was virtually zero, even at a 5000-fold excess over CK-3 concentrations. Cross-reactivity of CK-2, at B/B_o = 0.5, was between 3 and 17%, but binding was less avid than with CK-3. CK-2 purified from cardiac muscle showed very low cross reactivity compared with CK-2 obtained by *in vitro* hybridization. This may, or may not be, an important distinction, but further work is needed on this aspect, as well as on the effect of using a different tissue as the enzyme source. The authors concluded that this assay was probably specific for CK-3, especially since cross-reacting CK-2 is present in much smaller quantitites in serum than CK-3. In the RIA (sequential saturation, double-antibody system) for CK-1 (97) the detection range was approximately 1-150 µg/l of serum; the mean value in healthy persons (n = 209) was 3.4 µg/l, whereas in randomly selected patients it was

6.6 μg/l. Cross reactivity with CK-3 was virtually zero at a 14,000-fold excess over CK-1. CK-2 cross reacted, at $B/B_0 = 0.5$, at 3.1 to 17% (*in vitro* hybridized CK-2) and 1.1% (myocardial CK-2). This discrepancy between cross reactivity shown by the artificially hybridized CK-2 dimer and the (presumed) natural dimer from myocardium is a cause for some concern.

The low cross reactivities of CK-2 for both the CK-1 and CK-3 assay systems suggest that these assays are relatively specific for the dimeric association of the subunits and are relatively insensitive to the monomeric subunit. Nonetheless, the cross reactivity causes some confusion (i.e., either assay detects CK-2), and it seems that some colateral assay is needed since CK-1 is apparently (almost) ubiquitous in hospital populations.

Creatine Kinase Isoenzymes: Clinical Aspects

One of the most interesting abnormal CK to be reported is the so-called "macro-CK" of Yuu et al (98). This macro-CK was shown to migrate toward the cathode and to have a relative molecular mass of about 325,000 daltons (normal CK ~ 80,000 daltons). The macro-CK was found to elute from a Sephadex G-200 column between the IgM and IgG peaks. This macro-CK activity was shown to be different from mitochondrial CK, which migrates in the same zone (i.e., it is more cathodic than CK-3), by its resistance to elevated temperature, by its mass, and by the fact that it was thought, on the basis of lack of inhibition by anti-M subunit serum, to contain B subunit activity, although no CK-1 or CK-2 activity could be detected by electrophoresis or DEAE-Sephadex CL-6B chromatography. Finally, no IgG, IgA, IgM, or β-lipoprotein was detected in the macro-CK. Yuu and his coworkers speculated that the macro-CK may be a complex consisting of four CK molecules.

Another interesting abnormal CK was described by Ljungdahl and Gerhardt (99) in eight patients. They routinely use an anti-B inhibitor technique to detect the presence of B subunits, which are assumed to be the result of activity in the serum. Of approximately 800 patients studied, 8 showed the combination of normal, or slightly increased, total serum CK activity with an increased B subunit activity. The sera from these patients were examined by agarose electrophoresis. In each case one, or two, abnormally migrating enzyme bands were found between the CK-3 and CK-1 zones. By the use of anti-M serum these abnormal species were shown to contain the B-subunit. No common clinical denominator was discerned in the eight cases, and the authors surmised that the abnormal forms were altered CK-1 isoenzymes. This conclusion was based on the findings of Cho et al (100) and Bayer et al (101) that incubation of CK-1 in human serum made the isoenzymes electrophoretically more cathodic, although there was no change in the immunoprecipitability of these, less mobile, enzyme species (101).

The occurrence of CK-3 heterogeneity (102) adds to the possibility that extra CK "zones" exist. Soons et al (103-105) have investigated this aspect in considerable detail. In their first paper (103), they presented evidence for the occurrence of three CK-3 isoenzyme forms in sera, which were designated MM_1 (subunit structure M_1M_1), MM_2 (subunits M_1M_2), and MM_3 (subunits M_2M_2), MM_1 being most anodic (Fig. 11). These forms can most easily be

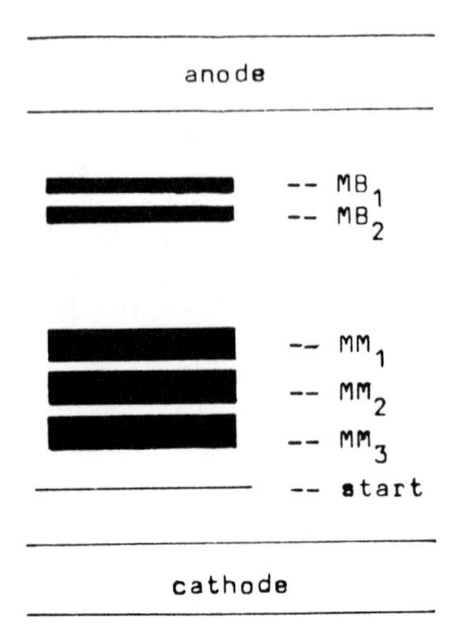

Figure 11. CK electrophoretogram of serum from a patient with total CK activity of 4600 units/l. Agarose electrophoresis 120 minutes, 85 V. (Reproduced from Wevers et al: *Clinica Chimica Acta* 75:377, 1977, by permission.)

obtained by prolonged electrophoresis. The CK-3 isoenzymes in the serum of a patient with myocardial infarction was studied by this technique during the course of the disease. It was shown that MM_3 predominated early in the disease but, with time, first MM_2 and then MM_1 predominated as the principal enzyme in the CK-3 zone.

Hybridization of the serum of a patient showing multiple CK-3 zones with an extract of brain produced CK-2 of two types (designated MB_1 and MB_2 with, presumably, the structure M_1B and M_2B, respectively). The existence of two CK-2 types was previously suggested, by the findings of Takahashi et al (106). Soons et al (104) then proceeded to demonstrate that the M_1 and M_2 hybrids could be isolated by isoelectric focusing (pI values: 6.24, 6.49, and 6.86), an example of which is shown in Figure 12. CK-2 had a pI value of 5.16. Another CK (X) was observed at pI 6.07, but this form could not be detected after electrophoresis of serum. The identity of this CK form remains a mystery.

A later paper (105) reports on an *in vitro* model for the *in vivo* findings (103). Extracts of heart muscle or skeletal muscle showed only MM_3 and MB_2 (i.e., M_2M_2 and M_2B). Incubation with serum produced a change to, successively, MM_2 (i.e., M_2M_1) then MM_1 (i.e., M_1M_1) and, for CK-2, MB_1 (i.e., M_1B). The authors regarded these changes in isoenzyme migration as being the result of postsynthetic process caused by the presence of a heat-labile material (incubation for 60°C for 30 minutes stops the change). The clinical implications of this work are considerable since this progression may be followed by estimating the different species and allow dating of the infarct and an indication that continuing damage is proceeding. Such techniques are likely to be superior to measurement of gross enzyme levels in monitoring fine changes. Further reports from Soons' laboratory are eagerly awaited.

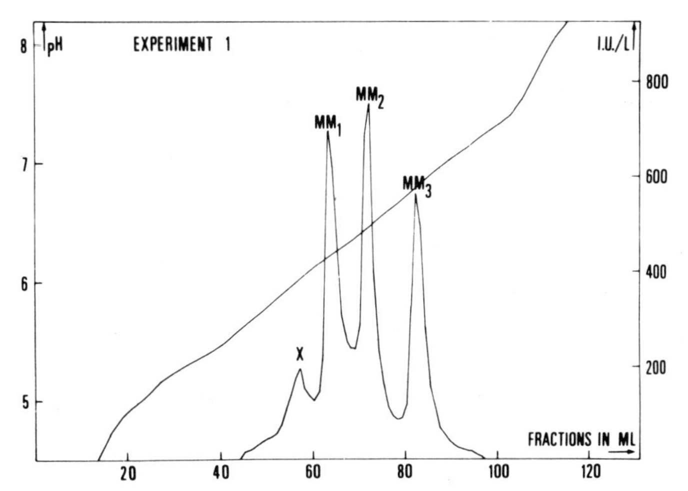

Figure 12. CK activity after isoelectric focusing in a sucrose density gradient. (Reproduced from Wevers et al: *Clinica Chimica Acta* 78:271, 1977, by permission.)

The relationship between the different forms of CK-3 described above, and creatine kinase Z, described by Leroux et al (107), is uncertain. CK-Z was found in heart muscle extracts and in sera of patients with myocardial infarction or with skeletal muscle trauma. In patients with infarction this enzyme species paralleled changes in serum CK-2 levels. CK-Z was located, electrophoretically, between CK-3 and CK-2. Leroux et al thought that CK-Z was similar to the atypical sub-band of CK described by Lim (108) and by Sax et al (109), and the splitting of CK-3 and CK-2 bands, observed by Smith (110), may be included in the same category, as is the CK-X of Madsen (111). Progress is so fruitful in this area of investigation that it is expected that these anomalous bands will soon be definitively described with an explanation for their mechanism of appearance.

I referred earlier to the ubiquitous nature of CK-1 (BB-CK) in serum. As is often the case, an examination of the tissue distribution of the isoenzymes gives a clue to their occurrence in the serum in certain pathological conditions. Several authors have presented data, quantitative or qualitative, on the distribution of CK isoenzymes in tissues (112-116). The two most useful compilations (114,115) are shown in Tables 4 and 5. Tsung (115) used tissue specimens obtained during surgery, whereas Jockers-Wretou and Pfleiderer (114) obtained their specimens 12 to 24 hours after death. There were slight differences in prostate, uterus, pancreas, and intestine and major differences in thyroid, liver, kidney, and lung; and these differences may be attributed to postmortem autolysis. However, the most important aspect of these data is the wide distribution of CK-1 throughout the body. Obviously, Tables 4 and 5 tend to give a simplistic idea of tissue distributions. For example, many workers have consistently reported the existence of small quantities of CK-3 in brain tissue (117-119) although it has recently been suggested that much more is present than those earlier reports had indicated.

Table 4. Creatine Kinase Concentration and Isoenzyme Patterns, Determined by Immunotitration, in Human Tissues Obtained at Autopsy.[a]

Tissue	Total CK Activity (units/gm frozen tissue)	Distribution of CK Isoenzymes MM (%)	MB (%)	BB (%)
Skeletal muscle	860-1310	96-100	0-3	0-1
Tongue	225-292	90-99	1-5	0-5
Heart (adult)	100-280	71-96	4-27	0-2
Heart (infant)	78-250	96-100	0-4	0
Aorta (arcus aortae)	3-7	81-88	8-14	3-5
Kidney	0-1	70-100	0	0-30
Spleen	0-1	65-75	0	25-35
Thyroid	13-34	64-79	6	15-30
Adrenal	0-1	35-60	0-10	25-40
Lung	2-9	27-72	0-4	18-69
Liver[b]	0-1	50	0	50
Prostate	7-10	34-39	2-6	59-60
Uterus	3-9	5-16	2-20	64-93
Pancreas	0-1	21-29	5-9	66-73
Intestine (mesentery)	5-6	11-13	7-9	78-80
Bladder	14-35	2-7	3-5	89-93
Stomach	15-23	3	2-6	91-95
Spinal cord	23-27	0	0	100
Cerebellum	50-87	0	0	100
Cerebrum	55-90	0	0	100

[a] Reproduced from Jockers-Wretou and Pfleiderer: *Clinica Chimica Acta* 58:223, 1975, by permission.

[b] From 6 specimens assayed only one showed CK activity.

Table 5. Creatine Kinase Concentration and Isoenzyme Patterns, Determined by Electrophoresis, in Surgical Specimens of Human Tissues[a]

Tissue	CK Activity units/gm wet tissue	MM (%)	MB (%)	BB (%)
Skeletal muscle (Gastrocnemius)	3281	100	0	0
Skeletal muscle (external intercostal)	1894	99	<1	<1
Heart (right atrium)	356-402	76-78	22	0-2
Brain	157	0	0	100
Urinary Bladder	162	2	6	92
Lung	9-14	16-35	0-1	69-84
Prostate	8-10	3-4	2-4	93-94

Table 5. (Continued)

Uterus	30-47	2	1-3	95-97
Thyroid	28-36	4-26	0-1	73-96
Pancreas	3	14	1	85
Stomach	86-120	2-4	1-4	94-95
Colon	125-148	3-4	0-1	96
Ileum	161	3	1	96
Kidney	15-21	8-12	0	88-92
Liver	3.8	90	6	4
Spleen	7	74	0	26
Placenta	1.7-2.8	15-48	6-22	46-63

[a] Reproduced from Tsung: *Clinical Chemistry* 22:173, 1976, by permission.

However these are minor criticisms of data, which are, on the whole, extremely valuable.

Since CK-1 is widely distributed throughout the body, it is not surprising that certain pathological states (e.g., carcinoma) may cause CK-1 to appear in serum if the affected tissue contains high concentrations of that isoenzyme. Thus, carcinoma of prostate (86,120,121), stomach (86,121,122), and lung (121,123) has been reported to produce increased CK-1 activity in serum. The gravid uterus and/or the placenta have been shown to produce elevations of CK-1 activity in serum during pregnancy and labor (124-126). Multisystem involvement, as in patients in intensive care units, might also be expected to cause increased serum CK-1 activity (127,128).

Several other sources of serum CK-1 have been suggested. One has been mentioned previously, namely, chronic renal failure (86, 89,129). Another is bone marrow. Reference was made earlier to the possible value of bone marrow ACP assays to determine the existence of metastases of prostatic carcinoma. Since prostatic tissue contains considerable CK-1 activity (see Tables 4 and 5, it has been argued (130) that this isoenzyme might also be a marker for metastases. However, it was found in all patients sampled that CK-1 was detected in the bone marrow, whereas only four patients (one with carcinoma of the lung and three with prostatic carcinoma) had CK-1 in their sera. It was concluded, therefore, that the CK-1 activity was probably derived from the marrow due to the presence of tumor.

Cardiac surgery has been noted to produce CK-1 activity in serum. Vladutiu et al (131) detected CK-1 intraoperatively in 22 of 25 patients undergoing aortocoronary bypass surgery in both the coronary sinus (the main vessel for venous drainage of myocardium into the left atrium) and the mixed venous blood. It was suggested that the source of this CK-1 may be the myocardium itself, although lung tissue, which is handled during cardiac surgery, has appreciable amounts of CK-1 (see Tables 4 and 5). It was noted that CK-1 activity appeared as early as 15 minutes after cardiopulmonary bypass was accomplished and disappeared within 6 hours after the surgery was completed.

The most obvious source for CK-1 in serum is brain tissue. It was shown in 1975 that surgical manipulation (119) or injury (132) of the brain can cause CK-1 to appear in serum, and malignant hyperpyrexia (133) and Reye syndrome (134) are also known to produce the same effect. Several groups (135-137) have made a systematic study of the occurrence of serum CK-1 activity after a variety of neurological insults. These disorders are listed in Table 6. Often only one case has been described.

Table 6. Neurological Disorders Associated With Increased Serum CK-1 Activity

Acute cerebral disorders	
Local and Diffuse brain damage	(132,136,137)[a]
Cerebral hemorrhage	(127,136,137)
Cardiac arrest with brain damage	(136)
Bacterial and viral infections	(136,137)
Seizures with and without alteration of consciousness	(127,137)
Migraine	(137)
Drug overdosage with intubation	(137)
Delirium tremens	(137)
Chronic cerebral disorders	
Optic atrophy	(137)
Spinocerebellar atrophy	(135)
Floppy-infant syndrome	(135)
Werdnig-Hoffman disease	(135)

[a] Reference numbers

Bell et al (137) also examined CSF for the presence of CK-1 activity. They found elevations in cases of acute cerebrovascular accident, seizure with prolonged alteration of consciousness, viral infection, drug overdosage with intubation, trauma, and acute multiple sclerosis. Another interesting aspect of these studies (136,137) was the suggestion that clinical outcome and the elevation of serum (or CSF) CK-1 activity seemed to be related. Obviously, a greater number of cases must be studied to establish this association with certainty.

The association of elevated CK-2 activity in serum with myocardial damage is now well documentated. For a recent review the reader is referred to Roberts and Sobel (138).

REFERENCES

1. Expert Panel on Enzymes, Committee of Standards of IFCC: Definitive Recommendation on IFCC Methods for the Measurement of Catalytic Concentration of Enzymes. Part 1. General Considerations Concerning the Detection of the Catalytic Concentration of an Enzyme in the Blood Serum or Plasma of Man, 1978.

2. *Clin Chem* 23:709, 1977.

3. Sostman HE, Manley KA: *Clin Chem* 24:1331, 1978.

4. Van Etten RL, Saini MS: *Biochim Biophys Acta* 484:487, 1977.

5. Choe BK, et al: *Prep Biochem* 8:73, 1978.

6. Choe BK, et al: *Arch Androl* 1:221, 1978.

7. Van Etten RL, Saini MS: *Clin Chem* 24:1525, 1978.

8. Vihko P, et al: *Clin Chem* 24:466, 1978.

9. Cooper JF, Foti A: *Invest Urol* 12:98, 1974.

10. Foti AG, et al: *Cancer Res* 35:2446, 1975.

11. Chu TM, et al: *Invest Urol* 15:319, 1978.

12. Lee C.-L, et al: *Cancer Res* 38:2871, 1978.

13. Foti AG et al: *Clin Chem* 23:95, 1977.

14. Foti AG et al: *N Eng J Med* 297:1357, 1977.

15. Catalona WJ, Scott WW: *J Urol* 119:1, 1978.

16. Carroll BJ: *N Eng J Med* 298:912, 1978.

17. Galen RS, Gambino SR: *Beyond Normality* in: *The Predictive Value and Efficiency of Medical Diagnosis*. New York, Wiley, 1975.

18. Gittes R: *N Engl J Med* 298:1398, 1978.

19. Editorial: *Brit Med J* 2:719, 1978.

20. Cooper JF, et al: *J Urol* 119:388, 1978.

21. Boehme WM, et al: *Cancer* 41:1433, 1978.

22. Cooper JF, et al: *J Urol* 119:392, 1978.

23. Catane R, et al: *N Y State J Med* 78:1060, 1978.

24. Choe BK, et al: *Arch Androl* 1:227, 1978.

25. Vikho AG, et al: *Clin Chem* 24:1915, 1978.

26. Foti AG et al: *Clin Chem* 24:140, 1978.

27. McDonald I, et al: *Arch Androl* 1:235, 1978.

28. Chu TM, et al: *Oncology* 35:198, 1978.

29. Ladenson JH, McDonald JM: *Clin Chem* 24:129, 1978.

30. Vihko P: *Clin Chem* 24:1783, 1978.

31. Choe BK, et al: *Invest Urol* 15:312, 1978.

32. Lantz RK, Eisenberg RB: *Clin Chem* 24:486, 1978.

33. Sanders GTB, et al: *Clin Chim Acta* 89:421, 1978.

34. Chambers JP, et al: *Metabolism* 23:801, 1978.

35. Lam WKW, et al: *Clin Chem* 24:309, 1978.

36. Lam WKW, et al: *Clin Chem* 24:1105, 1978.

37. Khan AN, et al: *Brit J Urol* 50:182, 1978.

38. Huseby N-E: *Biochim Biophys Acta* 483:46, 1977.

39. Tate SS, Meister A: *Proc Nat Acad Sci USA* 73:2599, 1976.

40. Miller SP, et al: *J Biol Chem* 251:2271, 1976.

41. Tate SS, Ross ME: *J Biol Chem* 252:6042, 1977.

42. Shaw LM, et al: *Clin Chem* 24:905, 1978.

43. Rosalki SB: *Adv Clin Chem* 17:53, 1975.

44. Wenham PR, et al: *Ann Clin Biochem* 15:146, 1978.

45. Burlina A: *Clin Chem* 24:502, 1978.

46. Kok PJMJ, et al: *Clin Chim Acta* 90:209, 1978.

47. Huseby N-E: *Biochim Biophys Acta* 522:354, 1978.

48. Freise J, et al: *J Clin Chem Clin Biochem* 14:589, 1976.

49. Smith AF, et al: *J Mol Med* 2:265, 1977.

50. Rej R: *Clin Chem* 24:1971, 1978.

51. Sampson EJ, et al: *Clin Chem* 24:1805, 1978.

52. Matsuda I, et al: *Clin Chim Acta* 83:231, 1978.

53. Bergmeyer HU, et al: *Clin Chem* 24:58, 1978.

54. Expert Panel on Enzymes, Committee on Standards of IFCC: *Clin Chem* 24:720, 1978.

55. Bergmeyer HU: In *Principles of Enzymatic Diagnosis*. H.U. Bergmeyer (ed): New York, Verlag Chemie, 1978, p 61.

56. Hørder M, Bowers GN: *Clin Chem* 23:551, 1977.

57. Rosalki SB, Bayoumi RA: *Clin Chim Acta* 59:357, 1975.

58. Moss DW: *Clin Chim Acta* 67:169, 1976.

59. Jung K, Böhm M: *Enzyme* 23:201, 1978.

60. Ratnaike S, Moss DW: *Clin Chim Acta* 74:281, 1977.

61. Jung K, et al: *Clin Chim Acta* 90:143, 1978.

62. Burger FS, Potgieter GM: *Clin Chem* 24:841, 1978.

63. Burtis CA, et al: *Clin Chem* 24:916, 1978.

64. Saks VA, et al: *Can J Physiol Pharmacol* 56:691, 1978.

65. Bickerstaff GF, Price NC: *Int J Biochem* 9:1, 1978.

66. Jacob et al: *Clin Chim Acta* 85:299, 1978.

67. Warren WA: *Clin Biochem* 8:247, 1975.

68. Szasz G, et al: *Clin Chem* 23:1888, 1977.

69. Szasz G, Gruber W: *Clin Chem* 24:245, 1978.

70. Szasz G, et al: *Clin Chem* 22:650, 1976.

71. Morin LG: *Clin Chem* 23:1569, 1977.

72. Szasz G: Proceedings of Eighth International Symposium on Clinical Enzymology, Venice, April, 1978.

73. Szasz G, et al: *Clin Chem* 24:1557, 1978.

74. Gruber W: *Clin Chem* 24:177, 1978.

75. Rollo JL, et al: *Clin Chem* 23:119, 1977.

76. Rollo JL, et al: *Clin Chim Acta* 87:189, 1978.

77. Roberts R, et al: *Amer J Cardiol* 33:650, 1974.

78. Sandifort CRJ: *Clin Chem* 23:2169, 1977.

79. Gerhardt W: Proceedings of Eighth International Symposium on Clinical Enzymology, Venice, April, 1978.

80. Gruber W, et al: *Clin Chem* 24:988, 1978.

81. Waldenström J, et al: *Clin Chem* 24:1004, 1978.

82. Hetland O, Lund PK: *Scand J Clin Lab Invest* 37:563, 1977.

83. Pruitt CA: *Amer J Med Tech* 44:500, 1978.

84. Feld RD, et al: *Clin Chem* 24:2039, 1978.

85. Coolen RB, et al: *Clin Chem* 24:1636, 1978.

86. Van Lente F, Galen RS: *Clin Chim Acta* 87:211, 1978.

87. Aleyassine H, et al: *Clin Chem* 24:492, 1978.

88. Aleyassine H, Tonks DB: *Clin Chem* 24:1850, 1978.

89. Chuga DJ, Bochner P: *Clin Chem* 26:1286, 1978.

90. Nicholson GA, O'Sullivan WJ: *Proc Aust Assoc Neurol* 10:105, 1973.

91. Roberts R, et al: *Science* 194:855, 1976.

92. Roberts R, et al: *Lancet* 2:319, 1977.

93. Willerson JT, et al: *Proc Nat Acad Sci USA* 74:1711, 1977.

94. Fang VS, et al: *Clin Chem* 23:1898, 1977.

95. Neumeier D, et al: *Clin Chim Acta* 79:107, 1977.

96. Van Steirteghem AC, et al: *Clin Chem* 24:414, 1978.

97. Zweig MH, et al: *Clin Chem* 24:422, 1978.

98. Yuu H, et al: *Clin Chem* 24:2054, 1978.

99. Ljungdahl L, Gerhardt W: *Clin Chem* 24:832, 1978.

100. Chu HW, et al: *Clin Chim Acta* 73:257, 1976.

101. Bayer PM, et al: *Clin Chim Acta* 79:261, 1977.

102. Kumuduvalli I, Watts DC: *Biochem J* 108:547, 1968.

103. Wevers RA, et al: *Clin Chim Acta* 75:377, 1977.

104. Wevers RA, et al: *Clin Chim Acta* 78:271, 1977.

105. Wevers RA, et al: *Clin Chim Acta* 86:323, 1978.

106. Takahashi K, et al: *Clin Chim Acta* 38:285, 1972.

107. Leroux M, et al: *Clin Chim Acta* 80:253, 1977.

108. Lim F: *Clin Chem* 21:975, 1975.

109. Sax SM, et al: *Clin Chem* 22:87, 1976.

110. Smith AR: *Clin Chim Acta* 36:351, 1972.

111. Madsen A: *Clin Chim Acta* 36:17, 1972.

112. Van Der Veen KJ, Willebrands AF: *Clin Chim Acta* 13:312, 1966.

113. Smith AF: *Clin Chim Acta* 39:351, 1972.

114. Jockers-Wretou E, Pfleiderer G: *Clin Chim Acta* 58:223, 1975.

115. Tsung SH: *Clin Chem* 22:173, 1976.

116. Jockers-Wretou E, et al: *Histochem* 54:83, 1977.

117. Lindsey GG, Diamond EM: *Biochim Biophys Acta* 524:78, 1978.

118. Klein MS, et al: *Cardiovasc Res* 7:412, 1973.

119. Nealon DA, Henderson AR: *Clin Chem* 21:1663, 1975.

120. Feld RD, Witte DL: *Clin Chem* 23:1930, 1977.

121. Hoag GN, et al: *Clin Chem* 24:1654, 1978.

122. Lederer WH, Gerstbrein HL: *Clin Chem* 22:1748, 1976.

123. Coolen RB, Pragay D: *Clin Chem* 22:1174, 1976.

124. Laboda HM, Britton VJ: *Clin Chem* 23:1329, 1977.

125. McNeely MDD, et al: *Clin Chem* 23:1878, 1977.

126. Bayer PM, et al: *J Clin Chem Clin Biochem* 15:349, 1977.

127. Coolen RB, et al: *Clin Chem* 21:976, 1975.

128. Mercer DW: *Clin Chem* 23:611, 1977.

129. Galen RS: *Clin Chem* 22:120, 1976.

130. Silverman LM, et al: *Clin Chem* 24:1423, 1978.

131. Vladutiu AO, et al: *Clin Chim Acta* 75:467, 1977.

132. Somer H, et al: *J Neurol Neurosurg Psy* 38:572, 1975.

133. Zsigmond EK, et al: *Anaesth Analg* 51:827, 1972.

134. Rock RC, et al: *Clin Chim Acta* 62:159, 1975.

135. Jockers-Wretou E, et al: *Clin Chim Acta* 73:183, 1976.

136. Kaste M, et al: *Arch Neurol* 34:142, 1977.

137. Bell RD, et al: *Ann Neurol* 3:52, 1978.

138. Roberts R, Sobel BE: *Amer Ht J* 95:521, 1978.

10
Non-Polypeptide Hormones

Beverley E.P. Murphy

A computer search of new methods for nonpolypeptide hormones published in 1978 recorded 170 articles relevant to clinical chemistry. Of these, 46% were about thyroid hormones (more than half of these, which all described commercial kits for T4, T3, and T3 uptake, appeared in the Japanese literature and were not available for review); 32% discussed steroids; 12% were about catecholamines; and 3% concerned hydroxy-vitamin Ds. Forty percent of the steroid articles dealt with estrogens in pregnancy. Although this list is incomplete, it gives some idea of the relative emphasis being placed on these assays.

Probably the most important recent advance in the assay of the nonpolypeptide hormones is the enzyme-linked immunoassay (EIA). In theory it has many advantages compared with radioimmunoassays (RIA) not the least of which are increased reagent stability, simpler equipment, and no hazard from radioactivity. Recently EIAs for many hormones have been developed, including thyroxine, progesterone, cortisol, and estrogens.

IODOTHYRONINES

Over the past year several EIAs have been described for thyroxine (T4). That of Riesen et al (1) describes the adaptation of the EMIT assay to an automatic analyzer (Type Kem-O-Mat). Results compared with an RIA in 92 samples correlated well ($r = 0.90$). Day-to-day analysis gave similar coefficients of variation (CV) for the two methods (6 to 10%). Another evaluation, in 392 patients, of the EMIT assay by Finley et al (2) using an ABA-100 bichromatic automatic analyzer gave similar results ($r = 0.94$, CV = 6 to 8%). These methods are rapid; 24 to 32 samples can be analyzed in 15 minutes. A manual EIA for T4 has also been described by Schall et al (3). Thus, results of the EIA obtained by these authors were almost the same as those of an RIA, except that hemolysis was likely to interfere with the EIA since the final step is spectrophotometric.

Although many relatively rapid and efficient manual radioassay methods are available for thyroid function evaluation, the urge to automate such methods completely remains strong. When very large numbers of samples are run by radioassay, counting time becomes an important factor. A novel approach to this problem for T4 RIA has been used by Luner (4), who suggested that quantitative autoradiography of multiple-well plates would permit the assay of more than 3,000 samples a day. After the plates had been in contact with x-ray film overnight, the optical density of the spots was scanned. The CV was 11% for the film-spot optical-density technique compared with 5% for gamma counting - - some loss of precision, but one that might be acceptable for screening purposes.

A continuous-flow RIA system for many of the principal indices of thyroid function (T4, T3, thyroglobulin antibodies, T4 binding) has been described by Nye et al (5). Antibodies are covalently linked to a magnetizable solid-phase support so separation of bound and free antigen can be achieved by applying an external magnetic field. The system operated at a rate of 30 samples per hour. Precision was good (4 to 5% at midrange for between-assay CV). Further details of the T4 assay have been given by one of the authors in another article (6).

A second continuous-flow system is the "Southmead" system (7). Here the antibody is covalently linked to a solid-phase porous particle (Sepharose). The reagents (serum sample, labeled antigen, antibody) are mixed in an air-segmented stream, incubated if necessary, and then passed through a separation block where all but the large particles of antibody are filtered out through a porous filter and counted. The antibody stream may also be counted. Between-assay precision was satisfactory (CV = 8.8%). This system has the advantage of using conventional automatic equipment with the addition of the relatively simple separator block.

In the third and most intriguing ARIA II system, the same antibody, covalently linked to a solid support, is used repeatedly (8). The sample, or standard, flows over the antibody, binds to it, and is followed by the tracer (^{125}I-thyroxine), which attaches to the residual antibody sites. The unbound tracer passes into a gamma-counting flow cell and is counted. Then the bound tracer is eluted from the antibody and is counted (Fig. 1). The antibody is finally washed with buffer and is ready for the next sample. The whole cycle takes 2.2 minutes. This technique appears to be efficient and is economical since the same antibody is reused many times. Precision (CV = 6%) was good, and the results compared closely with those using conventional assays (r = 0.94). This technique could not, of course, be used in RIAs that require longer incubation times for sufficient binding of antigen to antibody.

Only a few authors have examined the use of urine rather than plasma samples for the investigation of thyroid function. Rogowski et al (9) have reconsidered this possibility quite thoroughly but feel it is of limited practical importance.

In recent years attention has been directed to the inactive metabolites of T4, that is, reverse T3 and diiodothyronine (T2) (1), and several methods have appeared for their measurement (11-16). Although studies of these thyronines are of considerable physiological interest, to date they have no clear-cut clinical application.

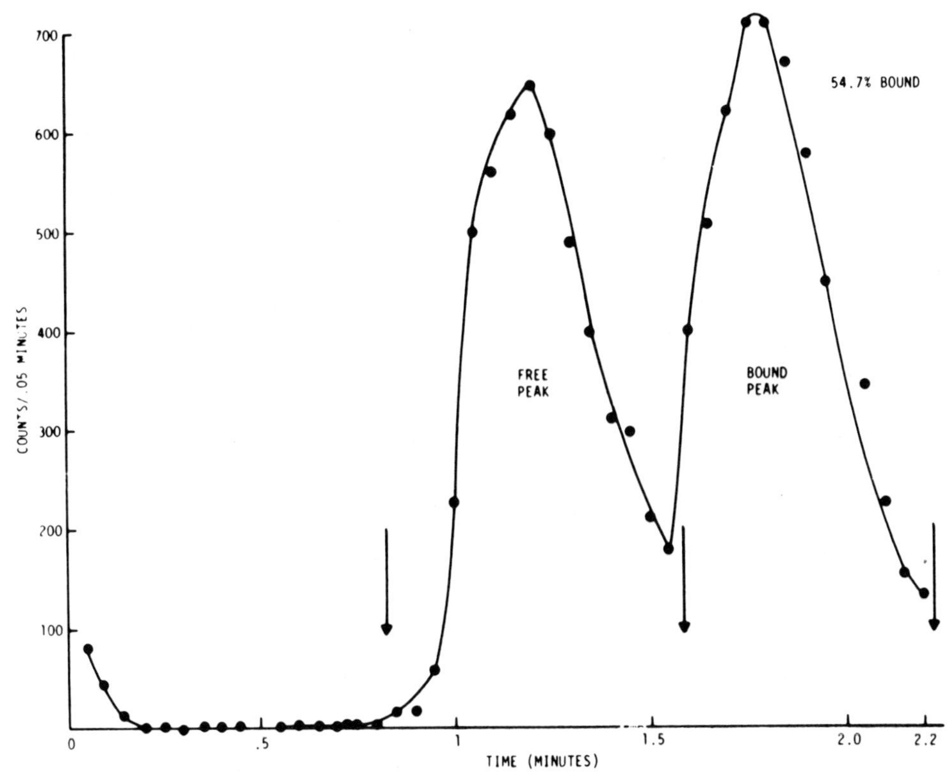

Figure 1. Typical ARIA elution curve, showing the bound and free peaks, the 2.2-minute cycle, the percentage bound, and the regions of count accumulation. (Reproduced with permission from Reese and Johnson, ref. 8)

While the "free T4 index," that is, the total T4 x T3 uptake ratio, is commonly used, little attention has been paid to the analogous entity, the "free T3 index." Sawin et al (17) have investigated the use of this index but it appears not to be as clinically useful as the free T4 index and is of little value as a routine test. This also applies to the measurement of thyroxine-binding globulin (TB) by RIA, since its measurement appears to have no advantage over that of the T3 uptake ratio to estimate thyroxine binding in serum (18,19).

For those laboratories that measure "free T4" directly, it is imperative that the tracer used be as pure as possible. Impurities may be present as by-products of preparation or may be generated in storage. One method of removing these, by binding known contaminants to specific antisera and then separating them on a column, has been described by Cooper and Burke (20).

A method for determining the level of free T4 in serum using polyacrylamide gel electrophoresis has been described (21). The results, while they correlated closely (r = 0.97), were about 40 times higher than those obtained by dialysis.

The clinical usefulness in preventing mental retardation resulting from neonatal hypothyroidism by screening of serum from newborn infants has been clearly established (22). Several techniques have been investigated, and perhaps the best approach at present is thyroxine screening with confirmation by TSH determination (23, 24). Most assays are carried out on dried blood spots collected on filter paper cards. Sadler et al (25) have used this technique in New Zealand, and as might be expected, they found that nonuniformity of the samples on the cards is the single greatest source of error. Jones and Oei (26) have also discussed this problem. Because TSH is more sensitive than T4 in detecting hypothyroidism, but technically more difficult and expensive, Miyai et al (27) have suggested using paired samples of infant blood for TSH determinations.

Ljunggren et al (28) have compared results of an RIA for T3 and T4 using capillary serum with those of an RIA using samples taken simultaneously from a cubital vein. An excellent correlation (r = 0.98) was obtained for both tests. Capillary sampling for T3 and T4 thus appears to be a useful technique, particularly for small children. It could also be used for neonatal screening but is less convenient than the filter paper cards.

Low T3 and high rT3 values have been reported in old age; however, Olsen et al (29) have found that this is an effect of disease rather than of age itself.

STEROIDS

While there is a growing tendency to develop EIAs to measure steroids most steroid assays are currently being done by RIA, using liquid scintillation counting of tritiated tracers. A major expense of these RIAs is that of the scintillator. Although almost all biochemical procedures have been scaled down to smaller and smaller volumes over the years, the size of counting vials for liquid scintillation counting has not changed since their introduction in the late 1950s. For most samples 10 ml of scintillator is used for each assay. This volume is quite unnecessary much of the time, as pointed out by Grothan and Sternberger (30), who use only 1 ml in a small tube (10 x 50 mm) placed in an adaptor within a conventional vial. Loss of counting efficiency was slight (42.8 ± 0.2% compared with 45.0 ± 0.1%).

Another approach to this problem, the substitution of iodinated tracers for tritiated tracers, is one which has been applied to virtually all the steroid hormones. While the much shorter half-life and greater biological hazard of the iodinated steroids are distinct disadvantages, iodinated compounds are much less expensive to count, and if large numbers of samples are being processed, this may be an important consideration.

Estrogens

The determination of estrogens in one form or another appears to be the most commonly performed endocrinological test in the latter half of pregnancy. This is perhaps surprising, since the physiological basis for its use is uncertain; its variability and expense are considerable; and its usefulness is poorly substantiated. Unfortunately at the present time, this test is the best method for assessing fetal well-being. In recent years several authors have questioned whether estrogen determinations are really helpful in managing pregnancy. Duenholter et al (31) have pointed out that there is little evidence to indicate that the reduction of fetal and neonatal mortality brought about by intervention based on estrogen assays exceeds the increased mortality that may result from inappropriate premature delivery based on misleading estrogen assays. Their prospective study consisted of two groups of high-risk pregnancies: in 315 cases (group A) the estrogen results (serum immunoactive estrogens) were reported; in 307 cases (group B) they were measured but were not reported. Nine perinatal deaths occurred in group A and 10 in group B. The authors concluded that estrogen determinations are not helpful in decreasing perinatal mortality. Bashore and Westlake (32), in an analysis of 321 high-risk pregnancies where unconjugated plasma estriol levels were determined, concluded that low plasma estriol values were not always ominous, and that therefore they should not be used as the sole criterion for early delivery. However they felt that estrogen levels might be helpful especially if measured daily. By contrast, Goebel (33) feels that estimation of serum estriol is a good test of fetal well-being. He measures both 24-hour total urinary estrogen excretion and serum estriol levels and has found an "excellent" correlation between the two.

However, plasma steroid levels in late pregnancy are extremely variable on a minute-to-minute basis. Buster et al (34) studied the variability in samples drawn every 5 or 15 minutes on one day and then again at similar times two days later from five pregnant women at 32 to 37 weeks' gestation. Levels of estradiol, estriol, progesterone, 17α-hydroxyprogesterone, and 16α-hydroxyprogesterone were measured in quadruplicate by radioimmunoassay. Although the experimental error of the assays themselves was reduced to negligible values by the quadruplicate determination, very large fluctuations were found in all the steroid levels, as shown in Figures 2 and 3. The single-sample 95% confidence limits were ± 36% for estradiol and ± 42% for estriol. To improve the clinical value of plasma measurements these authors suggested the use of multiple sampling every five minutes for half an hour or longer using an indwelling, heparinized, scalp-vein needle.

A study attesting to the variability of clinical estriol and total estrogen determinations in urine in late pregnancy was carried out by 27 clinical laboratories in Europe (35). Large interassay variations for the same sample were shown even in the same laboratory. Thus, urine determinations although they have the advantage of a prolonged collection period do not appear to be much more reliable than the serum estriol values. They also present the inherent problem of incomplete collection.

Because the collection of 24-hour urine samples is a nuisance and is often incomplete, some authors have suggested using the estrogen:creatinine ratio. This, of course, automatically adds the error of the creatinine determination to the already sizable one of the estrogen determination itself. Mukherjee et al (35), comparing results of first morning samples with those of either the previous 24-hour collection or the following 24-hour collection samples, found only a poor correlation (r = 0.45 and 0.39, respectively) even when corrected for creatinine (r = 0.54). Phillips et al (37) found that the estriol:creatinine (E/C) ratio in the early morning provided slightly more consistent results for paired within-week values (± 13.9 % SD for early morning E/C vs. ± 18.0% SD for 24-hour urine E/C and

± 18.9% for 24-hour estrogen excretion).

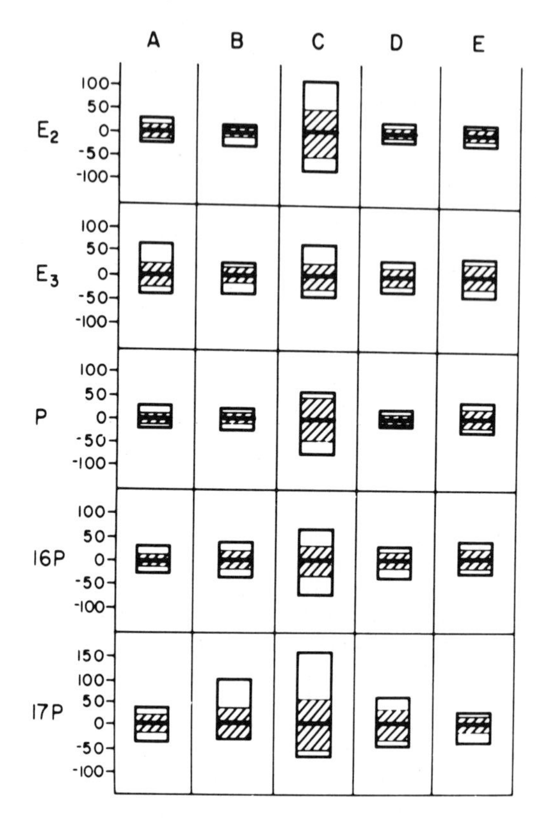

Figure 2. Variability expressed as % CV and the percentage range of all samples obtained from each subject. The mean concentration is expressed as the central solid line and plotted as %, the hatched area indicates the % CV; the open area the percentage range. E2: estradiol, E3: estriol, P: progesterone, 16P: 16α-hydroxyprogesterone, 17P: 17α-hydroxyprogesterone. Although subject C exhibited the greatest variability, each of the other subjects exhibited marked short-term variability for at least one steroid. (Reproduced with permission from Buster et al, ref. 34)

Distler et al (38) compared unconjugated plasma estriol, total plasma estriol, 24-hour urinary estriol, and the 24-hour urinary E/C ratio for their ability to monitor fetal distress. Their data suggested that the most reliable results were obtained using the unconjugated plasma estriol. In another study Kundu et al (39) compared the usefulness of sequential determinations of serum human placental lactogen, estriol, and estetrol levels for the assessment of fetal morbidity in 11 patients (6 diabetic, 5 toxemic). In this very limited series they felt that the estetrol levels fell earlier than the estriol levels and was therefore a more sensitive and reliable indicator of fetal morbidity in toxemic pregnancies. Placental lactogen levels remained low and were not considered to be helpful.

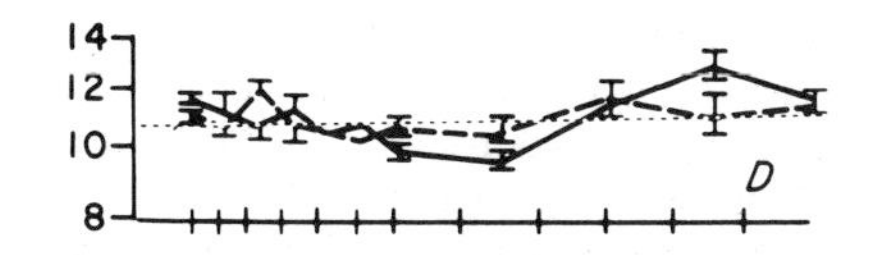

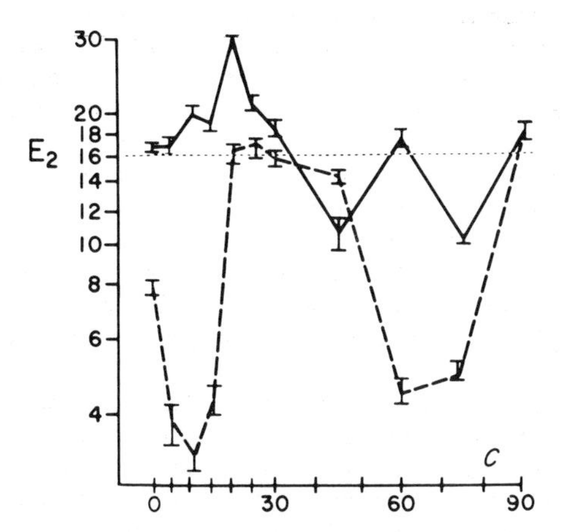

Figure 3. Examples of variability found in serum estradiol (E2). Maximum consecutive between-sample variability occurred on day 2 for subject C with an increase from 4.4 to 16.2 ng/ ml in a single 5-minute segment (a 3.7-fold increment). Subject C displayed the greatest overall variability and subject D the least overall variability. Each point is the mean of quadruplicate determinations ± SEM; the solid line is for day 1; and the dotted line is for day 2. (Reproduced with permission from Buster et al, ref. 34.)

In still another study the determination of total estrone (conjugated plus unconjugated) levels appeared to have no advantage over other estrogen assays (40). Determinations of estrone levels after an IV injection of dehydroepiandrosterone sulfate (DHAS) were not helpful.

Methodological variations have included the substitution of enzymes for radioactive tracer in EIAs for total plasma estrogens (41,42), the use of ^{125}I-estriol as tracer with polyethylene glycol (PEG) to separate free from bound antigen in an otherwise conventional RIA (43), and the direct RIA in plasma and urine of estriol-16-glucosiduronate, quantitatively the most important estrogen present in urine in late pregnancy (44).

However, no matter which estrogen method is used, the fact remains that the regulation of intrauterine steroid production is poorly understood, and the influence of obstetrical complications on these processes is even less well understood. Very low estrogen levels are observed in anencephaly, where in the latter half of pregnancy the adrenals atrophy presumably because of lack of fetal ACTH, and in placental sulfatase deficiency, where the enzyme that removes the sulphate from the major available precursor, DHAS, is largely lacking. Estriol itself has no discernible function in pregnancy, and infants with placental sulfatase deficiency are apparently normal. Since the precursors are stimulated by fetal ACTH, estrogen production might be expected to rise rather than fall in states of fetal stress. On the other hand, if large amounts of glucocorticoid are given to the mother, or if she receives large amounts of ACTH resulting in high cortisol levels, increased amounts of maternal corticoid cross the placenta and inhibit fetal ACTH production, thereby reducing estrogen excretion. Exogenous administration of synthetic steroids such as betamethasone is particularly effective in reducing estrogen production since they are not metabolized to as great an extent as cortisol when they cross the placenta. Altered maternal hepatic function and altered maternal renal excretion are other factors that may influence plasma and urinary levels.

While in general it may be said that low estrogen levels are often associated with fetal problems, interpretation of values in individual cases is extremely difficult. Since it appears that to be of any real use, plasma estrogen determinations must be done serially - - preferably daily rather than biweekly - - and preferably using multiple sampling techniques as suggested by Buster et al (34). It seems that the value of these tests must be more firmly established before they become a major commitment for a clinical laboratory.

Although they have negligible estrogenic activity, the role of the 2-hydroxyestrogens is coming under close scrutiny since it has been realized that they are quantitatively very important metabolites. New RIAs for serum (45) and urinary (46) catechol estrogens have been described. Gelbke et al (47) have reviewed the chemistry, formation, measurement, and significance of this group of steroids. Estetrol (2-hydroxyestriol) has been investigated as an index of fetal well-being (see above), but most studies suggest that estetrol determinations are about as useful as estriol determinations in ascertaining fetal well-being.

Levels obtained by RIA for estradiol compared well with those determined by mass fragmentography, suggesting that mass fragmentography may provide reliable reference methods for validating RIAs (48).

Cortisol

Since unbound cortisol is believed to be the physiologically active fraction, some authors have advocated measuring it rather than total cortisol. In effect both determinations must be performed since the unbound cortisol is usually derived as the percentage unbound multiplied by the total. Because of the considerable diurnal variation of total cortisol and the extra serum required for determining the unbound fraction - - 2 ml in a method recently published by Robin et al (49) - - the marginal increase in diagnostic accuracy is not worth the extra effort, particularly since urinary unconjugated cortisol is a much more accurate indicator of cortisol reaching the tissues.

Serum cortisol determinations are much more useful when used for function studies; for these total cortisol levels are entirely adequate. A particularly useful test is the short ACTH test. Lindholm et al (50) showed that not only is this test useful for establishing the ability of the adrenal cortex to produce cortisol but it also serves as a reliable indicator of the status of the hypothalamic-pituitary-adrenal axis (50). They compared the cortisol concentration at the peak after insulin injection and at 30 minutes after the administration of 250 μg of synthetic corticotropin. An excellent correlation was observed (r = 0.92) (Fig. 4). The ACTH test is preferred over the insulin hypoglycemia test, since there are no hazards or side effects, the peak is more predictable, and, for the latter reason, only two samples are required (0 and 3 minutes). The short test is also preferable to longer ACTH tests, such as the commonly used eight-hour ACTH infusion for three days or the four-hourly IM injection of ACTH for 48 hours. These longer tests require hospitalization, have some risk for the patient, are uncomfortable, and rarely give more information than does the short test, which can readily be used for outpatients.

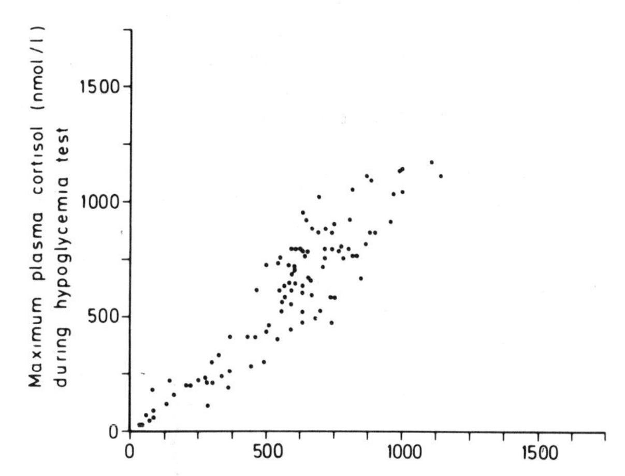

Figure 4. Relationship between peak plasma cortisol concentration during insulin-induced hypoglycemia and at 30 minutes after IV injection of 250 μg of corticotropin 1 - 24. A normal response in both tests is a plasma concentration of 500 nmole/l (18 μg/100 ml) or more. (From ref. 50.)

The assay of cortisol itself can be done simply and rapidly without extraction or chromatography for most clinical purposes either by radiotransinassay (RTA) or by RIA (52-54). Although greater specificity is claimed for RIA procedures, most are quite inadequate for the determination of cortisol in the sera of patients with congenital adrenal hyperplasia or in newborn infants sera that contain large amounts of unusual steroids (e.g. 17α-hydroxyprogesterone and 21-desoxycortisol), which cross-react with most antisera to cortisol. The most commonly used antiserum at present is to the 21-hemisuccinate conjugate, which invariably cross-reacts with 21-sulfates and 21-desoxycortisol. Since not all the steroids in serum and urine during pregnancy and in neonatal serum and urine have been identified, it is impossible to validate the results of RIAs for cortisol without chromatography (51,55).

The same problem of specificity applies to the recently described EIAs for cortisol, such as that of Kobayashi et al, which also uses an antiserum conjugated at the 21 position (56). Again, in many RIA procedures iodinated tracers may be substituted for tritiated ones (54,58,59). Although sampling for cortisol is usually done in serum or urine, it is also possible to use saliva (57), which may be related more closely to unbound cortisol. Such samples might be particularly useful in studying children.

Furuya et al (60) have modified the Porter-Silber method for determining 17-hydroxycorticoid (17OHCS) levels in urine. However, with the availability of assays for unconjugated cortisol, which is a more sensitive indicator of glucocorticoid production and is unaffected as are the 17OHCS by obesity and liver disease, little seems to be gained by measuring them at all.

Androgens

The use of iodinated testosterone has been investigated by Hampl et al (65) and has been found to be a satisfactory alternative to the tritiated tracer for RIA. Corker and Davidson (67) claimed to have developed an RIA for testosterone without chromatography; however, the values in women are "seriously overestimated" when assayed without chromatography so this method is quite unsatisfactory for routine use.

Because of the problem of specificity for testosterone, Rao et al (68) have turned to the use of the 19-O-carboxymethyl ether derivative of testosterone conjugated to bovine serum albumin. The antiserum to this hapten raised in rabbits had excellent specificity, thus allowing measurement in extracts from both men and women without prior chromatography. Cross-reaction with dihydrotestosterone was reduced to 7%.

Since the serum binding protein for estradiol and androgens (TeBG) varies inversely with the state of androgenicity, the evaluation of unbound testosterone in serum is advantageous. Wiest et al (66) have compared two procedures for this purpose: one using dextran-coated charcoal to separate protein-bound from unbound ligand and othe other using DEAE-cellulose filtration to isolate the TeBG-ligand complex directly. Although the authors felt that either method was suitable for routine clinical use they preferred the DEAE-cellulose method because of "its greater ease of manipulation."

Progesterone

Progesterone determinations are being used more and more widely to detect ovulation. The determination of progesterone levels, however, still requires preliminary extraction of the sample to remove the strong endogenous binding by plasma proteins. One RIA, that of Khadempour et al (61), which uses unextracted plasma, cannot distinguish levels below 3 ng/ml and is therefore only useful for samples during pregnancy and in the luteal phase. Since between-assay precision was rather poor (14.8%) this assay has little to recommend it.

Iodinated progesterone has been used as tracer by Scott et al (62) and by Allen and Redshaw (63) with results similar to those obtained when using tritiated tracer. Although decreased expense is one of the advantages claimed, this is only true if a large number of samples are being processed, and this is not the case in most laboratories. An EIA has been described for progesterone by Joyce et al (64) which has a sensitivity of 10 pg and appears to be suitable for clinical use.

Aldosterone

Al-Dujaili and Edwards (69) have described direct RIA for aldosterone in serum that allows assay without extraction or chromatography. Aldosterone linked to ^{125}I-histamine was used as ligand.

CATECHOLAMINES

Measurement of plasma and urinary catecholamines is important in the detection of pheochromocytoma and neuroblastoma, but their estimation still remains rather difficult. The available methods include fluorometry, radioenzymic assay (REA), and high-pressure liquid chromatography (HPLC) with electrochemical detection.

The widely used fluorometric methods are relatively insensitive. A system for measuring urinary catecholamines and their metabolites combining semiautomated fluorometric quantitation of dopamine (DA), epinephrine (E), and norepinephrine (NE), and manual methods for determining levels of homovanillic acid, normetanephrine (NMN), metanephrine (MN) and vanilmandelic acid (VMA) have been described by De Schaepdryver and Moerman (70).

A procedure for the preparation of 5 to 20-ml samples of urine for fluorometric assay of 3,4-dihydroxyphenylalanine (DOPA), 3,4 - dihydroxyphenylacetic acid (DOPAC), E, NE, and DA has been outlined by Dalmaz and Peyrin (71); purification on an aluminum oxide column precedes separation by ion-exchange chromatography and ether isolation of DOPAC from DOPA.

The radioenzymic methods are relatively more sensitive, but most are rather time consuming and cumbersome. They depend on the attachment of a radiolabeled methyl group enzymatically to the catechol. A single-isotope enzymic derivative assay was compared with a double-isotope technique by Evans et al (73). The single-isotope assay required half the time (1.5 days) and one-two hundredth of the sample volume, but it was judged to be equally reliable. A very sensitive REA for NE in 0.1 ml of deproteinized plasma has been described by Falke et al (72). As little as 2.5 pg of NE could be detected. Samples were added to partly purified phenylethanolamine-N-methyltransferase (PNMT) without prior extraction. A rapid REA for E, NE, DA, and DOPAC has been described by Saller and Zigmond (74). Deproteinized plasma (10 µl) was incubated with enzyme for 30 minutes and chromatographed on thin-layer plates, after which the spots were counted. With this technique it was possible to process more than 100 samples a day. Sensitivity was 15-25 pg.

Another approach is based on using HPLC with electrochemical detection. Hallman et al (75) have provided information on preliminary studies of this technique for human plasma, using 0.75 ml samples. After extraction on alumina, the catecholamines were desorbed and passed through a strong cation-exchange resin, and the effluent was passed over a detector electrode. An analogous procedure for separation of catecholamines and their O-methylated metabolites has been described by Borchardt et al (76). These methods appear very promising.

A gas chromatographic high-resolution mass fragmentographic method for E and NE in human plasma has been reported by Jacob et al (77), and it might be useful as a reference method.

VITAMIN D

To become biologically active, vitamin D must first be hydroxylated in the liver to 25OH-D and then in the kidney to 1,25-dihydroxy D, the most potent form known (78). The major metabolite in plasma is 25OH-D. The naturally occurring binding proteins of rats and humans provide good binding agents for 25OH-D, since they have affinity constants about 10^{10} l/mole for both 25OH-D2 and 25OH-D3. However, since there is some interference by other substances in serum a chromatographic step must be included (79).

Recently an attempt has been made to produce an antibody with specificity for 1,25-diOH-D3, the more potent form of the vitamin. 1,25-diOH-25-hemisuccinate-D3 was coupled to bovine serum albumin for use as an immunogen (80). However, other sterols cross-reacted, so preliminary separation using Sephadex LH-20 chromatography and HPLC was necessary, and the procedure required 10-20 ml of plasma.

The most sensitive assay for 1,25-diOH-D is a bioassay described by Stern et al (81). This assay measures as little as 1 pg of 1,25-diOH-D.

This field is currently a very active one and promises to yield much clinically applicable material in the near future.

REFERENCES

1. Riesen WF, Muacevic B, Jaggi M: *J Clin Chem Clin Biochem* 16:387, 1978.

2. Finley PR, Williams RJ: *Clin Chem* 24:165, 1978.

3. Schall RF Jr, et al: *Clin Chem* 24:1801, 1978.

4. Luner SJ: *Anal Biochem* 84:332, 1978.

5. Nye L, et al: *Clin Chim Acta* 87:307, 1978.

6. Anderson MJ: *Med Lab Sci* 35:173, 1978.

7. Ismail AA, West PM, Goldie DJ: *Clin Chem* 24:571, 1978.

8. Reese MG, Johnson LR: *Clin Chem* 24:342, 1978.

9. Rogowski P, Siersbaek-Nielsen K, Hansen JM: *Acta Endocrinol (Kbh)* 87:525, 1978.

10. Burman KD: *Metabolism* 27:615, 1978.

11. Faber J, et al: *Acta Endocrinol (Kbh)* 87:313, 1978.

12. Faber J, et al: *Clin Endocrinol* 9:279, 1978.

13. Grussendorf M, Hufner M: *Clin Chim Acta* 88:81, 1978.

14. Laurberg P: *Scand J Clin Lab Invest* 38:537, 1978.

15. Burman KD, et al: *J Clin Endocrinol Metab* 47:1059, 1978.

16. Premachandra BN: *J Clin Endocrinol Metab* 47:746, 1978.

17. Sawin CT, et al: *Ann Intern Med* 88:474, 1978.

18. Lecureuil M, et al: *Clin Chim Acta* 87:373, 1978.

19. Bisset JP, et al: *Nouv Presse Med* 7:2841, 1978.

20. Cooper E, Burke CW: *Clin Chim Acta* 86:51, 1978.

21. McDonald LJ, Robin NI, Siegel L: *Clin Chem* 24:652, 1978.

22. Dussault JH, Coulombe P, Laberge C: *J Pediatr* 86:670, 1975.

23. Walfish PG: *Lancet* 1:1208, 1976.

24. Bell AB, Coleman LH: *Clin Chem* 24:1755, 1978.

25. Sadler WA, Lynskey CP, Legge M: *Aust Paediatr J* 14:154, 1978.

26. Jones K, Oei T: *Clin Chem* 24:836, 1978.

27. Kiyoshi M, et al: *J Clin Endocrinol Metab* 47:1028, 1978.

28. Ljunggren JG, et al: *Scand J Clin Lab Invest* 38:593, 1978.

29. Olsen T, et al: *J Clin Endocrinol Metab* 47:1111, 1978.

30. Grothan HE Jr, Steinberger E: *Clin Chem* 24:1183, 1978.

31. Duenhoelter JH, Whalley PG, MacDonald PC: *Amer J Obstet Gynecol* 125:889, 1976.

32. Bashore RA, Westlake JR: *Amer J Obstet Gynecol* 128:371, 1978.

33. Goebel R: *Hormone Res* 9:339, 1978.

34. Buster JE, et al: *J Clin Endocrinol Metab* 46:907, 1978.

35. Huis int Veld LG, et al: *J Clin Chem Clin Biochem* 16:119, 1978.

36. Mukherjee TK, et al: *Obstet Gynecol* 51:166, 1978.

37. Phillips SD, et al: *Clin Chim Acta* 89:71, 1978.

38. Distler W, et al: *Amer J Obstet Gynecol* 130:424, 1978.

39. Kundu N, et al: *Obstet Gynecol* 52:513, 1978.

40. Axelsson O, Nillsson BA, Johansson EFB: *J Steroid Biochem* 9:1119, 1978.

41. Bosch AM, et al: *Clin Chim Acta* 38:59, 1978.

42. Exley D, Abuknesha R: *FEBS Lett* 91:162, 1978.

43. Liedtke RJ, et al: *Clin Chem* 24:1100, 1978.

44. Wright K, et al: *Amer J Obstet Gynecol* 131:255, 1978.

45. Ball P, et al: *Steroids* 31:249, 1978.

46. Chattoraj SC, et al: *Steroids* 31:249, 1978.

47. Gelbke HP, Ball P, Knuppen R: *Adv Steroid Biochem Pharmacol* 6:81, 1977.

48. Zamecnik J, Armstrong DT, Green K: *Clin Chem* 24:627, 1978.

49. Robin P, Predine J, Milgrom E: *J Clin Endocrinol Metab* 46:277, 1978.

50. Lindholm J, et al: *J Clin Endocrinol Metab* 47:272, 1978.

51. Murphy BEP: *J Clin Endocrinol Metab* 41:1050, 1975.

52. Ellis G, Morris R: *Clin Chem* 24:1954, 1978.

53. Connolly TM, Vecsei P: *Clin Chem* 24:1468, 1978.

54. Seth J, Brown LM: *Clin Chim Acta* 86:109, 1978.

55. Sulon J, et al: *J Steroid Biochem* 9:671, 1978.

56. Kobayashi Y, et al: *Steroids* 32:137, 1978.

57. Walker RF, Riad-Fahmy D, Read GF: *Clin Chem* 24:1460, 1978.

58. Brock P, et al: *Clin Chem* 24:1595, 1978.

59. Hampl R, et al: *J Steroid Biochem* 9:771, 1978.

60. Furuya E, Graef V, Nishikaze O: *Anal Biochem* 90:644, 1978.

61. Khadempour MH, Laing I, Gowenlock AH: *Clin Chim Acta* 82:161, 1978.

62. Scott JZ, et al: *Steroids* 31:393, 1978.

63. Allen RM, Redshaw MR: *Steroids* 32:467, 1978.

64. Joyce BG, et al: *Clin Chem* 24:2099, 1978.

65. Hampl R, et al: *J Clin Chem Clin Biochem* 16:279, 1978.

66. Wiest WG, et al: *Amer J Obstet Gynecol* 130:321, 1978.

67. Corker CS, Davidson DW: *J Steroid Biochem* 9:373, 1978.

68. Rao PN, et al: *J Steroid Biochem* 9:539, 1978.

69. Al-Dujaili AS, Edwards CR: *J Clin Endocrinol Metab* 46:105, 1978.

70. De Schaepdryver AF, Moerman EJ: *Clin Chim Acta* 84:321, 1978.

71. Dalmaz Y, Peyrin L: *J Chromatogr* 145:11, 1978.

72. Falke HE, Punt R, Birkenhager WH: *Clin Chim Acta* 89:111, 1978.

73. Evans MI, Halter JB, Porte D Jr: *Clin Chem* 24:567, 1978.

74. Saller CF, Zigmond MJ: *Life Sci* 23:1117, 1978.

75. Hallman H, et al: *Life Sci* 23:1117, 1978.

76. Borchardt RT, Hegazi MF, Schowen RL: *J Chromatogr* 152:255, 1978.

77. Jacob K, et al: *J Chromatogr* 146:221, 1978.

78. DeLuca HF: *Amer J Med* 57:62, 1974.

79. Aksnes L: *Scand J Clin Lab Invest* 38:677, 1978.

80. Clemens TL, et al: *Clin Sci Mol Med* 54:329, 1978.

81. Stern PH, et al: *J Clin Endocrinol Metab* 46:891, 1978.

11 Biochemistry of Pituitary Hormones

Janakiraman Ramachandran
Choh H. Li

This chapter will cover the significant developments in the biochemistry of the adenohypophyseal hormones during the past year. The hormones of the anterior pituitary gland fall into three groups on the basis of structural similarities. Corticotropin (ACTH), the melanotropins (α-MSH and β-MSH), lipotropin (β-LPH), and endorphin (β-EP) comprise the first group of closely related flexible polypeptides. The discovery that these five hormones are derived from a single common precursor is perhaps the most interesting development of the past year and will be considered in detail in this chapter. Somatotropin, or growth hormone (GH), and prolactin (PRL) exhibit considerable structural homology and are compact globular proteins possessing a characteristic three-dimensional structure. The placental hormone, chorionic somatomammotropin (CS), is closely related to this second group of pituitary hormones. The gonadotropins, lutropin (LH), and follitropin (FSH), as well as thyrotropin (TSH), are closely related glycoproteins forming the third group.

The glycoprotein hormones are composed of two nonidentical subunits of comparable size. The α-subunit is common to the three pituitary glycoprotein hormones. The human chorionic gonadotropin (hCG) is also a glycoprotein composed of two subunits and is closely related to LH.

HORMONES OF GROUP 1

ACTH

The amino acid sequences of the hormone isolated from ostrich pituitary glands (1) and from three species of whale (2) have been elucidated. The primary structure of whale ACTH is identical to the structure of the human hormone. The first 26 residues of ostrich ACTH are also identical to the human sequence, except for the replacement of lysine at position 15 by arginine in the avian hormone. There are six differences between ostrich and human ACTH in the region 27-39. These results further confirm that the biological activity of this hormone is derived from the highly conserved amino terminal half of the molecule (3).

It was recently recognized that ACTH inhibits the proliferation of normal rat adrenocortical cells in primary culture, rather than stimulating the growth of these cells as commonly believed (4). That ACTH is not the adrenal mitogenic factor has now been firmly demonstrated by investigating the effects of ACTH antiserum on adrenal growth and function in intact rats (5). Administration of antiserum to ACTH in male Sprague-Dawley rats ranging in age from 3 to 12 weeks for a period of 6 to 10 days had no effect on adrenal weight but caused a highly significant decrease in serum corticosterone levels. The ability of adrenocortical cells isolated from antiserum-treated animals to produce corticosterone in response to exogenous ACTH was also significantly lower than that of adrenocortical cells derived from normal rabbit serum-treated animals. The administration of ACTH antiserum to unilaterally adrenalectomized animals did not prevent the increase in weight, cell number, and DNA content of the remaining adrenal gland. However, the steroidogenic capacity of adrenocortical cells from antiserum-treated animals was significantly lower in comparison with the capacity of the cells obtained from the adrenals of unilaterally adrenalectomized rats treated with normal rabbit serum or from unoperated controls. Quantitative evaluation of the ultrastructural changes by morphometry showed that neutralization of endogenous ACTH with antiserum resulted in significant decreases in the absolute volumes of mitochondria, cytoplasmic matrix, and nuclei, although the changes were not as large as after hypophysectomy. These results and earlier studies of the action of ACTH on the proliferation of adrenocortical cells in monolayer culture show that the trophic action of ACTH is concerned solely with the induction and maintenance of the steroidogenic capacity of the adrenocortical cells.

The metabolism of ACTH has been investigated by administering tritiated hormone to adult male rats (6). Synthetic human ACTH, tritium labeled by reductive methylation (7), was injected intravenously, and the disappearance of radioactivity, bioactivity, and immunoreactivity was determined at various times. Total radioactivity was cleared from plasma in a multiphasic manner, and the disappearance of immunoreactivity closely paralleled that of radioactivity. In contrast, bioactivity decreased at a much more rapid rate. Gel exclusion chromatography of plasma showed that the major component up to 30 minutes after injection chromatographed with control radiolabeled ACTH. This component retained essentially full immunoreactivity, but its bioactivity decreased below the assay sensitivity within 10 minutes. These results suggest that the biotransformation of ACTH must depend on a rather subtle structural alteration probably involving the removal of one or a few N-terminal residues or side-chain modification. Maximum uptake of the radioactive hormone occurred in the kidneys and liver, with muscle and adipose tissues also accounting for appreciable removal of radioactivity. Urinary radioactivity never exceeded more than 2% of the injected dose up to several hours after injection and seemed to represent mainly oligopeptide fragments with no appreciable immunoreactivity.

Corticotropin-Inhibiting Peptide (CIP)

Sixteen years ago, Pickering et al (7) reported the isolation of a basic peptide from sheep pituitary extracts, the amino acid composition of which corresponded to residues 7-38 of ACTH. This peptide was recently isolated from human pituitary glands and characterized (8). The amino acid sequence of this peptide was found to be identical to residues 7-38 of human ACTH. This peptide, named *corticotropin-inhibiting peptide* (CIP), has been synthesized and found to be devoid of adrenal stimulating activity (8,9). CIP was found to act as a competitive inhibitor of the actions of ACTH on isolated rat adrenocortical cells. The presence of CIP in the pituitary raises the possibility that enhanced conversion of ACTH or its precursor to CIP may be responsible for clinical states associated with hypoadrenal function.

Melanotropins (MSH)

The pigmentary action of the melanotropins is clearly important in lower vertebrates. With the evolution of hair in the mammal, adaptive changes in skin color are far less important, and it is suspected that the melanotropins may have lost their significance as pigmentary hormones. There is no evidence that MSH has any physiological role as a pigmentary hormone in humans. In fact, the adult human pituitary has apparently lost the ability to produce MSH (10). For the past several years investigators have tried to elucidate the extrapigmentary actions of the melanotropins, which may be physiologically relevant in mammals. Analysis of the structure-activity relationships of a number of peptide analogs of ACTH and MSH has indicated that α-MSH may function as a lipolytic hormone in rabbits (11). Thody, Shuster, and colleagues (12-14) have shown that α-MSH stimulates the sebaceous glands of mammals and propose that this may be an important function of the hormone.

During the past two years evidence has accumulated to suggest that α-MSH may play an important role during fetal life. Silman et al (15) found immunoreactive α-MSH in large amounts in human fetal pituitary. Challis and Torosis (16) reported that α-MSH stimulated cortisol secretion by the fetal adrenal gland of the rabbit at a time when ACTH was relatively ineffective, whereas in the newborn rabbit ACTH was potent and α-MSH was ineffective. Immunoreactive α-MSH has been detected in the pituitary (17) and in the plasma (18) of pregnant women. Significant amounts of α-MSH were detected in pituitary extracts of rhesus monkey fetuses, using gel filtration and radioimmunoassay (19). It was also reported that α-MSH stimulates fetal growth in rats (20). These results strongly suggest that α-MSH may regulate the development and function of the fetus.

Lipotropin (LPH)

Lipotropin was discovered fortuitously during the investigation of a simplified isolation procedure for ACTH (21). Two peptides exhibiting lipolytic activity were isolated from ovine pituitaries and were designated β-LPH and γ-LPH. A linear polypeptide, β-LPH, consists of 91 amino acid residues. The 58-residue γ-LPH is identical to the sequence region 1-58 of β-LPH. Residues 41-58 in LPH correspond to the sequence of β-MSH. The 51-residue peptide of ovine β-LPH corresponding to residues 41-91 (22), as well as the 91-residue peptide (23), has been synthesized by the solid-phase procedure. Synthetic ovine β-LPH was found to be identical to the natural peptide in a large number of physicochemcial and biological properties.

The physiological role of LPH has remained obscure. Recent evidence strongly suggests that the biological function of LPH may be that of a prohormone for the opioid peptide, β-endorphin. However, Liotta et al (24) report that β-LPH is the predominant opioid-like peptide in the human pituitary and the anterior lobe of rat pituitary. The concentrations of β-LPH were comparable to those of ACTH on a molar basis. In human subjects, acute anterior pituitary stimulation using either insulin-induced hyperglycemia or vasopressin administration was associated with increased plasma β-LPH and ACTH levels. At the time of peak concentrations, significant amounts of β-endorphin were not detected (24). Jeffcoate et al (25) found β-LPH in the plasma of normal persons and both β-LPH and γ-LPH in the plasma of patients with disease of the pituitary-adrenal system (and with raised levels of ACTH). The presence of significant amounts of intact β-LPH in the circulation raises the possibility that LPH may have some function besides that of a prohormone for β-endorphin and β-MSH.

Endorphins (EP)

In December 1975, Hughes, Kosterlitz, and collaborators reported the isolation, structure, and synthesis of two similar pentapeptides from porcine brain with potent opiate agonist activity (26). They recognized that the entire peptide sequence of one of these peptides, namely, methionine enkephalin, is present in residues 61-65 of the β-LPH molecule. In search of β-LPH from camel pituitaries, Li and Chung (27) isolated a 31-residue peptide, which was found to correspond to residues 61-91 of β-LPH. This peptide was found to be more potent than the enkephalins in its analgesic properties, as well as in its affinity for brain opiate receptor, and was named β-endorphin (β-EP). β-EP has been isolated from porcine (28,29), ovine (30), bovine (31), rat (32), and human (33) pituitaries. The structure of β-EP appears to be highly conserved (Fig. 1). Both camel (34) and human (35) β-EP have been synthesized, in addition to a large number of analogs (36-39). Analysis of the structure-function relationships, using the guinea pig ileum assay, showed that potency increases as the peptide chain length is increased from that of enkephalin to endorphin. β_h-EP-(1-29) was as active as β_h-EP in inhibiting electrically stimulated contractions of the guinea pig ileum *in vitro*. However, all 31 amino acids were found to be necessary for eliciting full analgesic activity *in vivo* as measured by the tail flick and hot plate tests. β_h-EP-(1-29) was only 20% as active as β_h-EP in producing analgesia.

```
                              5                   10
Human: H-Tyr-Gly-Gly-Phe-Met-Thr-Ser-Glu-Lys-Ser-
                        15                  20
          Gln-Thr-Pro-Leu-Val-Thr-Leu-Phe-Lys-Asn-
                        25                      31
          Ala-Ile-Ile-Lys-Asn-Ala-Tyr-Lys-Lys-Gly-Glu-
                                                  OH

             Porcine:        Val           His          Gln-OH

Camel, Ovine, Bovine:        Ile           His          Gln-OH
```

Figure 1. Amino acid Sequence of β-EP from various species.

Intraventricular administration of β_h-EP in three human patients resulted in relief of intractable pain (40). A single dose of 200 μg of the peptide was effective, with no observable side effects. The pain relief thus produced was accompanied by alterations of the acute pain threshold as measured by the thermal dolorimeter. Both effects were reversed by naloxone. β_h-EP administered intravenously to patients with cancer pain produced good analgesia and mild improvement in mood (41). In a patient undergoing withdrawal from methadone complete relief of symptoms was reported 25 minutes after initiating an infusion of β_h-EP (440 μg/kg/30 min). Similarly, severe abstinence symptoms were effectively suppressed after the intravenous administration of β_h-EP (80 μg/kg) to patients with heroin withdrawal syndrome (42).

Biosynthesis of ACTH-MSH-LPH-EP

The most significant development in the biochemistry of the group 1 hormones in the past two years is the recognition that all the hormones of this group are derived from a single high-molecular-weight precursor. High-molecular-weight forms of ACTH were first detected in the blood of patients with ACTH-producing tumors (43) and subsequently in rat and mouse pituitary extracts (44). Gel filtration of extracts of mouse pituitaries or clonal mouse pituitary tumor cells (AtT-20/D-16v) under strongly denaturing conditions (6M guanidine hydrochloride) and analysis by sodium dodecyl sulfate-polyacrylamide gel electrophoresis (SDS-PAGE) revealed precursors of ACTH with molecular weights of 31,000 daltons (31K) and 23,000 daltons (23K) (45). Subsequently, it was found that the 31K protein is the precursor of both ACTH and β-LPH (46). When mRNA from AtT-20 cells is translated in a reticulocyte cell-free protein synthesizing system and the immunoprecipitates obtained with ACTH or endorphin antisera are analyzed by SDS-PAGE, only one labeled protein with a molecular weight of 28,500 daltons (28.5K) is found (47,48). Similar results were obtained with beef pituitary mRNA (49).

The common precursor of ACTH and EP has also been isolated from rat pituitaries, and the name *proopiocortin* has been proposed for this protein (50). The structure of proopiocortin has been analyzed in detail (48,51). The entire structure of proopiocortin can be accounted for by three smaller peptides: β-LPH, ACTH, and a glycopeptide of unknown function (referred to as 16K fragment). β-LPH (and therefore β-EP) is located in the carboxyl terminal of the precursor, ACTH is in the middle, and the 16K fragment forms the amino terminal. The complete amino acid sequence of proopiocortin of bovine origin has now been deduced from the nucleotide sequence of a 1,091-base pair cDNA insert coding for bovine proopiocortin determined by Nakanish et al (52). Inspection of the amino acid sequence (265 residues) reveals a number of interesting features. The molecule appears to consist of four repetitive units based on the core sequence Met-Glu-His-Phe-Arg-Trp-Gly of ACTH and MSH. These units are separated by paired basic residues. ACTH and β-LPH exist as contiguous peptides at the carboxyl terminal and are separated by the Lys-Arg sequence. Thus, in common with other protein precursors (proinsulin, proPTH, etc.), the biologically active sequences in proopiocortin are flanked by paired basic residues at which proteolytic processing takes place. Nakanishi et al (52) point out that the proopiocortin sequence contains a new MSH sequence flanked by paired basic residues in the amino terminal region and have termed it γ-MSH. The arrangement of the various peptides in proopiocortin is shown schematically in Figure 2.

The steps in the processing of proopiocortin to ACTH and EP have been deduced from analysis of the precursor from AtT-20 cells in pulse chase and continuous labeling experiments using radioactive amino acids and sugars (53). The 31K protein has been resolved into three glycoproteins of molecular weights, 29K, 32K, and 34K. The three forms have the same peptide backbone but differ from each other in the carbohydrate content. The nascent polypeptide of molecular weight 28.5K is glycosylated to the 29K form, which is converted to 32K and 34K forms by the addition of carbohydrate. The 29K form is also converted directly to ACTH (4.5K) and β-LPH (11.5K). The 32K form of proopiocortin is processed proteolytically to yield β-LPH and the 23K ACTH intermediate, which is further processed to yield the glycosylated form of ACTH (13K). In the anterior lobe of the pituitary proopiocortin is converted primarily to ACTH and β-LPH. In the intermediate lobe, ACTH and β-LPH formed from proopiocortin, are further processed by proteolysis at paired basic residues to yield α-MSH and corticotropin-like intermediate lobe peptide (CLIP) from ACTH and β-MSH and β-EP from β-LPH (53).

Both the ACTH (54) and the EP (55) content of AtT-20 cells were reduced by glucocorticoids. Nakanish et al (56,57) showed that the intracellular level of the mRNA coding for proopiocortin is depressed by glucocorticoids in pituitaries of adrenalectomized rats.

HORMONES OF GROUP 2

Somatotropin (GH)

The regeneration of growth-promoting activity by the noncovalent recombination of inactive fragments of human GH has been one of the most exciting developments in the peptide hormone field. Following the observation that treatment with human plasmin does not cause any change in the biological properties of hGH (58,59) it was shown that plasmin removed a hexapeptide corresponding to residues 135-140 of hGH (60). The plasmin-modified hormone is held together by the disulfide bridge between residues 53 and 165. Reduction and alkylation of the plasmin-treated hormone results in the formation of two peptides corresponding to residues 1-134 and 141-191 of the native hormone (60). The purified fragments have very little biological activity. However, recombination of two fragments results in the formation of a noncovalent complex that has full biological activity (61). The recombinant molecule is identical to the plasmin-treated native hormone in its chromatographic elution position, circular dichroism spectra, and immunochemical properties. These findings opened the way for the semisynthesis of hGH. The combination of the synthetic peptide corresponding to the carboxyl terminal fragment (residues 141-191) with the natural 1-134 fragment resulted in the restoration of full biological activity and immunoreactivity (62,63). Synthetic analogs containing norleucine in place of methionine and alanine in place of cysteine in the 141-191 fragment and in the 145-191 fragment complemented with natural 1-134 peptide to generate full growth-promoting activity (64,65). Plasmin digestion of hGH after reduction and alkylation of the hormone results in more extensive cleavage of

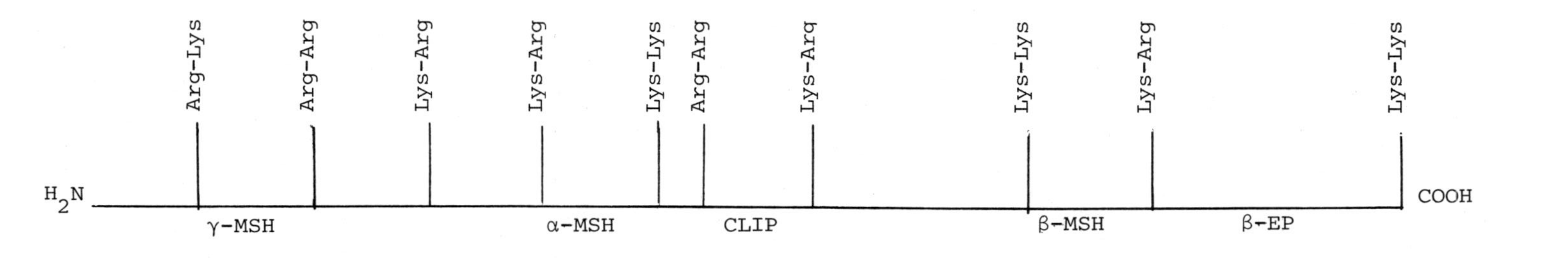

Figure 2. Structure of proopiocortin. The relative positions of the hormones of Group 1 are shown. The biologically active sequences are flanked by paired basic residues. (Adapted from S. Nakanishi et al, ref. 52.)

the peptide chain, compared with its action on native hGH (66,67). In addition to hGH 1-134, hGH 26-41, HGH 95-134, and hGH 42-134 are also formed.

Human chorionic somatomammotropin (hCS), the placental hormone is very similar to hGH in its primary structure (96% homology) yet has very low growth-promoting activity. Plasmin treatment of hCS results in the removal of a heptapeptide from the hormone (68). Reduction and alkylation of plasmin-treated hCS gave two fragments corresponding to residues 1-133 and 141-191 of hCS. The fragments were devoid of any biological or immunochemical activity (68). Complementation of the carboxyl terminal fragment of hCS (residues 141-191) with the amino terminal fragment of hGH (residues 1-134) yielded a product with 50% growth-promoting activity and nearly full immunoreactivity (69). The other hybrid, made up of hCS 1-133 and hGH 141-191, had very low growth-promoting activity ($< 10\%$). The immunochemical and receptor-binding properties of the two hybrids have been analyzed in greater detail (70). It was concluded that the determinants for binding hGH to hepatic GH receptors as well as to lactogenic receptors are present in the hGH 1-134 fragment. The antigenic determinants for both a monospecific antiserum to hCS and a monospecific antiserum to hGH also are on the amino terminal fragments of their respective antigens.

The dissociation kinetics of reduced, alkylated, plasmin-modified hGH has been studied in detail (71). The major products of the dissociation have been shown to be the monomeric form of hGH 141-191 and a trimeric form of hGH 1-134. The carboxyl terminal fragment appears to be a random coil, whereas the amino terminal fragment retains a considerable degree of secondary and tertiary structure (71).

The amino terminal heterogeneity, which is generated in bovine GH during its extraction from mildly acidic pituitary homogenates, has been traced to the action of a tripeptidyl amino peptidase in the pituitary, which is optimally active at pH 4 (72). This enzyme was found to be distinct from cathepsin D.

Significant progress has also been made toward the synthesis of GH in bacteria by the use of recombinant DNA technology (73,74). Polyadenylated mRNA obtained from rat pituitary tumor cells was used as the template for the preparation of growth hormone cDNA by reverse transcription. The amino acid sequences of rat GH and its precursor were deduced from the cDNA sequence (73). The cDNA was cloned and a hybrid gene was constructed between the cloned coding sequence of rat GH and the β-lactamase gene of plasmid pBR 322 (74). It was shown that this hybrid gene is expressed in *E. coli* by the immunological detection of the expressed GH sequences (74). These studies and the work on the expression of the proinsulin gene in *E. coli* (75) suggest that it would be possible to manufacture biologically important peptides by recombinant DNA technology. It should also be possible to produce new hormone analogues by expressing in bacteria genes that are designed specifically by substitution of the codons for specific amino acid residues through chemical synthesis.

Prolactin (PRL)

The existence of prolactin as a distinct pituitary hormone in humans was established only at the beginning of this decade. Human PRL was isolated in 1972, and recently the complete amino acid sequence was elucidated (76). Only 16% of the hPRL sequence is identical with the hGH sequence. Human PRL sequence identity with porcine, ovine, and rat PRL was found to be 77, 73, and 60%, respectively. PRL has also been isolated from chinchilla (77), salmon (78), and fin whale (79) pituitaries. PRL isolated from chinchilla pituitaries maintained in organ culture was found to be similar to other mammalian prolactins with regard to lactogenic properties, molecular weight, and susceptibility to deamidation (77). Salmon PRL also has a molecular weight similar to the mammalian prolactins (78). The amino acid composition and circular dichroism spectra of fin whale PRL were very similar to those of porcine PRL. The lactogenic potency of the whale hormone was 80% that of porcine PRL (79).

Ovine PRL was cleaved selectively between residues 53 and 54 by limited proteolysis with fibrinolysin (80). The two fragments, PRL-1-53 and PRL-54-199, were isolated and characterized. The fragments were devoid of lactogenic and immunochemical activities. Recombination of the two fragments resulted in full restoration of immunoreactivity but not of biological activity (80). The physicochemical properties of the recombinant molecule were very similar but not identical to those of the native hormone. Ovine PRL was cleaved at tryptophan residues 91 and 150 by reaction with 2-(2-nitrophenyl sulfenyl)-3-methyl-3'-bromoindolemine (81). A small fragment, PRL 92-150, and a large fragment containing PRL 1-91 and PRL 151-199 linked by a disulfide bridge were obtained. Neither fragment had bioactivity or immunoreactivity (81).

The biosynthesis of PRL has been investigated in monolayer cultures of normal rat pituitary cells (82). SDS-PAGE analysis of immunoprecipitates of [^{3}H]-leucine pulse-labeled extracts of pituitary cells revealed the presence of pre-PRL in addition to PRL. A chase with unlabeled leucine after the pulse-labeling resulted in the formation of labeled PRL only. Pre-PRL, which is the primary product formed in a translation system free of wheat germ cells and containing pituitary mRNA, was converted to PRL on the addition of a membrane fraction from dog pancreas. Partial sequence analysis showed that the processed product contained the correct amino terminus of PRL. These results strongly suggest that PRL is synthesized in the form of a precursor, pre-PRL, which is processed in a manner similar to that of other secretory proteins.

The molecular heterogeneity of immunoreactive hPRL in plasma and the pituitary has been studied (83). Three forms of hPRL corresponding to the monomer, dimer, and polymer have been detected by exclusion chromatography on Sephadex G-100. There was a significant increase in the proportion of PRL dimer in the blood of patients with PRL-secreting tumors. The PRL dimer is thought to be linked by intermolecular disulfide bonds and appears to be devoid of bioactivity as measured by radioreceptor assay, using rabbit mammary membrane preparations (83). The interassay variation of hPRL radioimmunoassay has been traced to the presence of varying amounts of polymeric forms in different batches of hPRL used for preparing the tracer (84). The actions of PRL, control of secretion, and pathological states have been reviewed (85).

HORMONES OF GROUP 3

A procedure for the isolation of six pituitary hormones, including the three glycoprotein hormones from a single batch of human pituitaries, has been described (86). The isolation of bovine pituitary glycoprotein hormones has been improved by incorporating affinity chromatography on concanavalin A-sepharose in the purification procedure (87). Immunochemical analysis of the unadsorbed fraction showed that less than 3% of the immunoreactive LH, FSH, and TSH are present as nonglycosylated forms in either crude pituitary extracts or concentrates.

Pregnant mare serum gonadotropin (PMSG) and its subunits have been further characterized (88). The subunits were found to have less than 1% of the activity of the parent hormone. The gonadotropin in pregnant mare endometrial cups (PMEG) has been purified and compared with PMSG (89). The amino acid composition of PMEG is similar to that of PMSG, but the PMEG preparation exhibited amino terminal heterogeneity. The carbohydrate content of PMEG was considerably less than that of PMSG. In bioassay for LH and FSH activity *in vitro* and *in vivo*, PMEG was less active (11 to 54%) than PMSG (89).

The structure of the carbohydrate moieties of the α-subunits of human LH, FSH, and TSH have been studied (90). A common structure of the carbohydrate moieties of three glycopeptides was deduced. These results suggest that the biosynthetic pathway for the carbohydrate chains is similar and may occur at the same site in the pituitary gland.

Lutropin (LH)

LH has been isolated from several nonmammalian species and has been shown to resemble mammalian LH in possessing two chemically distinct, noncovalently linked subunits (91). Hybrid LH molecules have been prepared by recombination of ovine LH subunits with sea turtle LH subunits (92). Assay of the hybrid hormones in both mammalian and reptilian assay systems showed that the β-subunit is primarily responsible for determining species specificity in responses to LH.

The interaction of the α and β subunits of ovine lutropin has been investigated by photochemical cross-linking (93) and with the aid of fluorescent probes (94). The beta subunit of ovine LH was reacted with 4-fluoro-3-nitrophenyl azide under select conditions to yield a product in which lysine 20 was modified (93). This derivative recombined stoichiometrically with the alpha subunit of LH to form a modified LH. Photolysis of the modified LH yielded a covalently coupled product in which the subunits are no longer dissociable. The modified LH was not attached significantly to rat testis homogenate preparations on photolysis. These results suggest lysine 20 in the alpha subunit of LH is located in the region of the binding interface between the two subunits. From measurements of the fluorescence of tyrosyl residues of ovine LH and the subunits it was inferred that most of the conformation changes that accompany dissociation of LH and recombination of the subunits occur in the α subunit (94).

Follitropin (FSH)

Follitropin and its subunits have been isolated from bovine (95,96) and porcine (97,98) pituitaries and have been characterized. The amino acid sequence of the beta subunits of porcine FSH has been elucidated (99). Porcine β-FSH is shorter than human β-FSH, at both the amino and carboxy ends. On the basis of the sequence homologies between the alpha and beta chains of the glycoprotein hormones and that of the three beta chains, it is suggested that the divergence of the alpha and beta chains preceded the divergence of the beta chains, which led to the three different hormones (99). Circular dichroism studies of mammalian follitropins showed that the tryptophan residue in FSH is not affected by N-bromosuccinimide treatment of the intact hormone, suggesting that the domain containing the tryptophan residue may be in or near a region of subunit-to-subunit contact (100).

Thyrotropin (TSH)

The complete amino acid sequences of the subunits of human TSH have been described (101), and the chemistry of human TSH has been reviewed (102). An age-dependent decrease in the ratio of bioactive to immunoreactive TSH in the serum of male Sprague-Dawley rats has been reported (103). Exclusion chromatography of whole rat serum and isotopically labeled pituitary incubates and extracts suggests that the decrease in the ratio of bioactivity to immunoreactivity is mainly due to the accumulation of high-molecular-weight material in the serum of the aging rat. This material cross reacts with antibodies to pituitary TSH and inhibits the production of thyroid hormones in intact mice (103).

REFERENCES

1. Li CH, et al: *Biochem Biophys Res Commun* 81:900, 1978.
2. Kawauchi H, Muramoto K, Ramachandran J: *Int J Peptide Prot Res* 12:318, 1978.
3. Ramachandran J: *Hormonal Proteins and Peptides* 2:1, 1973.
4. Ramachandran J, Suyama AT: *Proc Nat Acad Sci USA* 72:113, 1975.
5. Rao AJ, Long JA, Ramachandran J: *Endocrinology* 102:371, 1978.
6. Nicholson WE, et al: *Endocrinology* 103:1344, 1978.
7. Pickering BT, et al: *Biochim Biophys Acta* 74:763, 1963.
8. Li CH, et al: *Proc Nat Acad Sci USA* 75:306, 1978.
9. Lee CY, Ramachandran J, Li CH: *Endocrinology* 102(suppl):228, 1978.
10. Lowry PJ, Scott AP: *Gen Comp Endocrinol* 26:16, 1975.
11. Ramachandran J, et al: *Biochim Biophys Acta* 428:347, 1976.
12. Thody AJ, Shuster SJ: *Endocrinology* 64:503, 1975.
13. Thody AJ, et al: *J Endocrinol* 71:279, 1976.
14. Thody, AJ: *Front Hormone Res* 4:117, 1977.
15. Silman RE, et al: *Nature* 260:716, 1976.

16. Challis JRG, Torosis JD: *Nature* 269:818, 1977.

17. Visser M, Swaab DF: *Front Hormone Res* 4:42, 1977.

18. Clark D, et al: *Nature* 273:163, 1978.

19. Silman RE, et al: *Nature* 276:526, 1978.

20. Swaab DF, Visser M: *Front Hormone Res* 4:170, 1977.

21. Birk Y, Li CH: *J Biol Chem* 239:1048, 1964.

22. Lemaire S, Yamashiro D, Li CH: *Int J Pept Prot Res* 11:179, 1978.

23. Yamashiro D, Li CH: *J Amer Chem Soc* 100:5174, 1978.

24. Liotta AS, Suda T, Krieger DT: *Proc Nat Acad Sci USA* 75:2950, 1978.

25. Jeffcoate WJ, et al: *J Endocrinol* 77:27P, 1978.

26. Hughes J, et al: *Nature* 258:577, 1975.

27. Li CH, Chung D: *Proc Nat Acad Sci USA* 73:1145, 1976.

28. Bradbury AF, Smyth DG, Snell CR: *Biochem Biophys Res Commun* 69:950, 1976.

29. Graf L, Barat E, Patthy A: *Acta Biochem Biophys Acad Sci Hung* 11:121, 1976.

30. Chretien M, et al: *Biochem Biophys Res Commun* 72:472, 1976.

31. Li CH, Tan L, Chung D: *Biochem Biophys Res Commun* 77:1088, 1977.

32. Rubinstein M, Stein S, Udenfriend S: *Proc Nat Acad Sci USA* 74:4969, 1977.

33. Li CH, Chung D, Doneen B: *Biochem Biophys Res Commun* 72:1542, 1976.

34. Li CH, et al: *Biochem Biophys Res Commun* 71:19, 1976.

35. Li CH, et al: *J Med Chem* 20:325, 1977.

36. Yamashiro D, et al: *Int J Pept Prot Res* 10:159, 1977.

37. Yamashiro D, et al: *Int J Pept Prot Res* 11:251, 1978.

38. Yeung HW, et al: *Int J Pept Prot Res* 12:42, 1978.

39. Li CH, Tseng LF, Yamashiro D: *Biochem Biophys Res Commun* 85:795, 1978.

40. Hosobuchi Y, Li CH: *Commun Psychopharmacol* 2:33, 1978.

41. Katlin DH, et al: *Commun Psychopharmacol* 1:493, 1977.

42. Su C-Y, et al: *J Formosan Med Assoc* 77:133, 1978.

43. Yalow RS, Berson SA: *Biochem Biophys Res Commun* 44:439, 1971.

44. Orth DN, et al: *Endocrinology* 92:385, 1973.

45. Mains RE, Eipper BA: *J Biol Chem* 251:4121, 1976.

46. Mains RE, Eipper BA, Ling N: *Proc Nat Acad Sci USA* 74:3014, 1977.

47. Roberts JL, Herbert E: *Proc Nat Acad Sci USA* 74:4826, 1977.

48. Roberts JL, Herbert E: *Proc Nat Acad Sci USA* 74:5300, 1977.

49. Nakanishi S, et al: *FEBS Letters* 84:105, 1977.

50. Rubinstein M, Stein S, Udenfriend S:*Proc Nat Acad Sci USA* 75:669, 1978.

51. Eipper BA, Mains RE: *J Biol Chem* 253:5723, 1978.

52. Nakanishi S, et al: *Nature* 278:423, 1978.

53. Roberts JL, et al: *Biochemistry* 17:3609, 1978.

54. Watanabe H, Nicholson WE, Orth DN:*Endocrinology* 93:411, 1973.

55. Sabol SL: *Biochem Biophys Res Commun* 82:560, 1978.

56. Nakanishi S, et al: *Proc Nat Acad Sci USA* 74:3283, 1977.

57. Nakamura M: *Europ J Biochem* 86:61, 1978.

58. Chrambach A, et al: *Endocrinology* 93:848, 1973.

59. Mills JB, et al: *J Clin Invest* 52:2941, 1973.

60. Li CH, Graf L: *Proc Nat Acad Sci USA* 71:1197, 1974.

61. Li CH, Bewley TA:*Proc Nat Acad Sci USA* 73:1476, 1976.

62. Blake J, Li CH: *Int J Pept Prot Res* 7:495, 1975.

63. Li CH, et al: *Proc Nat Acad Sci USA* 74:1016, 1977.

64. Blake J, Li CH: *Int J Pept Prot Res* 11:315, 1978.

65. Li CH, Blake J, Hayashida T: *Biochem Biophys Res Commun* 82:217, 1978.

66. Mills JB, et al: *Endocrinology* 102:1366, 1978.

67. Mills JB, et al: *Endocrinology* 102:1377, 1978.

68. Li CH, Houghten R: *Int J Pept Prot Res* 12:114, 1978.

69. Li CH: *Proc Nat Acad Sci USA* 75:1700, 1978.

70. Burstein S, et al: *Proc Nat Acad Sci USA* 75:5391, 1978.

71. Bewley TA, Li CH: *Biochemistry* 17:3315, 1978.

72. Doebber TW, Divor AR, Ellis S:*Endocrinology* 103:1794, 1978.

73. Seeburg PH, et al: *Nature* 270:486, 1977.

74. Seeburg PH, et al: *Nature* 276:795, 1978.

75. Villa-Komaroff L, et al: *Proc Nat Acad Sci USA* 75:3727, 1978.

76. Shome B, Parlow AF: *J Clin Endocrinol* 45:1112, 1977.

77. Jibson M, Talamantes F: *Gen Comp Endocrinol* 34:402, 1978.

78. Idler DR, Shamsuzzaman KM, Burton MP: *Gen Comp Endocrinol* 35:409, 1978.

79. Kawauchi H, Tubokawa M: *Int J Pept Prot Res* 13:229, 1979.

80. Birk Y, Li CH: *Proc Nat Acad Sci USA* 75:2155, 1978.

81. Houghten R, Li CH: *Int J Pept Prot Res* 11:345, 1978.

82. Maurer RA, McKean DJ: *J Biol Chem* 253:6315, 1978.

83. Garnier PE, et al: *J Clin Endocrinol* 47:1273, 1978.

84. Fang VS, Armstrong J, Worlsey IG: *Clin Chem* 24:941, 1978.

85. Frantz AG: *N Engl J Med* 298:201, 1978.

86. Sairam MR, Chretien M, Li CH: *J Clin Endocrinol* 47:1002, 1978.

87. Bloomfield GA, Faith MR, Pierce JG: *Biochim Biophys Acta* 533: 371, 1978.

88. Papkoff H, Bewley TA, Ramachandran J:*Biochim Biophys Acta* 532: 185, 1978.

89. Papkoff H, Farmer SW, Cole HH: *Proc Soc Exp Biol Med* 158:373, 1978.

90. Hara K, Rathnam P, Saxena BB: *J Biol Chem* 253:1528, 1978.

91. Licht P, et al: *Rec Prog Hormone Res* 33:169, 1977.

92. Licht P, Farmer SW, Papkoff H: *Gen Comp Endocrinol* 35:289, 1978.

93. Burleigh BD, Liu W-K, Ward DN: *J Biol Chem* 253:7179, 1978.

94. Ingham KC, Bolotin C: *Arch Biochem Biophys* 191:134, 1978.

95. Becken JF, et al: *Biochimie* 59:825, 1977.

96. Cheng K-W: *Biochem J* 175:29, 1978.

97. Whitley RJ, Keutmann HT, Ryan RJ: *Endocrinology* 102:1874, 1978.

98. Closset J, et al: *Europ J Biochem* 86:105, 1978.

99. Closset J, et al: *Europ J Biochem* 86:115, 1978.

100. Guidice LC, et al: *Biochem Biophys Res Commun* 81:725, 1978.

101. Sairam MR, Li CH: *Can J Biochem* 55:755, 1977.

102. Sairam MR, Li CH: *Hormonal Proteins and Peptides* 6:1, 1978.

103. Klug TL, Adelman RC: *Adv Exp Med Biol* 97:259, 1978.

12
Biochemical Aspects of Genetic Disease

Per-Arne Öckerman
Arne Lundblad

A large number of rare disorders are inborn errors of metabolism (IEM). Each case of IEM is not of great importance to society, but it is important to the individual family. Taken together, however, cases of IEM represent quantitatively an important segment of human disease, especially in the infant population. Many of the disorders demand difficult diagnostic procedures. Very often the clinical symptoms and signs are so nonspecific that an IEM must be suspected and sought much more often than it is found. The use of diagnostic methods involves not only a highly varied setup of complicated techniques but also much work with many negative and few positive results. Cases of IEM consequently represent a large area of clinical and laboratory medicine. Everybody dealing with chemical diagnosis of IEM must specialize in certain topics and cannot cover all. We, therefore, have concentrated in this chapter on errors of the metabolism of simple and complex carbohydrates, certain cellular lipids, organic and amino acids, urea cycle defects, cystic fibrosis, and the muscular dystrophies. Large areas are not discussed, for example, hemoglobins, plasma proteins, coagulation, metals, erythrocyte enzymes, immunopathies, porphyrins, bilirubin, and hormones. A valuable review of IEM, with more than 500 references, has been published (1a).

METHODS OF GENERAL INTEREST

Clinical Identification

One main difficulty in identifying cases of IEM is that clinical symptoms and signs are often very uncharacteristic, or they are complex and hard to remember. They may also be nonspecific and may lead the clinician to think primarily of more common, nongenetic disorders. Good indicators of when to suspect an IEM are, therefore, needed. One such list has been compiled by Burton and Nadler (1b). Vomiting, diarrhea, jaundice, seizures, hepatomegaly, coarse facial features, abnormal urinary odor, and respiratory distress are the major clinical manifestations dealt with by these authors.

General Screening

Screening of whole populations, usually in the immediate postnatal period, has been carried out in many countries for several years. Only a small number of types of IEM have been suited to such screening because of very strict rules for qualification, that is, a simple, reliable, and cheap diagnostic method must be available, a suitable treatment must exist and must be cost-effective; and a delay in diagnosis should involve increased risks for the patient. No large changes occurred in this area during 1978, but experience is accumulating.

The practical difficulties in accomplishing an ambitious program can be great (2). Genetic screening is of interest not only to diagnose IEM but also to describe phenotypic characteristics that are important clues to the effect of exogenous toxic factors (3).

The pattern of genetic diseases can be very different among various countries. In publications from Finland during recent years specific national patterns of such diseases have been discussed (4). By analyzing urine from all newborns (in most countries only blood is sampled), it was found both in Slovakia (5) and in Wales (6) that alcaptonuria has a much higher incidence than earlier thought, 1:25,000 in Slovakia and 1:44,800 in Wales. The authors believe that the high incidence is not specific for the populations they studied but can be representative of many other countries.

Prenatal Diagnosis

Prenatal diagnosis has continually attracted much interest. Many centers now have gained good experience in this area, and informative surveys have been published during the past year (7-9). Both moral/ethical and practical/technical aspects are discussed.

Therapy

Therapeutic advances are described in some of the sections on specific diseases in this chapter. As an example, therapy for lysosomal storage diseases will be discussed here. One possible way to achieve a positive effect is to inject the missing enzyme entrapped in liposomes. These enzyme-containing liposomes are taken up by lysosomes in the liver, spleen, and other organs and can degrade the material stored in these organs. Such therapy was used in a case of type II glycogenosis (Pompe disease) by Ryman et al (10). Liver glycogen was reduced by this treatment, but no effect was noted on cardiac or muscle storage. If liposomes can also be persuaded to "home in" to other critical organs, for example, the heart and brain, this form of treatment might become more feasible and practical.

New Diagnostic Techniques

Improved methods for the diagnosis of individual IEM appear continuously (e.g., assay of neuraminidase for the diagnosis of sialidoses, see below). At the same time some techniques are developed that are more generally applicable for a group or even several groups of IEM.

An interesting new method to diagnose IEM has been presented by Kan and Dozy (11), who applied recombinant DNA techniques to diagnose successfully a fetus at risk for sickle cell anemia. This principle allows a very sensitive, rapid, and safe procedure and might be applicable to several types of IEM such as single gene defects, where the affected protein is known. The method, although as yet only used to a very limited extent, has been reviewed twice in the past year (12,13).

N.G. and N.L. Anderson have tried a different general approach (14,15). They obtain a very high resolution in the separation of serum, tissue proteins, and protein subunits by two-dimensional electrophoresis, using isoelectric focusing in one dimension and electrophoresis in sodium dodecyl sulfate in the second. This principle has yet to be more fully tested but could be of great future value for characterizing abnormality in protein composition in IEM.

Separation, identification, and quantitation of a whole group of substances, such as amino acids and organic acids, has long been possible. By the use of gas chromatography/mass spectrometry (GC/MS) and high pressure liquid chromatography (HPLC), new areas are successively being opened to exploration. This is true for oligosaccharides. Methods are now available (16-18) for the identification and quantitation of a series of oligosaccharides in urine. Such techniques may help to elucidate better the various oligosaccharidoses (see below).

The number of centers using sophisticated GC/MS equipment for multicomponent analysis of metabolites has been small but is now rapidly increasing. The use of more or less simplified versions of such techniques are now spreading all over the world (19).

For components that are not volatile or cannot be made volatile by some derivatisation, HPLC, rather than GC, may be applied for their separation. By coupling a mass spectrometer (MS) via a jet-separator to a HPLC, Takeuchi et al (20) were able to combine the separation of nonvolatile components by HPLC with unique analytical capacity of MS. Such a combination will no doubt increase the number of diseases that can be covered by analysis for the study of IEM.

GLYCOGENOSES

Glycogen Storage Disease Type I (GSD I)

Glucose-6-phosphatase (G6Pase) deficiency was the first IEM in which the enzyme defect could be characterized (21). The growth of knowledge of this disorder has long been relatively static as compared with the situation for most other types of IEM. Knowledge, however, is rarely stable for long and, as expected, important new information has also appeared for GSD I. The literature contains several case reports of type I glycogenosis where the G6Pase activity *in vitro* was normal but where no activity was detected in studies of glucose metabolism *in vivo*. For such cases the term type Ib glycogenosis has been used (22). Narisawa et al (23) recently described a case of an apparent deficiency in the transport of G6Pase from the cytosol to the lumen of the endoplasmic reticulum but with a normal G6Pase activity. They speculate that this is due to the absence of a specific protein that facilitates this translocation of the enzyme intracellularly. Rosenfeld et al (24) suggest the same deficiency in their case, although they give no direct positive evidence. To be able to separate these two GSD I variants, assays must be performed on fresh biopsy specimens. If frozen tissue is used, normal results for the G6Pase activity will be noted in the G6P transport-deficient type (1b).

Treatment of hepatic GSD continues to arouse much interest. Nocturnal feeding, specifically intragastric infusion, is now considered the method of choice for children with defective gluconeogenesis, and hypoglycemia (25). Such feeding normalizes blood glucose levels, improves secondary derangements of growth and of the metabolism of lactate, lipids, and urate (25,26), and corrects coagulation abnormalities. Much less therapeutic success was noted with kidney transplantation (27).

Glycogen Storage Disease Type II (GSD II)

GSD II, Pompe disease, was the first lysomal enzyme deficiency discovered (Hers 1963, ref. 28). One of the unsolved major problems in this disease is why some patients die in infancy while others have only a mild muscular disease compatible with adult life. Intermediate forms also exist. Up till now various explanations for this have been proposed. During 1978 several authors have attacked this problem, among them Beratis et al (29). Their results lend some support to the theory that infantile GSD II is the result of a structural gene mutation for α-glucosidase, which causes the synthesis of a catalytically inactive enzyme protein. In the adult form, the mutation causes a reduction in the amount of the enzyme protein present in the cells. An alternative mechanism, that of a total lack in infantile GSD II and some residual activity in the adult form, is supported by the results of Reuser et al (30). A third difference was noted by Bienvenu et al (31), who found a total deficiency in all tissues in infantile cases but a variable expression in different tissues in adult patients. These authors also noted a residual activity in adult cases.

Conflicting results were published by Murray et al (32), who could find no difference between the infantile and the adult form using activity assays and immunological studies. These authors failed to find any immunologically active α-glucosidase in both forms of the disease.

Pena et al used double-labeling (^{3}H-leucine and ^{14}C-leucine) of the proteins in cultured fibroblasts, a technique that may be generally applicable in IEM (33a). They found a 90% decrease in the amount of a protein with a molecular weight of 29,000 daltons in infantile GSD II (33b). In the adult form a similar but quantitatively less important deficiency was noted. This protein was not identical with α-glucosidase.

The enzymic diagnosis of GSD II is easily made on a muscle specimen. When blood cells are used there are more difficulties because of the existence of at least three isoenzymes of α-glucosidase of which only one is deficient. Dreyfus and Poënaru (34) have provided a good description of these isoenzymes, but they could find no difference between the forms of GSD II.

Other Glycogenoses

Although there are several other types of GSD, only one additional publication will be mentioned here. DiMauro and Hartlage (35) described a unique fatal infantile form of muscle phosphorylase deficiency. All other cases of myophosphorylase deficiency described so far have been much less serious, the first episodes of cramps and myoglobinuria appearing about the time of puberty in most patients. The authors noted a complete absence of the enzyme protein by immunodiffusion and speculate that this form is caused by a lack of both fetal and adult myophosphorylase.

GALACTOSEMIA

The publications in 1978 on the galactosemias expand earlier knowledge only slightly, giving additional examples of the results of screening (36), discussing improved diagnostic techniques (37), or describing various types or variants (37-40).

OTHER CARBOHYDRATE DISORDERS

Odièvre et al (41), working with a very large series of 55 patients with *hereditary fructose intolerance*, conclude that the long-term prognosis is extremely favorable if a fructose-free diet is given. However, persistent hepatomegaly and fatty changes in periportal hepatocytes seem to be the rule. According to the authors, these observations may support the hypothesis that only minimal amounts of fructose are essential for human beings.

In a study of a mother and her child Taunton et al(42) for the first time demonstrate a case of *fructose-1,6-diphosphatase deficiency* with mild symptoms starting in adult life. They found a relatively high residual activity of the enzyme and consider their cases to represent a new variant. Folate administration improved the condition in both mother and child.

There has been much discussion on the wisdom of giving milk to population groups with a high rate of *lactose intolerance*. In 1978 the American Academy of Pediatrics Committee on Nutrition stated (43): "The Committee's position remains unchanged. On the basis of present evidence it would be inappropriate to discourage supplemental milk feeding programs targeted at children on the basis of primary lactose intolerance . . . as long as milk continues to provide the best and cheapest source of high-quality protein."

MUCOPOLYSACCHARIDOSES (MPS)

The diagnosis of the various forms of MPS has been greatly helped by the characterization of the enzyme deficiencies. Much research in this area is being done, and new deficiencies are still being found (see below). The substrates used for the enzyme assays have mostly been very complex. It is, therefore, of practical interest that improved natural and synthetic substrates are becoming available and include, for example, those for α-iduronidase (44,45) and L-iduronic acid 2-sulfate-sulfatase (46) activity measurement. These findings have made the enzymatic diagnosis of Hurler, Scheie, and Hurler-Hunter syndromes easier.

Chemical assay methods for glycosaminoglycans (GAG) are also being successively improved. Some papers published during 1978 describe highly sensitive modified techniques, which, because of their high sensitivity, are important, especially in assays on amniotic fluid (47,48). In addition, new information on the technique used to measure GAG metabolism in cultured human fibroblasts has appeared (49), showing that cells must be trypsinized before GAG assay to avoid interference from pericellular GAG.

Corneal opacities are claimed to be one of the most valuable clinical criteria to differentiate MPS I from MPS II. Spranger et al (50) describe two cases of MPS II, one extremely severe and the other unusually mild. Both patients had corneal opacities. The authors conclude that a clear cornea no longer can be regarded as a hallmark of MPS II.

In addition to the well-known Sanfilippo syndromes A and B (MPS III A and III B) a Sanfilippo syndrome type C has been delineated by the Kresse-von Figura group (51). These authors not only differentiated this disease from the A and B forms but also found a new enzyme that acetylates α-glucosaminyl residues. This acetyl-CoA:α-glucosaminide N-acetyl-transferase was shown to be deficient in 11 patients with MPS III C.

The results of much important work on Morquio syndrome have been published. By studying cases of classical Morquio disease and a patient exhibiting a Morquio-Sanfilippo intermediate phenotype, it was shown that two different hexosamine-6-sulfate-sulfatases exist, one specific for the galactose configuration and deficient in Morquio disease and one specific for the glucose configuration and deficient in the new entity, Morquio-Sanfilippo intermediate phenotype (52-55).

Much of the present work being done on MPS is of the sophisticated enzyme chemistry type. However, important clinical studies are also being made. Gitzelmann et al (56) found an unusually mild course of β-glucuronidase deficiency in two brothers (MPS VII, cases 8 and 9) and concluded that MPS VII cases may be undetected in infancy.

Although about 25 cases of the Dyggve-Melchior-Clausen syndrome have now been described, the pathomechanism behind this disorder remains obscure (57,58).

EHLERS-DANLOS SYNDROME

Ehlers-Danlos syndrome is a relatively common heritable disorder of connective tissue. Eight different types have been described, and four of these have been shown to have defects in the biogenesis of collagen (59,60). It is believed that a defect in collagen metabolism is involved in all eight types.

OLIGOSACCHARIDOSES

The oligosaccharidoses constitute a rapidly developing area of clinical and laboratory medicine in which the basic knowledge is still incomplete. Despite the fact that *Aspartylglucosaminuria* (AGU) is easy to diagnose clinically and biochemically, only three cases have been reported outside Finland and northern Norway. Presently, more than 100 cases are known in Finland. A clinically related disorder, Salla disease, also found in Finland, is characterized chemically by a marked urinary excretion of free sialic acid (61).

Mannosidosis is known both in humans and in cattle. The human disease has been much studied with respect to its kinetic and immunological properties. Some recent papers (62-64) give conflicting evidence, and the significance of the fact that the amount of residual activity of acid α-mannosidase is quite variable remains unclear.

It is also still unclear whether some cases have an immunoreactive, enzymatically inactive protein and others do not.

Hair roots were tried with good success, instead of the more commonly used blood, fibroblasts, or liver for the diagnostic determination of α-mannosidase in a case of mannosidosis (64a).

In bovine mannosidosis the mutuant enzyme has been found to be immunologically reactive, enzymatically inactive, and present at a decreased concentration (65). Since bovine mannosidosis still is a problem in New Zealand and Australia, new modifications for heterozygote detection still appear (66).

It has long been clear that not only α-mannosidosis but also β-mannosidosis might exist, since there are two different mannosidases, one cleaving α-linked and one cleaving β-linked mannose. An improved synthesis of p-nitrophenyl-β-D-mannopyranoside (67) might make substrate for β-mannosidase analysis more easily available and might be used in the search for the first case of β-mannosidosis.

Modern analytical techniques for carbohydrates have allowed the identification of a series of fucose-containing oligosaccharides excreted in the urine in *fucosidosis* (68-70). In agreement with earlier evidence, the results of two recent studies (71-72) support the conclusion that there is only a very small production of a normal α-fucosidase in fucosidosis. However, there is no evidence to support the existence of a structurally altered enzyme.

MUCOLIPIDOSES, SIALIDOSES

Possible explanations for these disorders are still being discussed intensely. The most far-reaching systematic approach is by French authors. Strecker and Michalski (73) and Maroteaux et al (74) give a biochemical and clinical basis for six different types of sialidosis. They have assayed two different neuraminidases (sialidases), cleaving α-(2→6) and α-(2→3) linkages, respectively. Adding findings of urinary excretion of sialyl-oligosaccharides and results of clinical studies and assays of lysosomal hydrolases, they describe *sialidosis type A* (mucolipidosis II and III) with a total (ML II) or partial (ML III) deficiency of both α-(2→3) and α-(2→6)neuraminidase, a moderate excretion of urinary sialyl-oligosaccharides, and a leakage of lysosomal hydrolases into extracellular fluids. They also describe *sialidosis type B* (ML I), and nephrosialidosis and two neurological ocular types with α-(2→6) neuraminidase deficiency and massive excretion of urinary sialyl-oligosaccharides, but without lysosomal hydrolase exocytosis. The accumulated compounds show a ratio of α-(2→6) to α-(2→3)-sialyl linkages of 1 for sialidosis A and of 10-30 for sialidosis B.

Patients with neurological and ocular features (myoclonus and cherry-red spots) may belong to the same general disease category as the patients studied by O'Brien (75), Rapin et al (76), Thomas et al (77), Okada et al (78), and Koseki et al (79). Considering all these new studies, there seems to be much support now for the existence of sialidase deficiencies and for the existence of abnormal sialo-oligosaccharide excretion (80). However, other theories for I cell disease (ML II) have been proposed and include an abnormal assembly of carbohydrate residues in lysosomal enzymes, rather than a defective removal of sialic acid residues (81,82); a deficiency of a Golgi-associated α-mannosidase, possibly an endomannosidase (83); or a defect in a system phosphorylating carbohydrates on the acid hydrolase molecules, leading to a deficient receptor-mediated endocytosis of several lysosomal enzymes (82a).

AMINO ACIDS AND ORGANIC ACIDS

Among the large number of types of IEM described in amino acid metabolism, significant advances have been made only in a few and only these will be presented here.

Urea Cycle Defects

Five distinct disorders have been described so far, involving each of the five enzymes of the urea cycle: hyperammonemia due to carbamyl phosphate synthetase (CPS) deficiency or due to ornithine transcarbamylase (OCT) deficiency; citrullinemia due to argininosuccinic acid (ASA) synthetase deficiency; ASA-uria due to ASA lyase deficiency; and argininemia due to arginase deficiency. All five disorders have a similar clinical picture characterized by protein intolerance, vomiting, seizures, and various degrees of mental retardation associated with hyperammonemia.

Carbamyl Phosphate Synthetase Deficiency

Two new cases of lethal CPS deficiency have been described by Mantagos et al (84).

Argininosuccinic Aciduria

This much more well-known disorder of the urea cycle was described in a new case by Böhles et al (85), who used keto analogues of essential amino acids in treatment and achieved a remarkable clinical improvement. Such analogues have previously been used in the treatment of other defects of urea synthesis. The authors believe that these analogues open a new avenue in the treatment of inborn defects of urea synthesis.

Ornithine Transcarbamylase (OCT) Deficiency

Glasgow et al (86) concluded that low-protein diet plus oral keto acid supplementation resulted in the best metabolic control of OCT deficiency. They also noted that the OCT activity in the liver varied up to 40-fold in different biopsy specimens from a heterozygous female. This was thought to be caused by clusters of normal or abnormal hepatocytes that resulted from inactivation of either the abnormal or normal X chromosome in this X-linked disorder. Enzyme activity assayed on small liver biopsy specimens may, therefore, not be representative of the entire liver in female patients with OCT deficiency.

Hyperornithinemia

Hyperornithinemia is an obscure condition that has been described as part of various clinical entities, including gyrate atrophy of choroid and retina. Treatment has involved a drastic reduction in the dietary protein intake. By adding lysine, Giordano et al (87) could normalize plasma ornithine levels, thus allowing a normal protein diet and avoiding the deleterious effects connected with low-protein intake.

Familial Azotemia

Hsu et al (88) described the presence in five members of the same family of very high concentrations of blood urea. Creatinine levels were normal, and the patients were free of specific clinical symptoms. Decreased urea excretion despite otherwise normal renal function was responsible for the chronic azotemia in these patients.

Phenylketonuria (PKU)

Knox (89) in a editorial in the *New England Journal of Medicine* cites Murphy's Law applied to medicine: if anything can go wrong, it will. It has previously been stated that three different defects would allow phenylalanine to escape metabolic processes and to accumulate in and injure the immature developing brain in cases of phenylketonuria. These potential defects are *(1)* phenylalanine hydroxylase deficiency, known to be the lesion in most patients with classical PKU, and *(2)* deficiency of dihydropteridine reductase, an enzyme that regenerates biopterin, a cofactor needed in the hydroxylation process. In a child with phenylketonuria, who showed substantial neurological impairment despite early dietary control of blood phenylalanine levels, Kaufman et al (90) identified the third possible type of defect, deficient synthesis of tetrahydrobiopterin. The level of this hydroxylation cofactor was only 20% of normal in the liver. Until an easy way of getting this cofactor into the brain is found, the suggested treatment of this condition is the same as that of reductase deficiency, that is, the administration of dihydroxyphenylalanine (DOPA) and 5-hydroxytryptophan in addition to the restriction of dietary intake of phenylalanine.

Tyrosinemias

A series of disorders of tyrosine metabolism have been described (91) and include *(a)* neonatal tyrosinemia, a common transient condition that can be treated by ascorbic acid supplementation and a low-protein diet and that may be associated with perinatal cytomegalovirus infection (92); *(b)* hereditary tyrosinemia (tyrosinemia I), an autosomal recessively inherited severe liver disease with chronic liver and renal complications (the basic biochemical defect in this disorder is still unknown); *(c)* tyrosinosis, which is defined by only one case studied in 1932 and is accompanied by myasthenia gravis; and *(d)* tyrosinemia II (Richner-Hanhart syndrome) a symptom complex appearing with keratosis palmoplantaris, dendritic lesions of the cornea, photophobia, mental retardation, and autosomal recessive inheritance. In addition to these, some unclassifiable cases have been described.

Threoninemia

During a large-scale screening program in India, Reddi (93) detected an hitherto unreported disorder with threoninemia. The patient, an 8-month-old boy had periodic episodes of minor motor seizures and growth retardation and a markedly elevated level of threonine in serum and urine. No biochemical explanation can yet be given.

5-Oxoprolinuria (Pyroglutamic Acidemia)

In the so-called γ-glutamyl cycle, six enzymes catalyse the conversion of glutathione into γ-glutamyl amino acids and cysteinylglycine.

Two of these enzymes, glutathione synthetase and γ-glutamyl-cysteine synthetase, may be deficient in red blood cells, leading to hereditary hemolytic anemias. When there is a general deficiency of the activity of glutathione synthetase, a more serious disorder appears with chronic metabolic acidosis, usually of neonatal onset, increased rate of hemolysis, and in some cases neurological symptoms (94). An increase in the acetic acid concentration in the blood up to 10 times normal was found in the immediate postnatal period in two siblings with this condition (95). The hyperacetatemia could not be demonstrated later in life. The concentrations of thioredoxin, a potent reducing agent for both proteins and low-molecular-weight disulfides, and the activity of thioredoxin reductase were normal in 5-oxoprolinuria and cystinosis (96), indicating that the levels of glutathione and thioredoxin are regulated independently, and that the derangement in the conversion of cystine to cysteine in cystinosis is not due to decreased levels of either glutathione or thioredoxin in whole cells.

Carnitine Deficiency

Systemic carnitine deficiency is a rare syndrome; its major clinical manifestation is the development, usually in late childhood or adolescence, of a progressive lipid-storage myopathy. In new cases of this disorder, other symptoms and signs are manifested: increased blood lactate and pyruvate concentrations and transient ketoacidosis (97), hypoglycemia, hepatic dysfunction, and death from acute cardiac arrest (98). Since the disorder is responsive to specific replacement therapy with carnitine, the authors point out the importance of an early diagnosis, which can be made by carnitine assay on muscle biopsy specimens.

Cystathioninuria

Cystathioninuria was first described in a patient with severe mental deficiency. Later studies have demonstrated the condition in patients with a variety of clinical symptoms and even in healthy persons. It is presently unclear whether there is a causal relationship between the enzymic defect (cystathionase deficiency) and mental deficiency and between the defect and other clinical manifestations. Pascal et al (99), by studying lymphoid cell lines, found two or possibly three different modifications of the cystathionase molecule. According to these authors one form of cystathioninuria that is vitamin B_6-unresponsive appears to result from the failure to synthesize the enzyme protein. Vitamin B_6-responsive forms appear to result from the production of cystathionase molecules that are altered in their ability to combine with coenzyme but have maintained antigenic identity that is demonstrable on agar double-diffusion analysis. In some cases, there may be changes in antibody-binding capacity as well, although immunological cross-reactivity remains.

Maple Syrup Urine Disease (MSUD)

Screening for MSUD has been conducted since 1964. According to Naylor and Guthrie (100), more than 9.5 million newborns throughout the world have been tested for MSUD by the use of a bacterial inhibition assay for leucine on dried filter paper blood specimens. Forty-three cases have been detected, an incidence of approximately 1:240,000 newborns.

Taylor et al (101) described a patient with a combination of congenital lactic acidosis and high levels of branched chain amino acids. They demonstrated decreased activity of the dihydrolipoyl dehydrogenase component of the pyruvate, α-ketoglutarate, and branched chain keto acid dehydrogenase multienzyme complexes and speculated that all these dihydrolipoyl dehydrogenase activites, genetically and biochemically, are a single entity.

Nonketotic Hyperglycinemia (NKH)

Of 19 patients with NKH diagnosed in Finland from 1965 to 1977, 5 were alive in 1978. In this Finnish series described by von Wendt et al (102), plasma levels of branched chain keto acids were also elevated. The mechanism leading to such a poor prognosis is not known. De Groot et al (103) speculate on evidence from experiments on rats that glycine is neurotoxic. A few patients seem to have a milder form of the disease without life-threatening neonatal illness, severe mental retardation, or neurological deficits (104).

Analytical Techniques for Organic Acids

Assay of organic acids by GC/MS techniques has attracted much attention. The methods are being steadily improved. Gates et al (105, 106) provide data for 134 identified components in human urine and also present a very informative review of quantitative metabolic profiling based on gas chromatography (107). Technical aspects of their work are described in Chapter 2 of this volume. Data are now available for all ages, including newborns (108).

Methylmalonic Acidemia (MMA)

Several variants of MMA are known and are discussed by Morrow et al (109) and Matsuda et al (110). From their papers it can be deduced that deficiencies of the apoenzyme of methylmalonyl-CoA carbonylmutase, as well as of its coenzyme 5'-deoxyadenosylcobalamin, and of a racemase enzyme exist. Clinically, MMA can be classified into vitamin B_{12}-responsive and vitamin B_{12}-nonresponsive types. A mutase-deficient patient with severe hyperammonemia was described by Packman et al (111), who state that severe hyperammonemia can be seen not only in disorders of the urea cycle but also in methylmalonyl-CoA racemase and mutase deficiencies.

Propionic Acidemia

Shafai et al (112) report a case of propionic acidemia with severe hyperammonemia, thereby adding another condition to the list given by Packman et al (111). Previously, only moderate elevations of ammonia levels had been reported in propionic acidemia.

A number of previously unrecognized abnormal metabolites were identified and quantitated in the urine by Sweetman et al (113).

Isovaleric Acidemia

Isovaleric acidemia involves a high mortality risk if untreated. Evidence is now accumulating that glycine therapy has a positive beneficial effect (114,115).

SPHINGOLIPIDOSES

General

Sphingolipidoses are characterized by the accumulation of various lipids, resulting from deficiencies in specific lysosomal hydrolases. Some of the diseases are present in both cerebral and visceral forms, but the mechanisms for these diverse clinical manifestations are not known. Several cases of sphingolipidosis have a clinical course that does not correlate with the enzyme level determined by *in vitro* tests. One explanation for this has been the discovery of different protein and glycoprotein activators. Intensive research is now being done on such activators or modulators, as well as on the physiological role of the multiple forms of different sphingolipid hydrolases. An excellent review, with 110 references, has appeared (116). The field has also been reviewed by Adachi et al, who covered about 200 references (117).

Niemann-Pick Disease

Niemann-Pick disease (NPD) is an autosomal recessive disorder characterized by the accumulation of sphingomyelin in visceral and/or cerebral tissue. Five clinical forms (A to E) were previously distinguished, depending on the age at onset and on clinical and biochemical manifestations. A new type (F) of NPD has now been reported (118). This type has a childhood onset of splenomegaly and a lack of neurological involvement. Several other case reports have also been presented (119-121), and one case of successful prenatal diagnosis has been discussed (122). A few reports on the determination of sphingomyelinase may be mentioned (123-125). In one of these methods (123) it was shown that the enzyme can be measured reliably with methylumbelliferyl substrate in the presence of Triton X-100.

Gaucher Disease

Gaucher disease is an autosomal recessive disease that can appear in at least three clinical forms. Most common is the adult, or non-neuropathic, type 1. The infantile, or acute neuropathic, type 2, is the most severe form of Gaucher disease. The juvenile form, or subacute neuropathic, type 3, is less well defined. Two reports describe the prenatal diagnosis of Gaucher disease using natural and synthetic substrates (126-127). There is some conflicting information about the most reliable type of substrate. Thus, it is claimed (128) that both patients and carriers can be detected using leukocytes and the fluorogenic substrate 4-methyl-umbelliferyl-β-D-glucopyranoside in the presence of pure sodium taurocholate and Triton X-100. Butterworth and Broadhead (129), however, obtained variable amounts of residual activity using this substrate but were able to eliminate soluble and particulate components of β-glucosidase by preincubation of spleen or liver homogenates at pH 3.0, or with 100 mM sodium chloride at pH 4.0. A few new cases of the acute neuropathic form of Gaucher disease have also been reported (130-132).

Detailed studies on isolated mutant and normal glucocerebrosidase from human spleen tissue have been carried out (133). The enzyme was purified 26,000-fold. The results support previous data that the genetic basis of Gaucher disease is a structural mutation of glucocerebrosidase. Turner and Hirschhorn (134) studied β-glucosidase from cultured skin fibroblasts of control subjects and patients with

Gaucher disease. When compared with the enzyme from control fibroblasts, the enzyme from type 1 Gaucher disease was more rapidly inactivated at 50°C, had an altered pH activity curve, and was less effectively inhibited by deoxycorticosterone-β-glucoside and more effectively inhibited by deoxycorticosterone. They also conclude that cells from type 1 Gaucher disease contain a structurally altered form of β-glucosidase. The residual enzyme from patients with type 2 and type 3 Gaucher disease was indistinguishable from the enzyme from the control subjects in those parameters that were studied. A similar conclusion was reached by Choy and Davidson (135). Characterization of other hydrolases that have elevated (136) or decreased (137) concentrations have been performed. Chambers et al (138) studied the multiple forms of acid phosphatase and concluded that the major form of this enzyme in the serum of patients with Gaucher disease is similar to a minor component in spleen. The spleen in Gaucher disease contains relatively large quantities of a heat-stable activator of the normal glucocerebrosidase. The activator has been characterized as an acid glycoprotein with a molecular weight of 11,000 daltons. Chiao et al (139) have studied the subcellular localization of this factor and found it associated with the storage deposits in lysosomes of the Gaucher cell.

Lactosyl-Ceramidosis, Krabbe Disease, and GM_1-Gangliosidosis

In 1970 Dawson and Stein (140) described a child with a neurological disease characterized by an accumulation of lactosyl ceramide. They referred to this disorder as lactosyl-ceramidosis. Liver tissue and fibroblasts from this patient have been re-examined (141). A deficiency of neutral β-galactosidase was observed, and the authors suggest that this enzyme has an *in vivo* role in the cleavage of lactosyl ceramide. On the other hand, Cheetham et al (142) studied acid and neutral β-galactosidase activities in postmortem liver samples from people who died of nongenetic diseases. The neutral activities showed a bimodal distribution, suggesting the occurrence of two distinct populations of humans. No clinical symptoms were associated with the much lower neutral β-galactosidase activities.

Krabbe disease (globoid cell leucodystrophy) is one of the classical genetic leucodystrophies caused by a deficiency of galactosyl ceramidase. Detailed knowledge about this disease is available. The mutation in Krabbe disease appears to be structural and not regulatory; the defective gene produces altered galactosyl ceramidase, which lacks most of its enzymic activity but has retained most of its antigenic determinants (143).

GM_1-gangliosidosis is caused by a deficiency of GM_1-ganglioside-β-D-galactosidase. Four different variants have been described. A report of successful prenatal diagnosis has appeared (144). Three cases in adults with profound β-galactosidase deficiency but with lack of visceromegaly were reported (145). Further experiments with somatic cell hybridization have also been presented (146). The results obtained support the hypothesis that the gene mutation in types 1 and 2 is different from that in types 3 and 4. The residual liver acid β-galactosidase activity from a case of feline GM_1-gangliosidosis has also been studied (147). The results suggest a structurally altered enzyme. In contrast to the situation in humans, no immunological identity could be observed between normal and mutant enzyme.

The hepatic storage of oligosaccharides and glycolipids in a cat with GM_1-gangliosidosis was studied (148). Two oligosaccharides, both with terminal galactose, were partially characterized. They were shown to be distinct from the major oligosaccharides accumulated in the liver of patients with GM_1-gangliosidosis. Two structurally related oligosaccharides, an octasaccharide and a new pentasaccharide, were isolated from urine of a patient with GM_1-gangliosidosis (149).

Metachromatic Leukodystrophy (MLD)

MLD is characterized by the accumulation of galactosyl-3-sulphate ceramide and a deficiency of the enzyme arylsulphatase A (ASA). Several reports on new cases of this disease have appeared (150-154). One family has been found with low ASA activity but without any other symptoms of MLD (155). Christomanou and Sandhoff (156) have measured ASA activity in leukocytes and cultured fibroblasts from members of three families with late infantile and adult MLD. Values obtained for different heterozygotes were almost continuously scattered over the entire range from patient level to normal level. The possibility is discussed of a crucial minimal level of residual ASA activity necessary to avoid manifestation of the disease. Shapira et al (157) have compared the immunological cross-reactivity of ASA from the liver of normal persons, of patients with late infantile MLD, and of patients with juvenile MLD. Antisera were prepared against ASA from human liver, against a minor component with enzymatic activity, and against one of the two inactive subunits of the enzyme. In liver homogenate from patients with late infantile MLD, the active dimer and one of the subunits were absent. In contrast, in the juvenile form of the disease, both inactive subunits were immunologically detectable, whereas the active dimer could not be detected. The authors conclude that ASA is a dimeric enzyme. In cases of late infantile MLD, the molecular defect results in the absence of one of the subunits, and, therefore, no active dimer is formed. In contrast, in the juvenile form of the disease, the mutation does not cause the absence of either of the subunits, since they are immunologically detectable, but does cause the formation of an unstable dimer.

Fabry Disease

Fabry disease is an X-linked disorder of glycosphingolipid metabolism. α-Galactosidase A is completely missing in hemizygous males and is present at intermediate levels in heterozygous females. Female carriers can be detected by hair root analyses (158), and also by the presence of corneal dystrophy, angiokeratoma, or renal disease. Carriers with neurological manifestations have also been reported (159). The enzyme deficiency in Fabry disease results in the accumulation of, mainly, trihexosyl ceramide, a compound that also has been partly (160) and totally (161) synthesized. Johnson and Desnick (162) studied α-galactosidase A in cultured human endothelial cells and found physical and kinetic properties similar to the α-galactosidase A properties for other enzyme sources. They suggest that the preferential uptake of trihexosyl ceramide by the vascular endothelial cells is mediated by lipoprotein and involves a receptor on these cells. Poulos and Beckman (163) have studied α-galactosidase from human leukocytes, using radioactively labeled trihexosyl ceramide as a substrate. They found the enzyme inactive in the absence of detergent. Crude sodium taurocholate was the best detergent. Leukocyte extracts from a patient with Fabry disease had approximately 10% of the activity found in corresponding extracts obtained from normal people. The possibility of lowering the level of trihexosyl ceramide in the plasma of patients with Fabry disease was investigated (164). α-Galactosidase was immobilized by coupling to Sepharose 4 B. The immobilized enzyme was able to hydrolyze trihexosyl ceramide in an artificial system in which sodium taurocholate was used to solubilize the substrate, but no hydrolysis of the glycolipid could be detected with normal plasma or with plasma from a patient with Fabry disease.

Tay-Sachs and Sandhoff Disease

The hexosaminidase (hex) enzyme in man exists in two major forms, designated hex A and hex B. Two inherited diseases that are lethal in early childhood are associated with deficiencies of hexosaminidase. Hex A is deficient in Tay-Sachs disease, while hex A and hex B are deficient in Sandhoff disease. A few reports on new variants of these disorders have been published (165-170). Prenatal diagnosis of Tay-Sachs disease in amniotic cells (171,172) and in cell-free amniotic fluids (173,174) has been carried out. Additional reports on carrier detection have also been published (175-177). The specificity of human placental (178) and human liver (179) N-acetyl-β-D-hexosaminidase isozymes has been studied. Native hyaluronic acid could be hydrolyzed with placental hex A or hex B. Trisaccharides from hyaluronic acid and desulphated chondroitin-4-sulphate were hydrolyzed by liver hex A but not by liver hex B. Using antisera raised against cross-linked human placental hex A, cross-reacting material was detected in liver samples from eight unrelated patients with Tay-Sachs disease, suggesting a mutation in the structural gene coding for the α-subunit of hex A (180). Wood (181) studied the enzyme production after fusion between juvenile Sandhoff, Sandhoff and Tay-Sachs fibroblasts in various paired combinations. Only fusion of juvenile Sandhoff (or Sandhoff) with Tay-Sachs fibroblasts showed an increase of hex A, indicating that the genetic defect in juvenile Sandhoff disease probably represents an allelic mutation of the gene that is defective in Sandhoff disease.

Variant AB of infantile GM_2-gangliosidosis was studied by Conzelmann et al (182). Hex A was isolated from the tissue of a patient and was characterized biochemically and immunologically in comparison with an enzyme preparation from normal control tissue. No differences between hex A from normal and variant AB tissue could be detected, indicating that the defect involved in this disease is not at the genetic level of either α- or β-chains of hex A. In a later publication, Conzelmann and Sandhoff (183) were able to demonstrate that patients with the AB variant lack a stimulating factor necessary for the interaction of lipid substrates and the water soluble hydrolase. Finally, a report on the structure of minor oligosaccharides in the liver of a patient with Sandhoff disease has appeared (184).

CYSTIC FIBROSIS (CF)

Cystic fibrosis is the most common autosomal recessive disease among whites. The frequency generally accepted is 1:2,000 but varies in different populations from 1:500 to 1:80,000 (185). The basic biochemical disorder is unknown, but several hypotheses have been put forward (186). The diagnosis of CF is aided by the finding of abnormally high concentrations of sodium and chloride ions in the sweat (187,188), increased albumin levels in meconium, or decreased levels of pancreatic enzymes in feces. There are no generally accepted methods for making prenatal diagnoses or for detection of heterozygotes although strenuous efforts to fulfill these objectives are continuing.

Breslow et al (189) have developed a test based on the observation that cultured skin fibroblasts from CF patients are more resistant to dexamethasone than are normal cells. The method appears promising and may be useful for both heterozygote and prenatal diagnosis.

It has been suggested that cystic fibrosis protein (CFP) is a diagnostic marker (190), but it is claimed that the technique involved for the detection of CFP is too complicated for routine purposes (191). The relationship of CFP to the disease and to CF ciliary dyskinesia factors in serum is not clear. An attempt to fractionate different ciliary dyskinesia factors has been reported (192). A ciliostatic factor in saliva has also been purified (193). It is reported to have a low molecular weight and to be labile to variations in pH and temperature. Bauschbach et al (194) have observed a twofold increase in ^{45}Ca-labeling of normal human leukocytes incubated in the presence of serum of CF patients. A factor in CF serum obviously promotes the uptake of calcium by normal cells. It has also been shown that CF fibroblasts take up significantly more ^{45}Ca than do normal fibroblasts (195). It is not known if this calcium-uptake factor is related to the ciliary dyskinesia factors.

Hösli et al (196) recently have described the induction of alkaline phosphatase in skin-derived fibroblasts after stimulation with Tamm-Horsfall glycoprotein, isoproterenol, and theophylline. The reliability of this method to discriminate among normal people, CF patients, and CF heterozygotes has been tested. All cells from CF patients clearly show higher alkaline phosphatase activity as compared with those from heterozygotic and normal people. Cells of CF heterozygotes display a slight and hardly significant increase in alkaline phosphatase levels as compared with those of normal people.

Liver α-fucosidase (which was purified by affinity chromatography) from normal and CF patients had similar properties, except for a lower carbohydrate content in the CF α-fucosidase (197). Several other lysosomal enzymes were also studied in CF livers and found to be normal (198). CF liver sialyltransferase appeared more thermolabile than this enzyme from normal livers (199). It was suggested that the abnormalities observed with these liver enzymes may be secondary to general liver damage. It has also been reported that of 204 Swiss CF patients 9 had clinically overt liver disease (200).

Disaccharidases in small intestinal mucosa were also studied (201). Decreased activity was seen in CF and other patients with mucosal atrophy. In patients with pancreatic insufficiency, the disaccharidase levels were higher than those in normal subjects. Protease activity was measured in skin fibroblasts from CF patients (202). Reduced activity was observed, compared with normal subjects. The nature of the protease (measured with a 4-methyl-umbelliferyl-guanidino-benzoate substrate) remains to be clarified. Recent work has suggested that the α_2-macroglobulin of CF plasma has decreased ability to form a complex with the proteolytic enzymes trypsin, papain and thrombin (203). However, the ability of α_2-macroglobulin to form a complex with and inhibit plasmin was found to be completely normal.

Concentrations of spermidine in erythrocytes of CF patients are elevated as compared with those of controls. It has now been demonstrated that this elevation is caused by a defect in the conjugation of spermidine that inhibits excretion (204). Apart from the results briefly mentioned above, reports have been published on salivary amylase (205-207), cyclic nucleotides (208,209), prostaglandins and fatty acids (210-212), and hypergammaglobulinemic purpura (213), all related to CF. Gastrointestinal aspects of CF are further described in the Chapter by Forstner in this volume.

DUCHENNE MUSCULAR DYSTROPHY (DMD)

Duchenne muscular dystrophy is inherited as a sex-linked recessive trait. The primary cause of the disease is unknown. Review articles covering mainly the clinical aspects of DMD and related disorders have appeared (214-216). Thomson and Smith (217) have reported clinical improvements, such as increased physical strength, by oral administration of allopurinol. Since allopurinol inhibits xanthine oxidase, purines are retained and recycled to a larger extent. The results support the view that the cause of DMD may have some basis in defective muscle purine metabolism. An extensive review on diagnosis, carrier detection, and genetic counseling was published by Zatz (218). Determination of serum creatine phosphokinase (CK) levels is the most common method used in screening programs, and several articles covering this topic have been published (219-224). Sica and McComas (225) have reviewed experimental evidence for the neuronal hypothesis of DMD. Further evidence in favor of abnormal functioning of cellular membranes has been advanced (226-229). Pickard et al (226) have demonstrated a diminished percentage of lymphocyte-capping ability in DMD and some related disorders. Saturation transfer electron paramagnetic resonance spectroscopy has revealed distinct differences in the spectral intensities of erythrocytes from DMD patients as compared with controls (227). Several studies comparing the chemical composition of erythrocytes in normal people and in DMD patients have been carried out. No significant differences were seen in lipid and protein composition (230), membrane cholesterol (231), sialic acid (231), phospholipids (231,232), protein kinase (233), and adenylate cyclase (233). The activity of transketolase was investigated in the quadriceps muscles of patients with DMD and of patients with some related disorders (234). The activity was particularly high in muscles of DMD patients as compared with the muscles of normal people, indicating a high activity of the pentose phosphate shunt in dystrophic muscle cells.

REFERENCES

1a. Brock DJH, Mayo O (eds): *Biochemical Genetics of Man* ed 2. London, Academic Press, 1978, pp 469-560.

1b. Burton BK, Nadler HL: *Pediatrics* 61:398, 1978.

2. Grover R, et al: *Pediatrics* 61:740, 1978.

3. Kaback MM: *Pediat Clin N Amer* 25:395, 1978.

4. Nevanlinna HR: *Acta Neurol Scand* 57(suppl):37, 1978.

5. Srsen S, Varga F: *Lancet* 2:576, 1978.

6. Harper PS, Bradley DM: *Lancet* 2:576, 1978.

7. Miles JH, Kaback MM: *Pediat Clin N Amer* 25:593, 1978.

8. Holtzman NA: *Pediat Clin N Amer* 25:411, 1978.

9. Nelson MM, et al: *South African Med J* 54:305, 1978.

10. Ryman BE, et al: *Ann NY Acad Sci* 308:281, 1978.

11. Kan YW, Dozy A: *Lancet* 2:910, 1978.

12. Williamson B: *Nature* 276:114, 1978.

13. Marx JL: *Science* 202:1068, 1978.

14. Anderson NG, Anderson NL *Analyt Biochem* 85:331, 1978.

15. Anderson NL, Anderson NG: *Analyt Biochem* 85:341, 1978.

16. Purkiss P, Hughes RC, Watts RWE: *Clin Chem* 24:669, 1978.

17. Purkiss P, Hughes RC, Watts RWE: *Clin Chem* 24:714, 1978.

18. Chester MA, et al: *Monogr Hum Genet* 10:2, 1978.

19. Stantscheff P: *Fresenius' Z Anal Chem* 292:39, 1978.

20. Takeuchi T, Hirata Y, Okumura Y: *Anal Chem* 50:659, 1978.

21. Cori GT, Cori CF: *J Biol Chem* 199:661, 1952.

22. Senior B, Loridan L: *N Engl J Med* 279:958, 1968.

23. Narisawa K, et al: *Biochem Biophys Res. Commun* 83:1360, 1978.

24. Rosenfeld EL, et al: *Clin Chim Acta* 86:295, 1978.

25. Ehrlich RM, et al: *Amer J Dis Child* 132:241, 1978.

26. Benke PJ, Gold S: *Pediat Res* 12:204, 1978.

27. Emmett M, Narins RG: *JAMA* 239:1642, 1978.

28. Hers HG: *Biochem J* 86:11, 1963.

29. Beratis NG, LaBadie GU, Hirschhorn K: *J Clin Invest* 62:1264, 1978.

30. Reuser AJJ, et al: *Amer J Hum Genet* 30:132, 1978.

31. Bienvenu J, et al: *Clin Chim Acta* 84:277, 1978.

32. Murray AK, Brown BI, Brown DH: *Arch Biochem Biophys* 185:511, 1978.

33a. Pena SDJ, Wrogemann K: *Pediat Res* 12:887, 1978.

33b. Pena SDJ, et al: *Pediat Res* 12:894, 1978.

34. Dreyfus J-C, Poënaru L: *Biochem Biophys Res Commun* 85:615, 1978.

35. DiMauro S, Hartlage PL: *Neurology* 28:1124, 1978.

36. Levy HL, Hammersen G: *J Pediat* 92:871, 1978.

37. Schutgens RBH, Berntssen WJM, Pool L: *Clin Chim Acta* 86:301, 1978.

38. Beutler E, Matsumoto F: *Lancet* 1:1161, 1978.

39. Levy HL, et al: *J Pediat* 92:390, 1978.

40. Schapira F, et al: *Biomedicine Express* 29:136, 1978.

41. Odièvre M, et al: *Amer J Dis Child* 132:605, 1978.

42. Taunton OD, et al: *Biochem Med* 19:260, 1978.

43. Committee on Nutrition, American Academy of Pediatrics: *Pediatrics* 62:240, 1978.

44. Isemura M, et al: *J Biochem* 84:627, 1978.

45. Thompson JN: *Clin Chim Acta* 89:435, 1978.

46. Ginsberg LC, Di Ferrante DT, Di Ferrante N: *Carbohyd Res* 64:225, 1978.

47. O'Brien JF, Emmerling ME: *Analyt Biochem* 85:377, 1978.

48. Mitra SK, Blau K: *Clin Chim Acta* 89:127, 1978.

49. Fortuin JJH, Kleijer WJ: *Clin Chim Acta* 82:79, 1978.

50. Spranger J, et al: *Eur J Pediat* 129:11, 1978.

51. Klein U, Kresse H, von Figura K: *Proc Nat Acad Sci USA* 75:5185, 1978.

52. Ginsberg LC, et al: *Pediat Res* 12:805, 1978.

53. Di Ferrante N, et al: *Science* 199:79, 1978.

54. Horwitz AL, Dorfman A: *Biochem Biophys Res Commun* 80:819, 1978.

55. Glössl J, Kresse H: *Clin Chim Acta* 88:111, 1978.

56. Gitzelmann R, et al: *Helv Paediat Acta* 33:413, 1978.

57. Bonafede RP, Beighton P: *Clin Genet* 14:24, 1978.

58. Rastogi SC, et al: *Clin Chim Acta* 84:173, 1978.

59. Hollister DW: *Pediat Clin N Amer* 25:575, 1978.

60. Elsas LJ II, Miller RL, Pinnell SR: *J Pediat* 92:378, 1978.

61. Renlund M, et al: Personal communication.

62. Burditt LJ, Chotai KA, Winchester BG: *FEBS Lett* 91:186, 1978.

63. Bach G, et al: *Pediat Res* 12:1010, 1978.

64. Burton BK, Nadler HL: *Enzyme* 23:29, 1978.

65. Burditt LJ, et al: *Biochem J* 175:1013, 1978.

66. Healy PJ, Nicholls PJ, Butrej P: *Clin Chim Acta* 88:429, 1978.

67. Maley F: *Carbohyd Res* 64:279, 1978.

68. Lundblad A, et al: *Eur J Biochem* 83:513, 1978.

69. Strecker G, et al: *Biochimie* 60:725, 1978.

70. Nishigaki M, et al: *J Biochem* 84:823, 1978.

71. Alhadeff JA, Andrews-Smith GL, O'Brien JS: *Clin Genet* 14:235, 1978.

72. Thorpe R, Robinson D: *Clin Chim Acta* 86:21, 1978.

73. Strecker G, Michalski JC: *FEBS Lett* 85:20, 1978.

74. Maroteaux P, et al: *Arch Franc Pédiat* 35:819, 1978.

75. O'Brien JS: *Clin Genet* 14:55, 1978.

76. Rapin I, et al: *Ann Neurol* 3:234, 1978.

77. Thomas GH, et al: *Clin Genet* 13:369, 1978.

78. Okada S, et al: *Clin Chim Acta* 86:159, 1978.

79. Koseki M, Tsurumi K: *Tohoku J Exp Med* 124:361, 1978.

80. Dorland L, et al: *Eur J Biochem* 87:323, 1978.

81. Vladutiu GD, Rattazzi MC: *Biochim Biophys Acta* 539:31, 1978.

82. Vladutiu GD: *Biochem J* 171:509, 1978.

83. Kress BC, Miller AL: *Biochem Biophys Res Commun* 81:756, 1978.

84. Mantagos S, et al: *Arch Dis Child* 53:230, 1978.

85. Böhles H, et al: *Amer J Clin Nutr* 31:1808, 1978.

86. Glasgow AM, Kraegel JH, Schulman JD: *Pediatrics* 62:30, 1978.

87. Giordano C, et al: *Nephron* 22:97, 1978.

88. Hsu CH, et al: *N Engl J Med* 298:117, 1978.

89. Knox WE: *N Engl J Med* 299:715, 1978.

90. Kaufman S, et al: *N Engl J Med* 299:673, 1978.

91. Goldsmith LA: *Exp Cell Biol* 46:96, 1978.

92. Thoene J, et al: *J Pediat* 92:108, 1978.

93. Reddi OS: *J Pediat* 93:814, 1978.

94. Hagenfeldt L, Larsson A, Andersson R: *N Engl J Med* 299:587, 1978.

95. Porath U, Schreier K: *Deutsche Medizinische Wochenschrift* 103:939, 1978.

96. Larsson A, Holmgren A, Bratt I: *FEBS Lett* 87:61, 1978.

97. DiDonato S, et al: *Neurology* 28:1110, 1978.

98. Ware AJ, et al: *J Pediat* 93:959, 1978.

99. Pascal TA, et al: *Pediat Res* 12:125, 1978.

100. Naylor EW, Guthrie R: *Pediatrics* 61:262, 1978.

101. Taylor J, Robinson BH, Sherwood WG: *Pediat Res* 12:60, 1978.

102. von Wendt L, et al: *Neuropädiatrie* 9:360, 1978.

103. de Groot CJ, et al: In *Maturation of the Nervous System. Progress in Brain Research* vol. 48. Amsterdam, Elsevier, 1978, pp 199-205.

104. Frazier DM, Summer GK, Chamberlin HR: *Amer J Dis Child* 132:777, 1978.

105. Gates SC, Dendramis N, Sweeley CC: *Clin Chem* 24:1674, 1978.

106. Gates SC, et al: *Clin Chem* 24:1680, 1978.

107. Gates SC, Sweeley CC: *Clin Chem* 24:1663, 1978.

108. Alm J, Hagenfeldt L, Larsson A: *Ann Clin Biochem* 15:245, 1978.

109. Morrow G III, et al: *Clin Chim Acta* 85:67, 1978.

110. Matsuda I, et al: *Eur J Pediat* 128:181, 1978.

111. Packman S, et al: *J Pediat* 92:769, 1978.

112. Shafai T, et al: *J Pediat* 92:84, 1978.

113. Sweetman L, et al: *Biomed Mass Spectrom* 5:198, 1978.

114. Cohn RM, et al: *N Engl J Med* 299:996, 1978.

115. Yudkoff M, et al: *J Pediat* 92:813, 1978.

116. Pentchev PG, Barranger JA: *J Lipid Res* 19:401, 1978.

117. Adachi M, Schneck L, Volk BW: *Acta Neuropathol* 43:1, 1978.

118. Schneider EL, et al: *J Med Genet* 15:370, 1978.

119. Gatt S, Dinur T, Kopolovic J: *J Neurochem* 31:547, 1978.

120. Harzer K, et al: *Acta Neuropathol* 43:97, 1978.

121. Sogawa H, et al: *Eur J Pediat* 128:235, 1978.

122. Higami S, et al: *Tohoku J Exp Med* 125:11, 1978.

123. Besley GTN: *Clin Chim Acta* 90:269, 1978.

124. Jungalwala FB, Milunsky A: *Pediat Res* 12:655, 1978.

125. Seidel D, et al: *J Clin Chem Clin Biochem* 16:407, 1978.

126. Kitagawa T, et al: *Amer J Hum Genet* 30:322, 1978.

127. Chazan S, Zitman D, Klibansky C: *Clin Chim Acta* 86:45, 1978.

128. Wenger DA, et al: *Clin Genet* 13:145, 1978.

129. Butterworth J, Broadhead DM: *Clin Chim Acta* 87:433, 1978.

130. Forster J, et al: *J Pediat* 93:823, 1978.

131. Guibaud P, et al: *Arch Franc Pédiat* 35:949, 1978.

132. Grover WD, Tucker SH, Wenger DA:*Ann Neurol* 3:281, 1978.

133. Pentchev PG, et al: *Proc Natl Acad Sci USA* 75:3970, 1978.

134. Turner BM, Hirschhorn K: *Amer J Hum Genet* 30:346, 1978.

135. Choy FYM, Davidson RG: *Pediat Res* 12:1115, 1978.

136. Moffitt KD, et al: *Arch Biochem Biophys* 190:247, 1978.

137. Chiao Y-B, et al: *Proc Nat Acad Sci USA* 75:2448, 1978.

138. Chambers JP, et al: *Metabolism* 27:801, 1978.

139. Chiao Y-B, et al: *Arch Biochem Biophys* 186:42, 1978.

140. Dawson G, Stein AO: *Science* 170:556, 1970.

141. Burton BK, Ben-Yoseph Y, Nadler HL: *Clin Chim Acta* 88:483, 1978.

142. Cheetham PSJ, Dance NE, Robinson D: *Clin Chim Acta* 83:67, 1978.

143. Ben-Yoseph Y, Hungerford M, Nadler HL: *Amer J Hum Genet* 30:644, 1978.

144. Kudoh T, et al: *Hum Genet* 44:287, 1978.

145. Stevenson RE, Taylor HA Jr, Parks SE: *Clin Genet* 13:305, 1978.

146. de Wit-Verbeek HA, Hoogeveen A, Galjaard H: *Exp Cell Res* 113: 215, 1978.

147. Holmes EW, O'Brien JS: *Amer J Hum Genet* 30:505, 1978.

148. Holmes EW, O'Brien JS: *Biochem J* 175:945, 1978.

149. Lundblad A, Sjöblad S, Svensson S: *Arch Biochem Biophys* 188: 130, 1978.

150. Gordon N: *Postgrad Med J* 54:335, 1978.

151. Manowitz P, Ling A, Kohn H: *J Nerv Ment Dis* 166:500, 1978.

152. Luijten JAFM, et al: *Neuropädiatrie* 9:338, 1978.

153. Butterworth J, Broadhead DM, Keay AJ: *Clin Genet* 14:213, 1978.

154. Buonanno FS, et al: *Ann Neurol* 4:43, 1978.

155. Bosch EP, Hart MN: *Arch Neurol* 35:475, 1978.

156. Christomanou H, Sandhoff K: *Neuropädiatrie* 9:385, 1978.

157. Shapira E, DeGregorio RR, Nadler HL: *Pediat Res* 12:199, 1978.

158. Beaudet AL, Caskey CT: *Clin Genet* 13:251, 1978.

159. Bird TD, Lagunoff D: *Ann Neurol* 4:537, 1978.

160. Cox DD, Metzner EK, Reist EJ: *Carbohyd Res* 63:139, 1978.

161. Shapiro D, Acher AJ: *Chemistry and Physics of Lipids* 22:197, 1978.

162. Johnson DL, Desnick RJ:*Biochim Biophys Acta* 538:195, 1978.

163. Poulos A, Beckman K: *Clin Chim Acta* 89:35, 1978.

164. Schram AW, et al: *Biochim Biophys Acta* 527:456, 1978.

165. Lane AB, Jenkins T: *Clin Chim Acta* 87:219, 1978.

166. Felding I, Hultberg B: *Neuropädiatrie* 9:74, 1978.

167. O'Neill B, et al:*Neurology* 28:1117, 1978.

168. Momoi T, et al: *Pediat Res* 12:77, 1978.

169. Johnson WG, Chutorian AM: *Ann Neurol* 4:399, 1978.

170. O'Brien JS, et al: *Amer J Hum Genet* 30:602, 1978.

171. D'Azzo A, Hoogeveen A, De Wit-Verbeek HA: *Clin Chim Acta* 88:1, 1978.

172. Bladon MT, Milunsky A: *Clin Genet* 14:359, 1978.

173. Christomanou H, Cap C, Sandhoff K: *Klin Wschr* 56:1133, 1978.

174. Geiger B, Navon R, Arnon R: *Clin Chem* 24:1131, 1978.

175. Lowden JA, et al: *Amer J Hum Genet* 30:38, 1978.

176. Nakagawa S, et al: *Clin Chim Acta* 88:249, 1978.

177. Nakagawa S, et al: *J Lab Clin Med* 91:922, 1978.

178. Bach G, Geiger B: *Arch Biochem Biophys* 189:37, 1978.

179. Bearpark TM, Stirling JL: *Biochem J* 173:997, 1978.

180. Srivastava SK, Ansari NH: *Nature* 273:245, 1978.

181. Wood S: *Hum Genet* 41:325, 1978.

182. Conzelmann E, et al: *Eur J Biochem* 84:27, 1978.

183. Conzelmann E, Sandhoff K: *Proc Natl Acad Sci USA* 75:3979, 1978.

184. Ng Ying Kin NMK, Wolfe LS: *Carbohyd Res* 67:522, 1978.

185. Warwick WJ: *Helv Paediat Acta* 33:117, 1978.

186. Dann LG, Blau K: *Lancet* 2:405, 1978.

187. Bray PT, et al: *Arch Dis Child* 53:483, 1978.

188. Smalley C, Addy DP, Anderson C: *Lancet* 2:415, 1978.

189. Breslow JL, et al: *Science* 201:180, 1978.

190. Wilson GB, Fudenberg HH: *Pediat Res* 12:801, 1978.

191. Scholey J, et al: *Pediat Res* 12:800, 1978.

192. Wilson GB, Fudenberg HH: *J Lab Clin Med* 92:463, 1978.

193. Impero KE. Harrison GM, Nelson TE: *Pediat Res* 12:108, 1978.

194. Banschbach MW, et al: *Biochem Biophys Res Commun* 84:922, 1978.

195. Shapiro BL, et al: *Clin Chim Acta* 82:125, 1978.

196. Hösli P, Vogt E: *Hum Genet* 41:169, 1978.

197. Alhadeff JA, Watkins P, Freeze H: *Clin Genet* 13:417, 1978.

198. Alhadeff JA: *Clin Chim Acta* 89:469, 1978.

199. Alhadeff JA, Cimino G: *Clin Genet* 13:207, 1978.

200. Schwarz HP, et al: *Helv Paediat Acta* 33:351, 1978.

201. Antonowicz I, Lebenthal E, Shwachman H: *J Pediat* 92:214, 1978.

202. Platt MW, Rao GJS, Nadler HL: *Pediat Res* 12:874, 1978.

203. Choy H, et al: *Biochem Biophys Res Commun* 82:1325, 1978.

204. Rosenblum MG, et al: *Science* 200:1496, 1978.

205. Kenny D, et al: *Clin Chim Acta* 89:429, 1978.

206. Davidson GP, Koheil A, Forstner GG: *Pediat Res* 12:967, 1978.

207. Gillard BK, Markman HC, Feig SA: *Pediat Res* 12:868, 1978.

208. Davis PB: *Clin Chim Acta* 87:285, 1978.

209. Davis PB, Braunstein M, Jay C: *Pediat Res* 12:703, 1978.

210. Dodge JA, Hamdi I: *Lancet* 2:475, 1978.

211. Chase HP, Dupont J: *Lancet* 2:236, 1978.

212. Allué X, Sanjurjo P: *Pediatrics* 61:924, 1978.

213. Nielsen HE, et al: *Acta Paediat Scand* 67:443, 1978.

214. Furukawa T, Peter JB:*JAMA* 239:1537, 1978.

215. Furukawa T, Peter JB: *JAMA* 239:1654, 1978.

216. Appel SH: *Postgrad Med* 64:93, 1978.

217. Thomson WHS, Smith I: *Metabolism* 27:151, 1978.

218. Zatz M: *Pediat Clin N Amer* 25:557, 1978.

219. Emery AEH: *Lancet* 1:205, 1978.

220. Thompson CE: *N Engl J Med* 298:1479, 1978.

221. Beckmann R, et al: *Lancet* 2:105, 1978.

222. Gardner-Medwin D, Bundey S, Green S: *Lancet* 1:1102, 1978.

223. Drummond L, Veale AMO: *Lancet* 1:1258, 1978.

224. Dellamonica C, et al: *Lancet* 2:1100, 1978.

225. Sica REP, McComas AJ: *Canad J Neurol Sci* 5:189, 1978.

226. Pickard NA, et al: *N Engl J Med* 299:841, 1978.

227. Wilkerson LS, et al: *Proc Nat Acad Sci USA* 75:838, 1978.

228. Grassi E, et al: *Neurology* 28:842, 1978.

229. Lloyd SJ, Nunn MG: *Brit Med J* 2:252, 1978.

230. Kobayashi T, Mawatari S, Kuroiwa Y: *Clin Chim Acta* 85:259, 1978.

231. Godin DV, Bridges MA, MacLeod PJM: *Res Commun Chem Path Pharm* 20:331, 1978.

232. Koski CL, Jungalwala FB, Kolodny EH: *Clin Chim Acta* 85:295, 1978

233. Fischer S, et al: *Clin Chim Acta* 88:437, 1978.

234. Banerji AP, et al: *Clin Chim Acta* 86:307, 1978.

Editor's Note: Since going to press, the following additional references were received from the authors and are cited in the text.

64a. Phillips NC, Thorpe R: *Clin Chim Acta* 83:93, 1978.

82a. Ullrich K, et al: *Biochem J* 170:643, 1978.

13
Therapeutic Drug Monitoring

Charles E. Pippenger

INTRODUCTION

Drug therapy is usually aimed at abolishing or alleviating an acute or chronic pathological state. In chronic diseases with only occasional clinical manifestations, e.g. hypertension, asthma and the epilepsies, the clinician is largely dependent upon trial and error to ascertain the appropriate drug dosage for a particular patient that will produce a therapeutic drug concentration and thus the desired therapeutic response. The physician is never quite sure of success until some time has elapsed without the reappearance of symptoms. However, trial and error therapy places both the patient and the physician at the mercy of an unknown factor - - the utilization rate of the drug in a particular individual.

For hundreds of years, optimal drug dosages were established by trial and error. A patient was given a fixed quantity of drug and if the desired effect did not result, the dosage was increased until it was achieved or until clinical signs of toxicity appeared at which point the dosage was reduced. If the expected response still did not occur, a second drug was added to the patient's medication regimen or substituted for the original drug. If a response to the second drug was not forthcoming, the process was repeated until the desired effect was achieved or until all possible drugs were explored and exhausted. With many drugs it is no longer necessary to adjust therapy by trial and error. Today, therapeutic drug monitoring allows more accurate titration of dosage and thus greater individualization of drug therapy.

Only within the last three decades has a clearer understanding of the relationship between a drug's concentration within a biologic system and its pharmacologic activity become possible. Our current ability to monitor drugs and correlate their plasma concentrations, and by inference their tissue concentrations, with the observed therapeutic effects had to await the development of highly specific, reproduceable technologies. Monitoring serum drug concentrations allowed investigators to establish that a minimum effective concentration (MEC) of a given drug in plasma was necessary to elicit that drug's therapeutic effect. Furthermore, it is now accepted that there are optimum plasma concentration ranges within which therapeutic effects can be expected to occur in most patients receiving a particular drug. If plasma concentrations remain below the optimum therapeutic range the desired effect will not be achieved in most patients. Conversely, should plasma concentrations exceed the optimum therapeutic range, undesirable side effects or toxicity can be expected in most patients.

Because of problems in the preparation of this Chapter and the relative novelty of therapeutic drug monitoring compared with well-established areas of Clinical Biochemistry, this first article in the series will address broad issues of principle to provide the reader with a general background. Specific topics will be comprehensively covered in future volumes of this series. References as such will not be cited in the text, but a Bibliography based on recent key papers listed in alphabetical order of first authors, complete with titles of articles, is appended to assist the reader seeking more detailed information.

FUNDAMENTAL PRINCIPLES OF THERAPEUTIC DRUG MONITORING

General Remarks

It is the responsibility of any individual or laboratory engaged in routine therapeutic drug monitoring (TDM) to have a clear knowledge and understanding of the fundamental principles of clinical pharmacology. Failure to achieve this understanding results in a laboratory which, though capable of measuring serum drug concentrations, in reality is incapable of providing the necessary supporting information which is essential to ensure that the requesting physician and ultimately the patient achieves the maximum benefit from the assay.

The biological or pharmacological effect achieved following a given drug dose is a direct consequence of the formation of reversible bonds between the drug and tissue receptors controlling a particular response. For most drugs, the intensity and duration of a given pharmacological effect is proportional to the drug concentration at the receptor site. The exact mechanisms of drug-receptor interactions remain unclear. For a drug to exert the desired biological effect, it must reach and interact with the receptors regulating that specific response. Many factors can alter the concentration of drug ultimately achieved and maintained at a given receptor site.

Factors, such as age, sex, patient compliance with his prescribed drug regimen, individual differences in drug metabolism and excretion, and drug interactions (particularly during multiple drug therapy) contribute to the disposition of a drug within an individual patient, and thus, to the observed therapeutic response. Interactions between all the potential factors which influence drug therapy account for the broad inter-patient variability observed in serum drug concentrations following a single dose of a drug. Individual response to a given drug dose, however, remains constant because the factors which can alter drug utilization within the individual are relatively fixed.

Generally, inter-individual variations in response depend more on the relationship between total daily dose and plasma concentration than on the relationship between plasma concentration and intensity of the response. In other words, the probability of achieving a given blood level from a given drug dose is much less than the probability of obtaining a specific biological effect from a given plasma concentration of that drug. This is why drugs administered at fixed doses to large numbers of patients will produce marked variations in biological response. With a fixed or standard drug dosage, the desired therapeutic effect will be achieved in some patients; no therapeutic effect will occur in other patients; and clinical signs of drug intoxication usually associated with drug overdosage will be evident in still other patients. The titration of drug dosage to obtain a desired therapeutic plasma concentration can successfully eliminate the failure to achieve the desired biological effect which is a consequence of inter-individual variability.

Patient Compliance

Therapeutic drug monitoring has been well-established as a valuable tool for ensuring that the patient's serum drug concentrations are within the optimal therapeutic range. The most common cause of suboptimal drug levels and consequent failure to achieve the desired therapeutic response is patient noncompliance.

It has been suggested that over 60% of all patients do not take their drugs in the manner prescribed by the attending physician. Whenever a patient presents with consistently low serum drug concentrations when receiving an adequate drug dosage, noncompliance should always be considered as the probable cause. Noncompliance outside of hospital can usually be demonstrated by careful observation of the patient's daily drug intake over a specified interval of time (usually 5 drug half-lives) in a hospital setting, with frequent monitoring of serum drug concentrations. If the serum concentrations increase over the time interval selected, the patient was noncompliant outside of hospital. If the serum concentrations remain low other factors such as drug malabsorption or rapid drug metabolism should be suspected.

FUNDAMENTALS OF DRUG DISPOSITION

Drug Absorption

There are three major routes of drug administration: parenteral (intravenous or intramuscular), rectal, and oral. Parenteral administration is usually employed only in a hospital environment or physician's office. The entry of drugs into the general circulation by either the intravenous or the intramuscular route is generally rapid. Parenteral drug administration circumvents the problems associated with oral or rectal administration. Drug absorption following rectal administration is often erratic. Rectal administration is usually utilized only when patients cannot take drugs by mouth. Most drugs are administered orally. Following oral administration many factors can alter the amount of drug absorbed or the rate of absorption from the gastrointestinal tract into the general circulation. These include the solubility of the drug, the type of drug preparation, and simultaneous ingestion of food or other drugs.

Methods of drug manufacture can dramatically alter drug absorption. Brand-name drugs on which the patents have expired may be manufactured as generic drugs. A generic drug contains the same active chemical as the brand-name agent. For example, Dilantin is Parke-Davis' registered brand-name for sodium phenytoin. After patent expiration, manufacturers can market their own generic drug formulations. It does not necessarily follow that because a generic tablet contains the same drug, that the drug will be absorbed or distributed within the body in the same manner as a brand-name preparation of the drug. Any generic agent that does have the same absorption patterns as a brand-name compound is said to be bioequivalent; that is, the pharmacological responses to a given dose of the generic preparation are identical to those observed following the same dose of the brand-name preparation.

Slight changes in manufacturing procedures can dramatically alter a drug's bioavailability (absorption patterns). In 1968, a major outbreak of phenytoin intoxication occurred in Australia, when one manufacturer of phenytoin substituted lactose for calcium sulfate as a filler in its phenytoin preparation. This change was not originally considered significant by the manufacturer. Therefore, physicians prescribing the drug were not notified of the manufacturing change. When this new lactose formulation reached the marketplace and was ingested by patients, there was a massive outbreak of phenytoin intoxication. Subsequent investigation clearly established that phenytoin prepared with calcium sulfate is not as completely absorbed from the intestine as phenytoin prepared with lactose. Therefore, patients who had been titrated to appropriate serum concentrations of phenytoin on the calcium sulfate preparation actually absorbed significantly greater amount of phenytoin from the lactose preparation. Consequently, serum phentoin levels became elevated and the patients developed clinical signs of phenytoin intoxication.

Some patients on appropriate therapeutic doses of a drug will fail to achieve the desired pharmacological response. Generally, these patients have consistently low serum drug concentrations and are classified as either noncompliant or fast drug metabolizers. However, before one classifies a patient as a fast metabolizer, one should verify his or her ability to absorb the drug. Malabsorption of a drug administered orally can often be confirmed by measuring sequential plasma drug concentrations at given time intervals after the dose. The same procedure should be repeated following parenteral administration of the same dose of the drug as had previously been given by mouth. If the problem is malabsorption, the maximum plasma concentrations following an intravenous dose will be significantly higher than those achieved following the oral dose. Conversely, if the patient is a fast drug metabolizer, there will be no significant differences in the plasma concentrations achieved regardless of the route by which the drug was administered.

Plasma-Protein Binding

If a drug has entered the systemic circulation and is capable of binding to the plasma proteins, an equilibrium between free and bound drug will be established. By definition "bound drug" is that portion of a drug that is bound to plasma proteins. Bound drug is unable to cross cell membranes and consequently exerts no biological effect. Unbound or "free" drug is dissolved in the plasma water and can be transported across cell membranes. Thus, only free drug can cross biological membranes and interact with specific receptors to elicit the desired pharmacological response.

Each drug has its own characteristic protein-binding patterns, which are dependent on the physical and chemical properties of that drug. Drugs are either tightly or loosely bound, depending upon their affinity for the plasma proteins. A drug with greater affinity for plasma proteins can displace a weakly-bound drug from its plasma protein binding site. Protein-binding is also dependent on the physical characteristics of the plasma proteins, the presence or absence of other drugs or fatty acids, and the physical properties of the drug itself. It has been clearly demonstrated that fatty acids can interfere with protein-binding by displacing drugs from their protein-binding sites. Those drugs that are tightly bound will not be displaced quite so rapidly. Displacement of a drug from its plasma protein-binding site can, under certain circumstances, elevate free-drug concentrations at the tissue receptor sites with resultant clinical toxicity, even though the total plasma drug concentration remains unchanged.

Certain disease states can affect the protein-binding of a drug. Uremic patients lack the ability to completely bind phenytoin and other drugs to plasma proteins. This lack of binding capacity ranges from patients who can bind no phenytoin to patients who can bind only 10-70% of the total phenytoin present in plasma. If a patient could bind no phenytoin, levels of 1-2 μg/ml would be therapeutic and plasma concentrations above 2.5 μg/ml would result in toxic symptoms. This contrasts with normal patients where phenytoin is 90% bound and the therapeutic plasma concentration is 10-20 μg/ml.

When a patient presents with either clinical toxicity or a non-therapeutic response and the total plasma drug concentrations are known to be at the therapeutic level, one should consider the possibility of altered protein-binding. At present, the determination of protein-binding is a time-consuming and tedious procedure. The protein-binding status of a given patient can be assessed indirectly by measuring salivary drug concentrations, since only free drug is present in saliva. A word of caution is needed: for any drug that has an ionization constant (pK_a) that is significantly different from the pH of plasma (e.g. phenytoin), the salivary levels are a good indicator of free-drug levels. However, for any drug that has a pK_a that is close to the pH of plasma or saliva (e.g. phenobarbital), measurement of salivary concentrations will not reflect the true free-drug concentrations. Descriptions of the relationship between free-drug levels in saliva and plasma have been published.

Drug Metabolism

Any foreign compound that enters the body must be eliminated. As one studies the phylogenetic scale from fish to man, there is a progressive increase in the ability of the body to alter foreign compounds into compounds which are more water soluble and thus readily excreted. It is generally believed that the ability of the liver to metabolize drugs evolved as a mechanism for detoxification of substances ingested as food. Fish have no major drug metabolizing system; in contrast, man has an extensive and complex drug metabolizing system.

The liver drug metabolizing enzymes are nonspecific and interact with a wide variety of chemical structures. The entire purpose of hepatic drug metabolizing systems is to make compounds more water-soluble; therefore, the degradation of organic compounds leads to compounds that are less fat-soluble and more water-soluble. Metabolites of many drugs are conjugated within the liver to either glucuronic acid or sulfates, which increases their water-solubility and consequently the rate of renal metabolite excretion. For example, parahydroxyphenytoin, the major metabolite of phenytoin, is conjugated with glucuronic acid. This conjugation increases its water-solubility almost one hundred times.

The Hepatic Microsomal System

Most drug metabolism takes place within the microsomal fraction of the liver cell. The microsomes are also the site of metabolism of endogenous steroids. Since the enzymatic systems responsible for drug metabolism are not designed to recognize specific drugs, they act upon classes of compounds with similar structures. The same enzyme that is responsible for the hydroxylation of phenytoin is also responsible for the hydroxylation of many drugs which possess a phenyl ring. Therefore, when phenytoin is administered simultaneously with one of these drugs, there may be some clinically significant drug interactions as competition for active sites of the enzyme increases. Clinically, one would expect to see higher serum concentrations of the drug with the least affinity for the enzyme. Phenytoin has a very low affinity for microsomal enzymes. Thus, administration of a drug with a greater affinity for the enzyme than phenytoin will decrease phenytoin's rate of metabolism, and plasma phenytoin concentrations will become elevated.

One of the characteristics of the hepatic microsomal system is that it can be induced to metabolize drugs at a faster rate. Following chronic drug administration, the body, in its attempt to eliminate the drug, synthesizes new protein in the form of enzymes capable of metabolizing that agent. Unfortunately, many misconceptions about the clinical significance of enzyme induction abound. The activity of drug-metabolizing enzyme systems are not necessarily induced with every dosage increment or with the addition of a further drug to the patient's regimen. There is a maximum rate at which protein synthesis can occur. If a patient has been regularly receiving a drug with known enzyme-induction properties, he will achieve a maximum concentration of drug metabolizing enzyme. Therefore, it does not follow that a second drug of similar structure added to the patient's regimen will cause a marked increase in the rate of metabolism of both the first and second drugs, since the drug metabolizing system is already operating at its maximum capacity.

Genetic factors play a major role in determining the ability of a patient to metabolize drugs. It has been well-documented that individuals in certain families metabolize drugs (e.g. phenytoin, isoniazid) at a faster or slower rate than the general population. The importance of identifying fast and slow drug metabolizers cannot be overemphasized. A fast drug metabolizer will require a greater daily dose (mg/kg) of drug to achieve the same serum concentration (and the desired therapeutic response) than a normal individual. A slow drug metabolizer, if given standard drug dosages, will invariably exhibit drug toxicity. The absolute identification of fast and slow drug metabolizers is dependent upon examination of urinary drug metabolite excretion patterns, as well as on sequential determination of plasma drug levels.

The use of plasma drug concentrations alone to identify slow and fast metabolizers can be misleading. Generally speaking, the plasma drug concentrations of slow metabolizers will be significantly higher than would be observed in the general population receiving the same mg/kg/day dosage. Consistently high plasma concentrations on normal or low doses is suggestive of slow drug metabolism. However, a drug interaction that blocks drug metabolism will also result in elevated plasma concentrations.

Fast drug metabolizers usually exhibit consistently low plasma concentrations on standard therapeutic dosages. Often the desired therapeutic response, expected on the basis of the dose given, is not achieved. The most commonly encountered problem associated with patients on long-term therapy is their failure to take medications as prescribed. Since plasma drug levels in noncompliant patients mimic those observed in fast metabolizers, there is a tendency to identify noncompliant patients as fast metabolizers.

One way to distinguish between a fast metabolizer and a noncompliant subject is to hospitalize the patient and supervise intake of medication for a period of five to seven days while monitoring plasma concentrations of the drug. If the patient is indeed a fast metabolizer, there will be no significant change in his serum concentrations over the observation period. However, if the patient is noncompliant, there will be a consistent rise in the plasma concentrations over the time interval. It is advisable to collect 24-hour urine specimens for evaluation of metabolic patterns during this test period. If the patient is noncompliant, it is unnecessary to evaluate these urine specimens. However, if the plasma levels remain low, quantitation of drug metabolite concentrations in the urine specimens will confirm fast drug metabolism.

The clinical status of a patient can also dramatically alter drug utilization patterns. Hepatitis dramatically impairs drug metabolism. Generally, patients with hepatitis will become severely intoxicated when administered therapeutic doses of drugs which are dependent upon hepatic degradation for inactivation (such as phenytoin). Hepatitis leads to a marked decrease in the liver's capacity to metabolize drugs. Renal failure decreases urinary output and may cause elevation of plasma concentrations of drugs (such as phenobarbital) for which the major route of elimination is renal excretion.

Renal Excretion

Urinary excretion is the major pathway for the elimination of both drugs and drug metabolites. For those drugs which are not extensively metabolized any change in renal function will alter plasma concentrations of the drug. If renal function is impaired plasma drug concentrations can become elevated. Uremic patients and patients with congestive heart failure have decreased renal drug clearance. Most drug metabolites are conjugated to glucuronic acid or other compounds and are very water soluble. Thus, drug metabolites are rapidly eliminated even in the presence of a decreased urinary output.

DRUG ACTION

Sites of Action

The site at which a given drug acts to initiate the events which lead to a specific biologic effect is arbitrarily defined as that drug's "site of action" or "receptor site". A drug's biologic effect may occur by direct interaction with a receptor that controls a specific function, or by alteration of the physiologic process which controls that specific function.

Mechanism of Action

The "mechanism of action" of a drug is the means by which, at a given receptor site, it initiates its biological effect. The mechanism of action of most drugs depends upon their chemical interaction with a functionally viable component of the cell. However, since the exact mechanism of action for most drugs is unclear, a theoretical series of models has been developed to explain a given drug's mechanism of action. The fundamental concept upon which these models are based is that intracellular macromolecular receptors exist which, when activated, elicit a specific biologic effect. Drugs are postulated to combine reversibly with these receptors by means of ionic bonds, hydrogen bonds, and van der Waal forces. Interaction between drug and receptor is thought to form a drug-receptor complex of sufficient stability over a given time interval to alter the physiologic function of the target system and consequently to produce the observed pharmacologic effects.

PRACTICAL CONSIDERATIONS IN THERAPEUTIC DRUG MONITORING (TDM)

The Rationale for TDM

It is essential that the clinical pathologist or chemist doing routine drug monitoring understands the principles and techniques of clinical pharmacology as applied to patient care. Such an understanding will not only help him in his TDM work in the laboratory, but will enable him to assist the clinician more effectively in the interpretation of TDM data in order to optimize drug therapy. Clinical experience, supported by a wide variety of pharmacologic studies on dose-effect relationships, has clearly demonstrated that measurement of plasma drug concentrations yields a much better correlation with desired clinical effects than does total daily drug dosage.

For most drugs, the intensity of a pharmacologic effect tends to be proportional to the drug concentration in extracellular fluid. Drug present in extracellular fluid can enter tissues and interact with specific receptors to elicit a biologic effect. For example, antiepileptic drugs are believed to prevent seizures by binding to nerve membranes or by altering neurotransmitter release which is thought to stabilize neuronal membranes against the excessive electrical activity that generates seizures. Drug concentration in extracellular water is in equilibrium with drug concentration in plasma water. The latter, representing free drug concentration, is an indirect measure of drug concentration at the site of action. Many drugs are partially bound to plasma proteins and an equilibrium exists between the concentration of drug bound to plasma proteins and its concentration in plasma water as free drug. Only free drug is capable of crossing the various lipoprotein membranes that surround the receptor site.

Dosage-Concentration Relationships

For most drugs, there is a direct linear relationship between the dosage of drug and the plasma concentration achieved. However, one of the greatest misconceptions in this area is that such a linear relationship exists between all levels of drug dosage and the total plasma concentration, and that as the total dose of a drug increases there is a concomitant, directly proportional, linear increase in its plasma concentrations. This fallacy too often leads to dosage increments in patients unresponsive to lower doses in the misguided notion that an increase in plasma concentration is certain to occur and thus produce the desired effect. Unfortunately, this is not the case. In fact, TDM of certain drugs, such as phenytoin and amitriptyline, demonstrates a linear dose concentration relationship only over a limited range. Beyond this point, a marked elevation of plasma drug concentration completely disproportionate to the dose given can follow a negligible dosage increase. The phenomena operating in these instances, "saturation kinetics", reflect the limited capacity of some drug-metabolizing enzyme mechanisms. Clinically, saturation kinetics are evident when a patient rapidly develops toxicity after a small dosage increment.

One of the major advantages of TDM is that it can "predict" most pharmacologic responses. It does so by either assuring the clinician that plasma concentration is within the optimal therapeutic range in a given patient, or by informing him of subtherapeutic or toxic levels in the patient. Although TDM is not a panacea and the data generated by analyses must be interpreted in conjunction with the patient's clinical status, TDM as part of routine patient care gives the physician a valuable, more precise tool for assessing pharmacologic status, and helps him form a more rational basis for optimizing drug therapy.

TDM and Homeostasis

The ultimate objective of drug therapy is the maintenance of a therapeutic steady-state and/or the return of homeostasis to all systems within the organism. However, differences or alterations in all or any of such processes as drug absorption, plasma protein-binding, rate of drug metabolism, and renal excretion patterns of a drug and its metabolites can affect drug concentrations at receptor sites so as to intensify or reduce the expected pharmacologic effect.

THEORY OF PHARMACOKINETICS

General Remarks

The purpose of this section is to introduce certain fundamental concepts which have been established to explain the relationship between a given drug dosage and the plasma drug concentration attained. The concepts to be introduced represent only a brief overview. Therefore, the reader who wishes to pursue the subject in more depth is referred to the bibliography included at the end of this review.

Anyone who is engaged in routine therapeutic drug monitoring must constantly keep in mind that the plasma concentration achieved and maintained at any given moment in a given patient is a direct consequence of the interaction of a wide variety of processes including drug absorption, distribution, metabolism and drug excretion, as well as the physiological status of the patient over a given time interval. All these factors are inter-related and each plays a role in determining the steady-state concentration of drug which will be achieved on a fixed dosage regimen. The study of these inter-relationships forms the basis of pharmacokinetics.

Pharmacokinetics is the study of the time-course of drug and metabolite levels in different fluids, tissues and excreta of the body and of the mathematical relationships which can be utilized to develop models to interpret the blood level pattern observed in a given patient. In the practical sense, pharmacokinetics, as a discipline, represents an attempt to utilize mathematical models to predict the distribution and excretion patterns of a given drug (usually at steady-state concentrations) following a given dosage regimen. Practical clinical pharmacokinetics has been developed to the point where it is applicable to the study of patients receiving a given drug provided the theoretical limitations of the model are recognized.

Development of the analytical techniques to monitor drug concentrations in biological fluids resulted in early attempts to correlate the relationship between a given milligram per kilogram drug dosage and the observed plasma concentration in large patient populations. The fundamental assumption of these studies was that the patient was at a steady-state, that is, that the intake of drug was constant over a period of time and that drug elimination (as reflected in the rates of drug metabolism and excretion) was constant. Such studies represented early attempts to mathematically define the relationship between drug dose and plasma concentrations.

One of the unfortunate facts of pharmacokinetics is that most of the models do not take into account multiple-drug therapy and the consequent interaction between two drugs by virtue of which the kinetics of both drugs are altered, with subsequent changes in the plasma concentration of each drug. Therefore, the models serve only as a general guideline. A number of computer programs have been developed that predict the plasma concentration which will be achieved following a given dose from prior knowledge of concentration data with respect to time. Unfortunately, these programs and access to the information derived from them are not widely available in clinical chemistry laboratories or to practicing clinicians.

Definitions

For the purposes of most clinical laboratories, a detailed knowledge of pharmacokinetics is not essential although one should be cognizant of the terms and fundamental principles. Some of the relevant terms essential to an understanding of pharmacokinetic principles are listed in the following definitions which have been modified from previously published reviews.

First-Order Kinetics. The rates of any given process associated with drug utilization - - absorption, excretion and metabolism - - are dependent upon concentration such that the rate of the process involved (absorption, excretion, or metabolism) is directly proportional to the concentration of the drug in the system. A graphic plot of serum drug concentration (which represents the balance between the rates of each process) versus total dose (mg/kg) yields a straight line.

Zero-Order Kinetics. At a certain concentration point, drug absorption, excretion, and bio-transformation become independent of concentration. It is believed that all drug processes will convert from first-order to zero-order kinetics, since a point will be reached at which enzyme or transport mechanisms become saturated. Hence a graphic plot of serum drug concentration versus total daily dose (mg/kg) yields an initially straight line (first-order kinetics) that curves sharply upward as the saturation point is reached. The change in rate which occurs beyond the saturation point represents zero-order kinetics.

Fortunately, most drugs never achieve concentrations in the body which approach the transition point from first-order to zero-order kinetics. For most drugs, the serum concentration of a drug achieved at therapeutic dosage is low relative to the drug concentration necessary to saturate the particular system involved. Therefore, first-order kinetics are observed throughout the therapeutic range. There are notable exceptions to this rule since both phenytoin and aspirin exhibit saturation kinetics near the upper limits of the therapeutic range. For any drug that exhibits zero-order kinetics, a very small dosage increment may result in a clinically significant elevation of serum concentration.

Elimination Half-Time (Drug Half-Life). Drug half-life is the time required for elimination from the body of half the concentration of a drug present at the initial starting time. As an example, if the concentration of phenytoin were 20 μg/ml, the time required for the body to clear the drug to a concentration of 10 μg/ml might be 24 hours, provided that no further doses of the drug had been administered. Elimination half-time is commonly referred to as the drug half-life. Elimination half-time is a combined reflection of the individual rates of several different processes. The rates of drug metabolism and excretion are the primary determinants of the elimination half-time in any given patient.

DRUG DOSAGE RELATIONSHIPS

Single Drug Doses

Following drug administration, a peak plasma concentration is reached when the absorption phase is almost complete. The plasma concentration begins to decline, even as drug continues to be absorbed. This decline in plasma concentration is dependent upon the rates of absorption, metabolism and excretion of the drug. Once the absorption phase is complete, the rate of decline in the plasma concentration is a reflection of the elimination rate (the sum of the rates of renal excretion and metabolism of the drug), and the half-life of the drug can be determined by measuring the decline in plasma concentration over a fixed interval with respect to time.

Multiple Drug Doses - - Steady-State

By definition, steady-state is reflected by those plasma concentrations that recur with each dose once an equilibrium is reached between the dose of drug and the amount being eliminated.

When oral long-term drug therapy is initiated, the drug will continue to accumulate within the body until such time as the rate of elimination (which constitutes all tissue distribution, metabolic and renal processes involved in drug disposition) equals the rate of administration. Over a period of time, body and plasma drug concentrations will increase exponentially until they reach a steady-state plateau. It requires 7 half-lives of drug administration before the steady-state concentration is achieved and stabilized. Steady-state processes are 97% complete within 5 half-lives. As a rule of thumb for practical purposes, 5 times the half-life of any drug approximates the time required to achieve steady-state. For example, phenytoin, which has a half-life of 24 hours, requires a period of 24 x 5 or 120 hours (5-6 days) to achieve a steady-state level. In contrast, drugs such as primidone or valproate, which have half-lives of 6-8 hours, require only 30-40 hours to reach steady-state levels.

It is important to note, the same principles which govern the gradual accumulation of a drug to steady-state levels also apply when the drug is discontinued. That is, if plasma phenytoin is at a steady-state concentration, and drug administration is completely discontinued, there will be a period of 5-6 days (5-6 half-lives) before the drug is completely eliminated from the body. This is why drugs with prolonged half-lives can still be detected in serum for 3-4 weeks after administration of the last drug dose. Phenobarbital, whose half-life is four days, is an example of such a drug.

The drug concentration during steady-state is directly proportional to the patient's elimination half-time. Thus, if two patients differ in their respective rates of elimination of a drug, the daily drug dose necessary to maintain the same steady-state concentration in the two patients will differ by the same factor that the rates of drug elimination differ. This is why individualization of drug therapy is necessary to ensure appropriate steady-state concentrations. Every individual has a different elimination half-time on a fixed dosage regimen which is dependent on his overall physiological status at that point in time.

Steady-state drug concentrations are usually maintained in individual patients by various combination of total drug dosage and dosage intervals. As a rule of thumb, in order to maintain a smooth constant, steady-state drug concentration without excessive fluctuations, a drug dosage interval should be one-half of that drug's half-life. For example, phenytoin would be given every 12 hours, since its half-life is 24 hours, whereas primidone would be given every 3 hours since its half-life is 6 hours. Short dosage intervals are often impractical but drugs with short half-lives should at least be administered once each half-life. The important point is to maintain the valley drug concentration within the therapeutic range without the peak concentration reaching toxic levels. As long as the dosage schedule and the interval between doses are selected so that there are not significant fluctuations between peak (the highest drug concentration achieved) and valley (the lowest concentration achieved) concentrations of drug within the dosage interval, appropriate steady-state concentration will be maintained. If the dosage interval is too long with respect to the half-life of the drug, the serum concentrations just prior to the next dose (valley concentration) may be insufficient to provide the desired therapeutic effect. For example, a patient whose phenytoin level falls from 14 µg/ml to 9 µg/ml may have a seizure at the lower level but not at the higher. For any drug, it is possible to maintain plasma concentration within the therapeutic range at all times by adhering to appropriate dosage schedules.

If drug absorption is very rapid, or the drug dosage intervals are excessively short, it is possible for a patient to experience transient periods of drug intoxication, which can present with any of the side effects normally associated with the drug. At the time of clinical symptoms the plasma concentrations are also above the therapeutic range. These symptoms usually appear transiently at fixed intervals following drug administration throughout the day. This toxicity is attributed to a peak plasma concentration above the optimal therapeutic range shortly after drug administration. Such transient side effects can often be eliminated by increasing the drug dosage interval.

APPLICATION OF PHARMACOKINETIC PRINCIPLES

General Remarks

Laboratories engaged in routine therapeutic drug monitoring can be divided into two groups: *(1)* those which perform analytical assays, and simply report the value; *(2)* those which perform the analytical assay but also have the capability to interface with the attending physician with respect to the pharmacological status of the patient. Every laboratory engaged in routine therapeutic monitoring should fall into the second category. The ultimate responsibility of any laboratory engaged in therapeutic drug monitoring is to ensure that all available information relevant to the patient's pharmacological profile can be utilized to optimize his therapeutic regimen.

Clinical pharmacokinetics is a valuable tool for understanding and interpreting the response of an individual patient to a given drug regimen. A wide number of texts containing detailed mathematical derivations of the fundamental principles of clinical pharmacokinetics are available. Computer programs which apply these principles to dosage calculations for individual patients are also available. These programs can be utilized to calculate expected plasma concentrations of any given drug which will be achieved over a fixed time interval following a given dose. A series of simple rules exist which will generate approximately the same information as the computer programs without the necessity of complex mathematical formulas or computer programs.

Patient Information

Wide individual variability exists in patient utilization of drugs as a direct consequence of genetic factors, multiple drug therapy, age and weight. It is extremely difficult and dangerous to generalize about the relationship between plasma concentrations and drug dose. Rather, it is necessary to apply the fundamental principles of clinical pharmacology to achieve the desired pharmacological effect without introducing unwanted drug side effects. The importance of individualized drug therapy is particularly significant in children who utilize drugs at a faster rate than do adults. Conversely, geriatric patients generally utilize drugs at a slower rate than the general adult population.

In order to derive as much information as possible about the pharmacological status of the patient, each laboratory engaged in therapeutic drug monitoring should have the following information available: *(1)* the patient's age - - it is clearly established that there are marked differences in drug utilization which are age dependent; *(2)* weight - - the weight of the patient is essential for mathematical calculations of the relationship between drug dose and plasma concentration; *(3)* all drugs the patient is receiving - - knowledge of the drugs the patient is receiving, in addition to the drug being monitored, is essential for identification of potential drug interactions which have been reported to alter plasma concentrations; *(4)* total daily dosage of drugs - - knowledge of the total daily dosage for each drug administered is necessary to mathematically determine the patient's total daily drug dose in mg/kg. Without this information, it is impossible to correlate the patient's actual plasma drug concentration with his expected plasma drug concentrations; *(5)* the time the last dose of drug was administered and the time the blood specimen was drawn - - without this information it is impossible to know whether the actual plasma concentration represents a peak or trough level. This information is extremely important for accurate interpretation of plasma levels of drugs with short half-lives such as theophylline and lidocaine; and *(6)* the clinical status of the patient - - it is well established that acute or chronic diseases can alter drug utilization patterns dramatically. This is particularly important for regulation of drug-therapy in patients with hepatitis or renal failure. Without knowledge of the clinical status of the patient, it is impossible for those who are interpreting the drug levels to introduce the factors necessary to correct for altered drug utilization associated with a given disease state.

If information with respect to all of the components described above is available, a great deal of insight into the pharmacological status of the patient can be obtained which will allow individualization of drug therapy.

Dose-Response Ratio

For all drugs which exhibit first order kinetics there is a direct relationship between the total daily dose and the plasma concentration achieved at steady-state. By dividing the observed plasma concentration in a given patient by the patient's total daily dose in mg/kg the contribution of each mg/kg of drug administered to the total steady-state plasma concentration can be obtained.

$$\text{Dose Response Ratio} = \frac{\text{drug plasma concentration } (\mu g/ml)}{\text{total daily drug dose } (\mu g/ml}$$

Example: Dose Response Ratio for phenobarbital - $\text{DRR} = \dfrac{30\ \mu g/ml}{3\ mg/kg} = 10.$

Utilizing the DRR, it is possible to calculate a patient's expected plasma concentration at steady-state as follows:

Expected Plasma Concentration = DRR (μg/ml/mg/kg) X total daily drug intake (mg/kg).

DRR for many drugs have been established, based upon data obtained from routine monitoring of large patient population. Therefore, if one knows the DRR, the plasma concentration which would be expected in the average patient of a general population can be calculated on the basis of the patient's total daily drug intake (mg/kg). For example: a patient weighs 60 kilograms and receives a total daily dose of 180 mg. Total daily dose 180/60 kg = 3 mg/kg/day. If the patient is 26 years old then expected plasma concentration is as follows:

$$\frac{10\ \mu g/ml}{mg/kg} \times 3\ mg/kg = 30\ \mu g/ml$$

Thus a phenobarbital concentration of 30 μg/ml at steady-state would be expected. On the other hand, if a patient weighing 80 kilograms was receiving the same mg phenobarbital dose, one would expect to achieve a plasma concentration of 21 μg/ml on the basis of a total daily dose of 2.1 mg/kg/day. Since the drug utilization rates for children are higher than adults, different age dependent DRR factors are used. For example: the DRR for phenobarbital is 5 in children 6-10 years. Therefore, a 7-year old child receiving 3 mg/kg/day of phenobarbital would be expected to have phenobarbital steady-state concentration of 15 μg/ml (5 μg/ml/mg/kg x 3 mg/kg = 15 μg/ml).

All drugs should be administered on the basis of body weight. Instead of a patient receiving an arbitrary dose of drugs, such as 300 mg of phenytoin a day, a patient should be given a dose based on mg/kg/day; most epileptic patients should receive 5 mg/kg/day, in order to achieve and maintain optimal therapeutic drug concentrations at steady-state. The reason for the great concern about weight is that if one utilizes simply so-called therapeutic doses of a drug without considering patient weight, and the patient is excessively overweight, the plasma concentration achieved on the standard therapeutic dose will be below the therapeutic range, whereas plasma levels achieved in a patient who is under the normal weight for the general population will be elevated.

It has been demonstrated that approximately 50% of the patients receiving "the standard therapeutic dose of phenytoin" - - 300 mg/day - - achieved steady-state plasma concentration levels which were below the optimal therapeutic range. Modification of dosage regimens to take into account patient weights have resulted in a significant improvement in the number of patients who are maintained within the optimal therapeutic range with the desired pharmacological effect.

Plasma concentrations for drugs with first-order kinetics during the steady-state are linearly related to dose. For any change of dosage, after a steady state has been achieved, the principles regulating the time required to achieve a *new* steady-state still apply. To illustrate this important principle, if the maintenance dose of a drug were doubled, the new steady-state concentration would not double until the completion of 5 to 6 half-lives. If a plasma drug concentration is determined before achievement of a steady-state, for example after two half-lives, it will not reflect the true steady-state concentration of the drug. Plasma concentrations measured at steady-state provide the most valuable information enabling individualization of drug therapy. Measurement of plasma concentrations before steady-state is achieved does not yield as much information about the patient as levels measured at steady-state. Before steady-state, changes in the rate of elimination, distribution and utilization of a drug are constantly fluctuating. It is possible, if one knows the time of initiation of drug therapy to extrapolate predicted steady-state plasma concentrations by correcting for the number of half-lives before sampling.

Drug half-life by definition is the time required for elimination of half the plasma concentration of drug present at the initial starting time. It must be remembered that half-life is in reality the elimination half-time of a drug. The half-life of the drug is one of the most important pieces of information for anyone attempting to interpret blood levels. Multiplication of the half-life times 5 yields the time required to achieve steady-state drug concentrations.

Elimination Rate (Half-Life)

By definition the half-life of a drug is the time necessary for the plasma concentration to decrease by 50%. There are many factors which can alter the half-life of a drug. Since half-life is dependent upon the rates of drug metabolism and excretion, it is the sum total of these two processes which define a drug's elimination rate. Elimination rate does not necessarily refer to the actual elimination of the drug from the body, since drug metabolism converts drugs from pharmacologically active to pharmacologically inactive compounds. Thus the pharmacological activity of the drug may be eliminated even though the metabolite is still present in the body. Analytical techniques for routine therapeutic drug monitoring do not measure drug metabolites as parent compound. Therefore, therapeutic drug levels can serve as an indicator of altered drug disposition.

Any factor which alters the drug half-life will alter the drugs' steady-state concentration. During multiple drug therapy two drugs may compete for the same metabolic site. This competition will decrease the rate of metabolism of the drug which is excluded from the metabolic site and prolong its half-life. Since the half-life is prolonged a new higher steady-state drug concentration will be achieved. For this reason, it is necessary to be able to identify all drugs a patient is receiving which are potentially capable of altering a given drug half-life. Most clinically significant drug interactions are readily identifiable in the presence of elevated plasma concentrations of a given drug. Generally, it will be observed that the adjunctive drug has a metabolic pathway which is similar to that of the therapeutically monitored drug.

Drug half-lives are dramatically altered during renal and liver disease because the elimination rates of the drugs are changed. Consequently, new steady-state levels will be achieved which may differ significantly from those observed in healthy individuals. Genetic factors are also a major determinant of drug half-life. In a large population of patients, one could predict that if the entire population were given the same mg/kg dosage of a drug, there would be marked differences in their ability to utilize the drug. These genetic differences will be reflected in the steady-state plasma concentrations observed in this population. For example, in a population of patients receiving phenytoin at a standard therapeutic dose of 5 mg/kg/day, one would theoretically expect all patients to have a therapeutic drug level of 15 μg/ml. In reality, plasma concentrations will range from 5 μg/ml - - which represents drug malabsorption or patient non-compliance - - to levels of 40-50 μg/ml - - which repres-

ent patients who exhibit hepatic or renal disease or are genetically slow drug metabolizers.

The importance of genetics is reflected in the individual patient's drug utilization patterns. As an example, we may consider the incidence of fast and slow metabolizers in patients receiving isoniazid, a drug commonly used in the treatment of tuberculosis. Approximately 40% of all Caucasians are rapid acetylators of isoniazid. In contrast, over 90% of Japanese and Eskimos are rapid acetylators. This genetic variability requires individualization of therapeutic regimens to assure the maintenance of optimal isoniazid plasma concentrations.

Sampling Time

Following the administration of any drug, there is always a peak level which represents the point of maximum absorption, and a trough (valley level) which represents the lowest point achieved following a given dose of drug. The trough level occurs when the absorption process is almost complete, and is a consequence of the process of drug elimination (drug metabolization and excretion) which occurs during each dosage interval.

The object of all drug therapy is to ensure that a minimum effective concentration (MEC) of drug is present throughout a dosing interval. If plasma concentrations of the drug fall below the MEC during any given dosing interval an exacerbation of the patient's clinical status can be expected. Therefore, the blood specimens for routine therapeutic monitoring should be drawn immediately prior to the next drug dose. The actual plasma concentration observed at that time should be within the optimal therapeutic range.

Availability of appropriate information related to therapeutic regimens and physical characteristics of a given patient allows individualization of drug therapy. Calculation of dose response ratios allows prediction of a patient's plasma drug concentrations. Deviation from expected values is indicative of patient noncompliance or altered clinical status. Consideration of half-lives and steady-state is crucial to the proper utilization and interpretation of plasma drug concentrations.

PEDIATRIC CLINICAL PHARMACOLOGY

For several years, clinicians and investigators have recognized the existence of an age-related difference between children and adults in the utilization of antiepileptic drugs. Unfortunately, most of these investigations have been concerned with the effects of only one drug regardless of the number of agents the patient was receiving.

In the course of our own studies, serum drug concentrations were analyzed by gas-liquid chromatography or homogeneous enzyme immunoassay (EMIT) in 828 specimens from children and adolescents 1 to 20 years of age who were receiving multiple antiepileptic drugs. An age-dependent relationship between plasma drug concentration and drug dose was clearly evident. Children below the age of 11 years rapidly utilized phenytoin, phenobarbital, and primidone. Recently, other investigators have reported rapid utilization of carbamazepine and ethosuximide in children. Serum drug concentrations in a pediatric population tend to be significantly lower than those in an adult population on equivalent mg/kg dosages. Consequently, children should receive higher maintenance dosages of a given drug to achieve optimal antiepileptic drug concentration.

Others have reported and we have confirmed, a marked decrease in serum phenytoin concentrations and an elevation of serum phenobarbital concentrations after phenobarbital was added to the regimens of children on phenytoin therapy. Children at every age medicated with a combination of phenytoin and phenobarbital consistently exhibited lower serum phenytoin and higher phenobarbital concentrations than children receiving the same mg/kg dosage of these agents as single dose therapy. Children receiving primidone and phenytoin also exhibited decreased phenytoin and increased phenobarbital concentrations, suggesting that an interaction also occurs between phenytoin and the phenobarbital derived from the metabolism of primidone.

A marked decrease in antiepileptic drug utilization is observed as children enter the early stages of puberty. Pubescent children achieve serum drug concentrations that approximate more closely those expected in adults on the same mg/kg dosage. By the time children reach late adolescence (16 to 20 years), their drug patterns are essentially identical to those of adults. One of the major problems associated with the treatment of epilepsy or any other disease state in children is the clinical deterioration that often occurs as they approach puberty. This may be characterized by an exacerbation of seizures or increased lethargy and dullness. Many of the problems associated with seizure control in pubescent children may be directly related to the alteration of their drug-utilization patterns. The importance of this problem is demonstrated by an example based on our clinical experience.

A child who, at the age of 9, was placed on a standard pediatric dosage of phenobarbital (6 mg/kg) and achieved a serum phenobarbital concentration of 30 µg/ml, was seen by his pediatric neurologist once a year. Drug levels were carefully monitored, seizures were well controlled, and after his twelfth birthday, the parents and teachers became aware of gradually increased lethargy in the child. On his yearly return visit, the serum phenobarbital concentration was found to have increased from 30 to 60 µg/ml, reflecting the characteristic adult levels observed with a phenobarbital dosage of 6 mg/kg. This increased serum phenobarbital level was attributable directly to changes associated with puberty.

Chronic phenobarbital toxicity is subtle, often occurs without dramatic changes, and may go undetected by the casual observer. Because the child's potential ability to learn is diminished by increasing lethargy, serum antiepileptic drug concentrations should be monitored frequently (at least every 3 months) as he approaches early pubescence, regardless of whether or not the patient is seen by a physician. If serum drug concentrations increase, the appropriate dosage alteration should be implemented. If such a procedure is followed, unnecessary phenobarbital toxicity can be virtually eliminated.

GERIATRIC CLINICAL PHARMACOLOGY

The population of geriatric patients is consistently increasing. At the present time, approximately 11% of all Americans are over the age of 65. The extensive use of drugs in elderly patients has important consequences. Firstly, older patients take more medication because they have more disease. A variety of chronic disorders requiring therapy occur commonly in this age group, including osteoarthritis, nutritional deficiency, anemia, diabetes, malignancy, and atherosclerosis with its various manifestations. It is the rule rather than the exception for the elderly patient to have multiple diseases for which he takes multiple therapeutic agents. Thus the likelihood of drug interactions is greatly increased in this age group.

The second problem related to the use of drugs in the elderly is the greatly increased incidence of adverse reactions which have been documented in numerous studies. It is not clear to what extent altered handling of drugs by the body is responsible for the elderly patient's sensitivity to drugs, since other factors are undoubtedly important, such as non-compliance and mistakes in drug intake as a result of multiple therapy. Factors which may affect drug disposition in the elderly include: *(1)* changes in protein binding; *(2)* alterations in biotransformation; *(3)* reduced elimination related to decreased renal function; and *(4)* an altered drug distribution consequent upon changes in body composition. Unfortunately this problem has not been extensively studied and the information available on any given class of drugs administered to geriatric patients remains rather sketchy. It is anticipated that over the next few years, this deficiency will be corrected and more data will certainly become available.

BIBLIOGRAPHY

Allonen H, et al: Passage of digoxin into cerebrospinal fluid in man *Acta Pharmacol Toxicol (Kbh)* 41:192, 1977.

Anthony M, Hinterberger H, Lance JW: Plasma sodium valproate levels and clinical response in epilepsy *Proc Aust Assoc Neurol* 14:208, 1977.

Aronson JK, Graham-Smith DG, Wigley FM: Monitoring digoxin therapy *Quart J Med* 47:111, 1978.

Assael BM, et al: Gentamicin dosage in preterm and term neonates *Arch Dis Child* 52:883, 1977.

Barkley RA, Cunningham CE: Do stimulant drugs improve the academic performance of hyperkinetic children? *Clin Pediatr* 17:85, 1978.

Barot MH, et al: Individual variation in daily dosage requirements for phenytoin sodium in patients with epilepsy *Br J Clin Pharmacol* 6:267, 1978.

Bartels H, Oldigs HD, Günther E: Use of saliva in monitoring carbamazepine medication in epileptic children *Eur J Pediatr* 126:37, 1977.

Baruzzi A, et al: Plasma levels of di-no-propylacetate and clonazepam in epileptic patients. *Int J Clin Pharmacol Biopharm* 15:403, 1977.

Barza M, Lauermann M: Why monitor serum levels of gentamicin? *Clin Pharmacokinet* 3:202, 1978.

Bell T, Bigley J: Sustained-release theophylline therapy for chronic childhood asthma *Pediatrics* 62:352, 1978.

Benowitz NL, Meister W: Clinical pharmacokinetics of lignocaine *Clin Pharmacokinet* 3:177, 1978.

Birkett DJ, et al: Multiple drug interactions with phenytoin *Med J Aust* 2:467, 1977.

Callaghan N, et al: The effects of toxic and non-toxic serum phenytoin levels on carbohydrate tolerance and insulin levels *Acta Neurol Scand* 56:563, 1977.

Chien C, Solomon K, Platek TE: Macro-monitoring: A step toward rational psychopharmacotherapy *Am J Hosp Pharm* 35:397, 1978.

Christophidis N, et al: Fluorouracil therapy in patients with carcinoma of the large bowel: A pharmacokinetic comparison of various rates and routes of administration *Clin Pharmacokinet* 3:330, 1978.

Chow M, et al: Prediction of gentamicin serum levels using a one-compartment open linear pharmacokinetic model *Am J Hosp Pharm* 35:1078, 1978.

Cranford RE, et al: Intravenous phenytoin: Clinical pharmacokinetic aspects *Neurology* 28:874, 1978.

Crouthamel WG, Kowarski B, Narang PK: Specific serum quinidine assay by high-performance liquid chromatography *Clin Chem* 23:2030, 1977.

Dam M, et al: Interaction between carbamazepine and propoxyphene in man *Acta Neurol Scand* 56:603, 1977.

Dixon R, Crews T: An ^{125}I-radioimmunoassay for the determination of the anticonvulsant agent clonazepam directly in plasma *Res Commun Chem Pathol Pharmacol* 18:477, 1977.

Doering W, König E: Anstieg der digoxinkonzentration in serum unter chinidinmedikation *Med Klin* 73:1085, 1978.

Doherty JE: How and when to use the digitalis serum levels *JAMA* 239:2594, 1978.

Drayer DE, et al: Steady-state serum levels of quinidine and active metabolites in cardiac patients with varying degrees of renal function *Clin Pharmacol Ther* 24:31, 1978.

Eadie MJ, Lander CM, Tyrer JH: Plasma drug level monitoring in pregnancy *Clin Pharmacokinet* 2:427, 1977.

Edelbroek PM, De Wolff FA: Improved micromethod for determination of underivatized clonazepman in serum by gas-chromatography *Clin Chem* 24:1774, 1978.

Ericsson CD, Duke JH Jr, Pickering LK: Clinical pharmacology of intravenous and intraperitoneal aminoglycoside antibiotics in the prevention of wound infections *Ann Surg* 188:66, 1978.

Friedman RB, Young DS, Beatty ES: Automated monitoring of drug-test interactions *Clin Pharmacol Ther* 24:16, 1978.

Friis ML, Christiansen J: Carbamazepine, carbamazepine-10,11-epoxide and phenytoin concentrations in brain tissue of epileptic children *Acta Neurol Scand* 58:104, 1978.

Gadalla MA, Peng GW, Chiou WL: Rapid and micro high-pressure liquid chromatographic method for simultaneous determination of procainamide and N-acetylprocainamide in plasma *J Pharm Sci* 67:869, 1978.

Ginchansky E, Weinberger M: Relationship of theophylline clearance to oral dosage in children with chronic asthma *J Pediatr* 91:655, 1977.

Goldstein A, Horns WH, Hansteen RW: Is on-site urine testing of therapeutic value in a methodone treatment program? *Int J Addict* 12:717, 1977.

Halkin H, et al: Steady-state serum digoxin concentration in relation to digitalis toxicity in neonates and infants *Pediatrics* 61:184, 1978.

Hendeles L, Weinberger M, Johnson G: Monitoring serum theophylline levels *Clin Pharmacokinet* 3:294, 1978.

Hendeles L, Weinberger M, Wyatt R: Guide to oral theophylline therapy for the treatment of chronic asthma *Am J Dis Child* 123:876, 1978.

Hollister LE: Treatment of depression with drugs *Ann Intern Med* 89:78, 1978.

Holmstrand J, Anggård E, Gunne LM: Methadone maintenance: plasma levels and therapeutic outcome *Clin Pharmacol Ther* 23:175, 1978.

Holt DW, Williamson JD, Volans GN: Digoxin measurements in general practice *Br J Clin Pharmacol* 4:321, 1977.

Johnston GD, Kelly JG, McDevitt DG: Do patients take digoxin? *Br Heart J* 40:1, 1978.

Johnston GD, McDevitt DG: Digoxin compliance in patients from general practice *Br J Clin Pharmacol* 6:339, 1978.

Karlsson E: Clinical pharmacokinetics of procainamide *Clin Pharma-*

cokinet 3:97, 1978.

Katz RM, Rachelefsky GS, Siegel S; The effectiveness of the short- and long-term use of crystallized theophylline in asthmatic children *J Pediatr* 92;663, 1978.

Knudsen FU, Vestermark S: Prophylactic diazepam or phenobarbitone in febrile convulsions: a prospective, controlled study *Arch Dis Child* 53:660, 1978.

Koup JR, et al: Interaction of chloramphenicol with phenytoin and phenobarbital. Case report *Clin Pharmacol Ther* 24:571, 1978.

Koup JR, Brodsky B: Comparison of homogeneous enzyme immunoassay and high-pressure liquid chromatography for the determination of theophylline concentration in serum *Am Rev Respir Dis* 117:1135, 1978.

Lagerströn PO, Persson BA: Liquid chromatography in the monitoring of plasma levels of antiarrythmic drugs *J Chromatogr* 149:331, 1978.

Lang D, Hofstetter R, von Bernuth G: Plasma-digoxin-konzentration in verschiedenen altersstufen *Klin Wochenschr* 56:93, 1978.

Langslet A, et al: Plasma concentrations of diazepam and N-desmethyldiazepam in newborn infants after intravenous, intramuscular, rectal and oral administration *Acta Paediatr Scand* 67:699, 1978.

Latini R, et al: Kinetics and efficacy of theophylline in the treatment of apnea in the premature newborn *Eur J Clin Pharmacol* 13:203, 1978.

Laxdal OE, et al: Improving physician performance by continuing medical education *Can Med Assoc J* 118:1051, 1978.

Leahey EE Jr, et al: Interaction between quinidine and digoxin *JAMA* 240:533, 1978.

Lee B, Turner WM: Food and Drug Administration's adverse drug reaction monitoring program *Am J Hosp Pharm* 35:929, 1979.

Lichey J, et al: Human myocardium and plasma digoxin concentration in patients on long-term digoxin therapy *Int J Clin Pharmacol Biopharm* 16:460, 1978.

Lie KI, et al: Efficacy of lidocaine in preventing primary ventricular fibrillation within 1 hour after a 300 mg intramuscular injection *Am J Cardiol* 42:486, 1978.

Loes MW, et al: Relation between plasma and red-cell electrolyte concentrations and digoxin levels in children *N Engl J Med* 299:501, 1978.

Loughnan PM, McNamara JM: Paroxysmal supraventricular tachycardia during theophylline therapy in a premature infant *J Pediatr*

92:1016, 1978.

Mangione A, et al: Pharmacokinetics of theophylline in hepatic disease *Chest* 73:616, 1978.

Manion CV: Interpretation of digoxin blood levels by programmable calculator *Am J Hosp Pharm* 35:947, 1978.

Marks S, et al: Evaluation of three antibiotic programs in newborn infants *Can Med Assoc J* 118:659, 1978.

McKenzie SA, Baillie E, Godfrey S: Effect of practical timing of dosage on theophylline blood levels in asthmatic children treated with choline theophyllinate *Arch Dis Child* 53:322, 1978.

Mihaly GW, et al: Measurement of carbamazepine and its epoxide metabolite by high-performance liquid chromatography, and a comparison of assay techniques for the analysis of carbamazepine *Clin Chem* 23:2283, 1977.

Morselli PL, et al: Carbamazepine and carbamazepine-10,11-epoxide concentrations in human brain *Br J Clin Pharmacol* 4:535, 1977.

Moussa MA: Statistical problems in monitoring adverse drug reactions *Methods Inf Med* 17:106, 1978.

Nation RL, Peng GW, Chiou WL: High-pressure liquid chromatographic method for the simultaneous quantitative analysis of propranolol and 4-hydroxypropranolol in plasma *J Chromatogr* 145:429, 1978.

Newlands ES: The current role of cancer chemotherapy *Br J Radiol* 51:756, 1978.

Nielsen-Kudsk F, Magnussen I, Jakobsen P: Pharmacokinetics of theophylline in ten elderly patients *Acta Pharmacol Toxicol (Kbh)* 42:226, 1978.

Okada RD, et al: Relationship between plasma concentration and dose of digoxin in patients with and without renal impairment *Circulation* 58:1196, 1978.

Painter MJ, et al: Phenobarbital and diphenylhydantoin levels in neonates with seizures *J Pediatr* 92:315, 1978.

Paxton JW, Rowell FJ: A rapid, sensitive and specific radioimmunoassay for methotrexate *Clin Chim Acta* 80:563, 1977.

Pearson JR: Drug screening by enzyme immunoassay with the American Monitor KDA *Clin Chem* 24:1823, 1978.

Penchas S, Zajicek G: Plasma digoxin levels and the interbeat interval signal in atrial fibrillation *Z Kardiol* 67:104, 1978.

Peng GW, Gadalla MA, Chiou WL: High-performance liquid-chromatographic determination of theophylline in plasma *Clin Chem* 24:357, 1978.

Perucca E, et al: Water intoxication in epileptic patients receiving carbamazepine *J Neurol Neurosurg Psychiatry* 41:713, 1978.

Pfeifer HJ, Greenblatt DJ, Koch-Weser J: Adverse reactions to practolol in hospitalized patients: a report from the Boston Collaborative Drug Surveillance Program *Eur J Clin Pharmacol* 12:167, 1977.

Pollock J, et al: Relationship of serum theophylline concentration to inhibition of exercise-induced bronchospasm and comparison with cromolyn *Pediatrics* 60:840, 1977.

Pickering LK, DuPont HL, Satterwhite TK: Evaluation of spectinomycin and gentamicin in the treatment of hospitalized patients with resistant urinary tract infections *Am J Med Sci* 274:291, 1977.

Ponto LB, et al: Drug therapy reviews: tricyclic antidepressant and monoamine oxidase inhibitor combination therapy *Am J Hosp Pharm* 34:954, 1977.

Preskorn SH, Abernethy DR, McKnelly WV Jr: Use of saliva lithium determinations for monitoring lithium therapy *J Clin Psychiatry* 39:756, 1978.

Pynnönen S: The pharmacokinetics of carbamazepine in plasma and saliva of man *Acta Pharmacol Toxicol (Kbh)* 41:465, 1977.

Raposa T: Sister chromatid exchange studies for monitoring DNA damage and repair capacity after cytostatics *in vitro* and in lymphocytes of leukaemic patients under cytostatic therapy *Mutat Res* 57:241, 1978.

Reynolds EH: Drug treatment of epilepsy *Lancet* 2:721, 1978.

Rietbrock I, Streng H, Pesold R: Digoxinkonzentrationen im plasma von intesivpatienten eines anaesthesiologischen krankengutes *Klin Wochenschr* 56:503, 1978.

Rogers HJ, et al: Phenytoin intoxication during concurrent diazepam therapy *J Neurol Neurosurg Psychiatry* 40:890, 1977.

Rutherford DM, Okoko A, Tyrer PJ: Plasma concentrations of diazepam and desmethyldiazepam during chronic diazepam therapy *Br J Clin Pharmacol* 6:69, 1978.

Saccar CL: Drug therapy in the treatment of minimal brain dysfunction *Am J Hosp Pharm* 35:544, 1978.

Schneider J, Ruiz-Torres A: Digitalis effect and blood concentration *Int J Clin Pharmacol Biopharm* 15:424, 1977.

Shapiro W: Correlative studies of serum digitalis levels and the arrhythmias of digitalis intoxication *Am J Cardiol* 41:852, 1978.

Shenfield GM, Thompson J, Horn DR: Plasma and urinary digoxin in thyroid dysfunction *Eur J Clin Pharmacol* 12:437, 1977.

Shorvon SD, et al: One drug for epilepsy *Br Med J* 1:474, 1978.

Simons FE, et al: Pharmacokinetics of theophylline in acute asthma *J Med* 9:89, 1978.

Slaughter RL, Schneider PJ, Visconti JA: Appropriateness of the use of serum digoxin and digitoxin assays *Am J Hosp Pharm* 35:1376, 1978

Spector R, et al: Does intervention by a nurse improve medication compliance? *Arch Intern Med* 138:36, 1978.

Stern EL: Possible phenylethylmalondiamide (PEMA) intoxication *Ann Neurol* 2:356, 1977.

Tansella M, et al: Plasma concentrations of diazepam, noradiazepam and amylobarbitone after short-term treatment of anxious patients *Pharmakopsychiatr Neuropsychopharmakol* 11:68, 1978.

Tedeschi L, et al: Simultaneous gas liquid chromatographic dosage of phenobarbital and diphenylhydantoin in serum *Farmaco (Prat)* 33:83, 1978.

Venulet J: Methods of monitoring adverse reactions to drugs *Prog Drug Res* 31:231, 1977.

Vozeh S, et al: Changes in theophylline clearance during acute illness *JAMA* 240:1882, 1978.

Walther H, et al: Dosage monitoring through measurement of serum level -- a task of clinical pharmacology *Int J Clin Pharmacol Biopharm* 16:387, 1978.

Wareham DV, Bauer SP, Foster BJ: Review of gentamicin therapy based on pharmacokinetics *Am J Hosp Pharm* 35:317, 1978.

Washington University Case Conference: Theophylline toxicity *Clin Chem* 24:1603, 1978.

Weissman AM, Burleson KW: Advances in computer-based pharmacy systems *Med Instrum* 12:237, 1978.

Whyte J, Greenan E: Drug usage and adverse drug reactions in paediatric patients *Acta Paediatr Scand* 66:767, 1977.

Wilensky AJ, et al: Clorazepate kinetics in treated epileptics *Clin Pharmacol Ther* 24:22, 1978.

Windorfer A, Stünkel S, Weinmann HM: Bestimmung des serumknozentrationsverhaltnisses von primidon/phenobarbital bei antikonvulsiver primdonbehandlung als weiteres kriterium bei der therapiebeurteilung

Monatsschr Kinderheilkd 126:507, 1978.

Wisnicki JL, Tong WP, Ludlum DB: Analysis of methotrexate and 7-hydroxymethotrexate by high-pressure liquid chromatography *Cancer Treat Rep* 62:529, 1978.

Witherspoon LR, Shuler SE, Garcia MM: Spurious underestimation of results for digoxin radioimmunoassay with a commercial kit *Clin Chem* 24:494, 1978.

Wolf SM, Forsythe A: Behavior disturbance, phenobarbital, and febrile seizures *Pediatrics* 61:728, 1978.

Wyatt R, Weinberger M, Hendeles L: Oral theophylline dosage for the management of chronic asthma *J Pediatr* 92:125, 1978.

Yoshioka H, et al: Dosage schedule of gentamicin for chronic renal insufficiency in children *Arch Dis Child* 53:334, 1978.

Yourassowsky E: Indications, interpretation and applications of antibiotic assay *Infection* 6:140, 1978.

14
Plasma Proteins

Chester A. Alper
Alvin E. Davis, III

INTRODUCTION

Although the measurement of serum proteins commenced with determination of serum total protein concentration and the albumin/globulin ratio, the modern era was ushered in with the introduction of paper electrophoresis a quarter of a century ago. The most recent phase began with the measurement of specific proteins in patients' sera some 15-20 years ago, and today involves the automated or manual estimation of up to 25-30 proteins. It is now clear that serum contains in excess of 100 different kinds of protein molecules, each of which has a specific function and the synthesis of which is under precise genetic control.

Because the blood is such an accessible tissue and is used so extensively in the diagnosis of disease affecting all organ systems certain groups of proteins are of paramount interest to one particular clinical specialty. For example, the transcobalamins, transferrin, haptoglobin, and hemopexin are claimed by the hematologists, while immunologists have proprietary feelings toward the immunoglobulins and proteins of the complement system. Dermatologists, allergists, and nephrologists often have similar feelings about these same proteins. Endocrinologists have long since staked their claims to thyroxin-binding globulin, transcortin, and testosterone-binding globulin. Thyroxin-binding prealbumin, on the other hand, formerly in the realm of endocrinology, must now be shared with the nutritionists because of its function as the transport protein for vitamin A.

We point out these realities to emphasize the remarkably scattered character of the clinical literature on plasma proteins. The problem is further compounded by the fact that new knowledge of plasma proteins, even as it relates to disease, comes from the work of physical chemists, structural biochemists, physiologists, immunologists and others who work with proteins *per se*. The task of reviewing the literature relevant to the plasma proteins in clinical and laboratory medicine has, for the reasons outlined above, been a prodigious one. We make no claim to having reviewed all relevant articles published in 1978 but have sought to present only a few of the developments that seem to be of importance to us. We have used the cited papers to call attention to developments in a rather small number of areas related to serum proteins: protease inhibitors, the complement system, genetics, and measurement techniques. We have also included a potpourri of assorted publications which struck our fancy and which we hope will be of special interest.

PROTEASE INHIBITORS

It has become clear over the past several years that the protease inhibitors in plasma play important roles not only in the regulation of the coagulation, complement and kinin systems, but also in protection from autodigestion by gut and white cell proteases elaborated during normal physiological functions. The identification in 1963 by Laurell and his group in Sweden of α_1-antitrypsin deficiency and the definition of genetic polymorphism of this protein provided further insight and tools to study this central protease inhibitor. By convention, which will be followed in this Chapter, genes, alleles, and genetic loci are designated in italics. Gene products, proteins or phenotypes are designated in capital letters. In July, 1979, a new nomenclature for genes and gene products was promulgated at the International Society of Human Genetics in Edinburgh. This nomenclature will supersede the nomenclature outlined here. The genetic locus for α_1-antitrypsin termed *Pi* (for protease inhibitor), has over two dozen alleles, including a null gene and the hypomorphic alleles, Pi^Z and Pi^S. The latter two alleles are associated with products which appear in the circulation at about 10% and 25% respectively of the concentration of product of the common allele, Pi^M. The Pi Z protein accumulates as inclusions in the livers of individuals who carry Pi^Z, perhaps because the structural mutation that characterizes this variant prevents post-ribosomal release of newly synthesized protein. It is abundantly clear that persons who are Pi^Z, Pi^-, and Pi^{SZ} are especially prone to develop neonatal obstructive jaundice, childhood cirrhosis, and a progressive, fatal form of chronic obstructive pulmonary disease in young adulthood. The question of whether heterozygotes or carriers of the deficient Pi genes are more susceptible to these diseases is of considerable general interest since the frequency of Pi^{MZ} in most Caucasian populations is 2-4%. Current results suggest that fine measurements of pulmonary function and structure show abnormalities in subjects who are genetically Pi^{MZ}, but the latter do not have detectable clinical disease (1). It may be that cigarette smoke and otherwise polluted air accelerates aging in a lung already so predisposed by heterozygous α_1-antitrypsin deficiency. In two recent mass screening studies for severe α_1-antitrypsin deficiency (2,3), neonatal cholestasis occurred in about 10% of infants whose serum contained the Pi Z protein and approximately 3% had hepatic cirrhosis at 2 years of age. From other studies it appears that as many as 40% of babies with neonatal hepatitis have severe α_1-antitrypsin deficiency.

Earlier structural studies of α_1-antitrypsin had suggested that Pi Z, Pi S, and Pi M variant proteins differed by an amino acid substitution and also in carbohydrate composition. It is now known that Pi M protein contains glutamic acid in a position where Pi Z has lysine, and that in a different position Pi S has valine substituted for glutamic acid in the Pi M molecule (4,5). Neither of these substitutions should affect the attachment of carbohydrate. It is furthermore known that the relative charge differences between Pi M, Pi Z and Pi S proteins persist after desialidation, although virtually all of the microheterogeneity of α_1-antitrypsin is eliminated by this treatment (6,7). The bewildering array of genetic variants in the Pi system was increased by isoelectric focusing which revealed subtypes of the Pi M protein reported in 1976 and 1977, and these findings were recently extended (8). Chemical modification of lysyl and arginyl residues in α_1-antitrypsin, as previously applied to antithrombin III, yielded suggestive information about the structure of the active inhibitory site (9). Whereas arginyl residues were essential for antithrombin III, lysine appears to be critical for α_1-antitrypsin inhibitory function.

Antithrombin III slowly inhibits thrombin and some other coagulation proteases by forming an enzyme:inhibitor complex in a 1:1 molar ratio, and this complex-formation is markedly enhanced by heparin. Inhibition by antithrombin III of the non-coagulation protease, urokinase, and acceleration of inhibition by heparin has recently been described (10). Complex formation in a 1:1 molar ratic was demonstrated by sodium dodecyl sulfate-polyacrylamide gel electrophoresis, and by crossed immunoelectrophoresis. It has been recognized for some time that congenital deficiency of antithrombin III may lead to thrombotic disorders. Patients with nephrotic syndrome have a propensity for thrombosis. Kauffmann and colleagues have evaluated antithrombin III levels in 48 patients with proteinuria (11). Eight of nine patients with signs of thrombosis had antithrombin III levels which were less than 70% of normal. These authors further showed a significant negative correlation ($p < 0.001$) between the degree of proteinuria and plasma antithrombin III concentration. Antithrombin III was found in the urine of 32 out of 42 patients tested.

Alpha$_2$-plasmin inhibitor is a protease inhibitor which has been recognized and purified only during the past five years. It is the most potent and efficient plasmin inhibitor of human plasma. Its probable physiologic function has been clarified by the recognition of a patient with total deficiency of α_2-plasmin inhibitor (12). This patient presented with a severe hemorrhagic diathesis consisting of hemothorax, hemarthrosis, and prolonged bleeding following minor injuries. His clinical status was improved by therapy with the drug, tranexamic acid, which inhibits plasminogen activation. Coagulation studies in this patient were all normal, with the exception of shortened euglobulin and whole blood clot lysis times. His plasma contained no immunochemically detectable α_2-plasmin inhibitor. In addition, several family members including the patient's parents had approximately half-normal α_2-plasmin inhibitor levels, which suggest that they were heterozygous for the deficiency state. All other protease inhibitors were present in normal quantities in the propositus and his parents. The patient showed no evidence of fibrinogenolysis *in vivo*. It thus appears that α_2-plasmin inhibitor is required to maintain integrity of local hemostatic clots, while the other protease inhibitors are sufficient to protect from fibrinogenolysis.

THE COMPLEMENT SYSTEM

The complement system comprises a group of more than 20 proteins which are involved in several aspects of the inflammatory response. The components, when activated, are designed by an over-bar. Thus, for example, C1r when activated is designated $C1\bar{r}$. Two pathways for complement activation exist, the classical pathway and the alternative pathway. The classical pathway, consisting of C1 (and its components, C1q, C1r, and C1s), C2 and C4, is activated by interaction of C1 with immune complexes. The alternative pathway, on the other hand, does not require immunoglobulin but is activated by interaction of the components of the pathway, factor B, factor D, properdin and C3 on the surface of activator particles. Substances capable of activating the alternative pathway include complex polysaccharides (yeast cell walls, inulin, zymosan), bacterial lipopolysaccharides, aggregated immunoglobulins, and some mammalian cells (such as rabbit erythrocytes). Activation of both the classical and alternative pathways results in cleavage and activation of C3, C5 and the terminal cytolytic complement components. The early reactions of either pathway are characterized by sequential limited proteolysis of the components which generates proteins active in the cytolytic pathway, and causes the release of activation peptides, many of which have important biologic functions. As examples, a peptide derived from cleavage of C2 has kinin-like activity; the C3a fragment of C3 has anaphylatoxic activity; and the C5a fragment released from C5 has both anaphylatoxic and chemotactic activities.

Activation of the terminal components of the complement system is characterized by protein-protein interactions with complex-formation which does not depend upon enzymatic cleavage of the proteins. The physicochemical and functional characteristics of the C5b-9 complex in both the fluid phase and upon various surfaces are being examined in detail (13-16). It had previously been shown that complex-formation among the terminal components resulted in expression of a neoantigenic determinant which was not present on any of the individual proteins and that a unique serum protein, the S-protein, was capable of binding to the complex, and inhibiting its membranolytic activity (13). Podack and Müller-Eberhard demonstrated binding of the anionic detergent desoxycholate to intermediate complexes of the terminal components with increasing binding of the detergent paralleling further complex assembly (14). The amphiphilic nature of the terminal component complexes was demonstrated by charge-shift electrophoresis (14,15). Desoxycholate above the critical micellar concentration, but not nonionic detergents, exercised multiple effects upon complex-formation by these terminal components; it prevented aggregation of nascent C5b-7, dissociated the S-protein from its combination with C5b-9 in the SC5b-9 aggregate, and extracted C5b-9 complexes bound to complement-treated sheep erythrocyte membranes without dissociating the complexes. The C5b-9 complex during its formation could combine with either desoxycholate, the S-protein, lipoprotein, or lipid vesicles; if any of these substances were absent, it would self-aggregate. It was proposed that the C5b-7 complex overcomes the charge barrier of the membrane by ionic interactions and then associates with the membrane by secondary hydrophobic interactions. C8 and C9 binding further increase the hydrophobicity of the complex, resulting in transmembrane channel formation (14). The C5b-9 complex was extracted from liposomes with triton X-100, and both the native and proteolyzed complex could be reinserted into liposomes. The detergent-extracted complex maintained the same structure observed by electron microscopy as did the C5b-9 complex on membranes. This complex has been shown by Bhakdi and co-workers to be a hollow cylinder oriented vertically on the target membrane, with a height of approximately 15 nanometers and an internal diameter of 10 nanometers (16). Thus, insertion of the hydrophobic lipid binding region of the C5b-9 complex appears to result in the formation of a transmembrane pore as postulated earlier by Mayer in his "doughnut" theory.

The proteins controlling the complement system have been the subject of intense study during the past several years. The importance of the C3b inactivator and $\beta 1H$ in inactivation of C3b and of the alternative pathway convertase $\overline{C3bBb}$ has been well documented, as has the role of the C3b inactivator in C4b inactivation. Although the function of the $C\bar{1}$ inhibitor has been known for some time, its function and mechanism of action are now being evaluated in more detail. Ziccardi and Cooper have demonstrated that activation of C1 in fresh serum results in apparent disappearance of antigenic C1r and that this disappearance was due to masking of $C1\bar{r}$ antigenic determinants by complex formation with $C\bar{1}$ inhibitor (17). This did not occur in the serum of patients with hereditary angioneurotic edema unless exogenous $C\bar{1}$ inhibitor was added. Similar findings resulted using some antisera to C1s. It was further demonstrated that $C1\bar{r}$ and $C1\bar{s}$ were present in soluble complexes with $C\bar{1}$ inhibitor and were not simply bound to the surface of the activating agents used. Laurell and her colleagues evaluated C1 subcomponents immunochemically following treatment of serum with $C1\bar{r}$, $C1\bar{s}$, trypsin, heat aggregated IgG, immune complexes, cell-bound antibody and C-reactive protein-protamine complexes (18). C1 was shown to dissociate and $C1\bar{s}$ shifted in electrophoretic mobility to an α_2 position, while C1q became more cathodal. $C1\bar{s}$ in the α_2 position was further shown to be present in complex with $C1\bar{r}$ and $C\bar{1}$ inhibitor. This complex had a molecular weight of 5×10^5 daltons.

Previous clinical studies have shown that therapy of hereditary angioneurotic edema with the attenuated androgen, Danazol, results in disappearance of symptoms of the disease with an increase in plasma concentrations of C1 inhibitor, C4, and C2. A more recent clinical trial has shown that within one week of beginning therapy with Danazol (200 mg daily), functional and antigenic concentrations of $C\bar{1}$ inhibitor began to increase (19). When $C\bar{1}$ inhibitor concentrations remained less than 50% of normal, C4 concentrations correlated with those of $C\bar{1}$ inhibitor, but this correlation was lost when $C\bar{1}$ inhibitor rose to greater than 50% of normal. Further, $C\bar{1}$ inhibitor concentrations were shown to increase progressively with increasing Danazol dose up to a concentration of approximately 12 mg/dl (normal mean $C\bar{1}$ inhibitor concentration is 18 mg/dl) with the maximum dose of Danazol (600 mg/day).

Another control protein of the classical pathway, the C4-binding protein, which appears in many ways to act in a manner functionally analogous to the effects of $\beta 1H$ on C3b and the alternative pathway convertase, has been recognized during the past two years. This protein was first identified in mouse serum, but the human counterpart has now been isolated and characterized (20). Human C4-binding protein is a 10.7S glycoprotein with a molecular weight of 540-590,000 daltons. It is composed of disulfide-linked 70,000 dalton subunit polypeptide chains. It migrates on electrophoresis at pH 8.6 as a β globulin in the absence of divalent cations and as a γ globulin in the presence of calcium. C4-binding protein binds to purified C4b; this binding is not dependent on divalent cations and is not inhibited by diisopropylfluorophosphate. C4-binding protein is multivalent with each molecule binding four or five C4b molecules. One important function of this protein is that of an essential cofactor in the proteolysis of C4b by the C3b-inactivator (21). In the presence of both proteins, the α'-chain of C4b is cleaved into three fragments with molecular weights of 47,000 (α2), 25,000 (α3), and 17,000 (α4) daltons. The α3 and α4 fragments are disulfide linked to other chains of C4b, while the α2 fragment is not. These reactions do not occur in the absence of either C4-binding protein or the C3b-inactivator; neither protein could be replaced by β1H, and native C4 was unaffected by incubation with C4-binding protein and C3b-inactivator. Thus, β1H is a required cofactor for inactivation and proteolysis of C3b by the C3b-inactivator, and the C4-binding protein is a required cofactor for inactivation and proteolysis of C4b by the C3b-inactivator.

As alluded to earlier, the activation peptides produced on proteolysis of several complement components have important biologic functions. These phenomena began to be recognized about 15 years ago when Dias da Silva and colleagues first demonstrated the anaphylatoxic activity of the C3a fragment of C3. The past few years have resulted in a remarkable understanding of both the functional and structural characteristics of both the C3a and C5a peptides largely from the work of Hugli and colleagues. The amino acid sequences of both the C3a and C5a peptides have been determined (22). $C5a_{desArg}$, that is, C5a which has had its carboxyl terminal arginine removed by serum carboxypeptidase B, was purified from activated serum in the absence of inhibitors of carboxypeptidase. C5a consists of 74 amino acids (8200 daltons), while the carbohydrate portion of the molecule accounts for 3000 daltons. The carbohydrate portion of the molecule consists of a single oligosaccharide unit attached to an asparagine at position 64. More than 9% of the molecule consists of half-cystine residues, with two repeating cysteine sequences in the linear structure. Six of the seven half-cystines in C5a are located at positions nearly identical to those in C3a. The high degree of homology between C3a and C5a suggests a common genetic ancestry.

Although C3a and C5a both have anaphylatoxic activity, only C5a has chemotactic activity for polymorphonuclear leukocytes (23). Using a Boyden chamber assay, Fernandez and colleagues showed that C5a was chemotactically active at concentrations from 0.04 to 1.7×10^{-8}M; $C5a_{desArg}$ was active only in the presence of nonactivated serum. With a skin window assay, C5a was maximally active from 2 to 6×10^{-7}M, and $C5a_{desArg}$ was active from 0.2 to 1.2×10^{-6}M. C3a was inactive in both assays. These data suggest that the $C5a_{desArg}$ molecule is capable of functioning as a major chemotactic factor under physiologic conditions.

Previous studies have suggested defective opsonization in several groups of patients, including patients with nephrotic syndrome, sickle cell anemia and Leiner's Syndrome. The abnormalities underlying these defects remain incompletely understood. It was suggested, for example, that a defect in C5 function was responsible for the abnormality found in Leiner's Syndrome. However, the finding that patients with congenital C5 deficiency have normal opsonization of yeast makes this hypothesis extremely unlikely. Two new quantitative methods for the measurement of yeast opsonizing activity of serum have been reported (24,25). In one of these, free yeast cells are counted with a Coulter counter following incubation of opsonized yeast with white blood cells (24). The other assay takes advantage of the fact that Brewer's yeast incorporates ^{3}H-uridine, while phagocytes and phagocytosed yeast do not (25). In both these studies, it was found that approximately 7% of the normal human population have abnormal serum opsonizing activity. The reasons for these abnormalities are not clear. However, results such as these should be borne in mind when attempting to interpret reports of abnormal opsonization by sera of patients with various diseases.

In 1965, it was first reported that some patients with chronic glomerulnephritis had depressed serum C3 levels. This disease was subsequently morphologically defined as membranoproliferative glomerulonephritis, which has more recently been subdivided into at least two types. It was found that the sera of some of these patients contained a factor (termed C3 nephritic factor) which induced cleavage of C3 in normal serum, and that this C3 cleavage was mediated via the alternative complement pathway. Although an early report by Thompson suggested that C3 nephritic factor was an immunoglobulin, subsequent studies suggested that it was the activated form of an initiating factor of the alternative pathway. In 1977, several reports definitively showed that C3 nephritic factor is an IgG immunoglobulin, and that it is an autoantibody directed toward the alternative pathway convertase, C3bBb. It thus functions by prolonging the half-life of this complex enzyme, allowing more extensive cleavage of C3 and C5. Further studies of C3 nephritic factor have demonstrated its limited heterogeneity, its light chain composition, its subclass composition and its polypeptide chain composition (26,27). Nephritic factors may contain either kappa or lambda light chains, or both, and may consist of virtually any single subclass or combination of subclasses. In addition, there is a suggestion that the heavy chains of some nephritic factors may be slightly larger than the heavy chains of normal IgG.

GENETIC ASPECTS OF PLASMA PROTEINS

It is now clear that a number of proteins isolated and characterized from serum or plasma are not identical to the original transcription products. For example, albumin in the rat as assembled on ribosomes in the liver (pre-pro-albumin) is larger than that found in the circulation by two segments at the amino terminal portion of the molecule. The segment at the amino terminus (the pre-segment) remains attached to the ribosome and the pro-albumin is released, presumably by highly specific proteolytic cleavage. As the pro-albumin reaches the Golgi apparatus, a second cleavage occurs and the familiar serum albumin is ultimately released into the circulation. A variant human albumin designated $\text{Albumin}_{\text{Christchurch}}$ has recently been studied in detail (28). It is slightly larger in size than normal human albumin and has an extra six amino acids at the N-terminus. Remarkably, the sequence of this extra segment is identical to that of the pro-segment of rat proalbumin except for one residue: glutamic acid in place of arginine at the C-terminus. The next 10 amino acid residues of this variant were identical to those of the N-terminus of normal human albumin. The implication of these findings is that $Albumin_{Christchurch}$ has a single amino acid substitution (presumably GLU for ARG) at the point of proteolytic cleavage between the pro-segment and the rest of the albumin molecule such that cleavage cannot occur (most serine proteases cleave at Arg- or Lys- bonds).

A different kind of post-translational processing occurs in the generation of multichain proteins and consists of scission of a single subunit protein at one or more points. This was shown to occur in the generation of A and B chains from insulin, and the work of Colten and co-workers has shown it to be operative in the production of the two- and three-chain circulating forms of C3 and C4. The mechanism may apply to many multichain plasma proteins. In recent studies (29), about 5% of plasma C4 was found to be in the single chain "pro-C4" form, and pro-C4 had no functional hemolytic activity by the usual assays. The implication of these findings is that C4 is the product of a single cistron despite the fact that the bulk of C4 in plasma is a three-subunit structure.

Genetic polymorphism in human complement proteins is extensive and in most instances appears to involve single point mutations. Human C4 has been known for a decade to vary electrophoretically from individual to individual, but it has not been possible to explain this variation adequately by any genetic model. Important advances in this area were made by O'Neill and co-workers (30,31) who postulated that there are two closely linked genetic loci for human C4 and the product of one of these carries the previously described blood group specificity, Ch(a), and the other carries Rg(a). These two loci, in turn, are closely linked to the major histocompatibility complex. This model also postulates that there are common C4 haplotypes in which either the Ch^a or Rg^a loci are unexpressed.

Among the new genetic polymorphisms described was that for human C7 (32), the seventh component of complement. About 1% of random individuals show variant forms, of which there are two, when serum samples are examined by isoelectric focusing. Linkage studies revealed very close linkage between the loci for *C7* and *C6*. This proximity of the two loci may explain the abnormality in a patient studied by these same authors (33) who had a subtotal deficiency of both C6 and C7 proteins and no clinical disease. This defect was transmitted as a single autosomal co-dominant trait to a number of family members. Both C6 and C7 proteins had abnormal banding patterns on isoelectric focusing. The authors, Lachmann and colleagues, postulated that C6 and C7 proteins are coded by a single transcript such that the separate molecules result from post-translational cleavage. Of clinical interest is the observation that C7 deficiency, just as deficiency of other late-acting complement components, C5, C6 and C8, may be associated with increased susceptibility to bacteremic infections with neisseria (34), including gonococcal sepsis and meningitis.

Isoelectric focusing coupled with immunofixation and other specific techniques of pattern development already used with success in analysing the complement components has revealed more extensive genetic polymorphisms within already polymorphic systems. Thus, Gc^1 can now be divided into Gc^{1F} and Gc^{1S} (35). Gene frequencies for these differ strikingly among different populations. These findings considerably enhance the use of Gc typing in forensic medicine, gene localization and population genetic studies. Similarly, after some initial confusion about which protein was being examined (36), widespread genetic polymorphism was found in human transferrin (37): Tf^C consists of two common alleles, Tf^{C1} and Tf^{C2}.

The use of radioactive ligands by Daiger and associates had earlier demonstrated that Gc-globulin was the transport protein for vitamin D. Using this approach, the same group showed two common alleles of a locus for transcobalamin II (38).

A number of reports have appeared from Europe on a positive association between the $C3^F$ gene and atherosclerotic heart disease. In a recent study, the incidence of C3 F protein in such patients was over twice that of the general population. Among treated hypertensive patients, the relative risk of coronary heart disease was increased by a factor of 10 ($p < 0.002$) for C3 F carriers (39).

ASSAY TECHNIQUES

Electroimmunoassay

The value of measuring specific proteins in clinical medicine is now well established. Serum protein electrophoresis has gradually come to play its proper role as a screening procedure, without the pseudoquantification of α, β, and γ globulins that dominated the field earlier. Because of its simplicity, radial immunodiffusion in agar or agarose gel, first introduced by Mancini and co-workers, has been the technique most widely used to measure individual proteins by means of monospecific antibodies. Since the method as originally devised took several days to reach the endpoint, a number of modifications, such as the "kinetic" approach (in which ring diameters were measured as they grew) were introduced. The electroimmunoassay of Laurell provided two important advantages: it shortened the time to reach the end result to several hours; and it reduced the contribution of antigen molecular size to the final answers. The latter point is of great importance since the molecular size of antigens leads to considerable errors when complex-forming or polymer-forming proteins such as haptoglobin, or fragmenting proteins such as C3, C4 or factor B, are measured by radial immunodiffusion. A recent micromodification of Laurell's electroimmunoassay (40) appears to have more research interest than clinical application. Attempts to use Laurell's crossed-immunoelectrophoresis with multivalent antisera have had limited success in the measurement of large numbers of individual proteins in single serum samples.

Immunonephelometry

Immunonephelometry, primarily because of its speed and the ease with which it can be automated, was reintroduced after a quarter of a century of dormancy. The principle is extraordinarily simple. Early in the course of antigen-antibody interaction, microaggregates are formed which scatter light. Certain agents, such as polyethylene glycol, accelerate this reaction. One can wait until there is relatively little change in light scattering (20 minutes to 1 hour without polyethylene glycol or several minutes with this reagent) and perform endpoint determinations, or one can analyze the very early portions of the reaction kinetically (rate nephelometry) and obtain an answer within a minute or so. Sensitivity is in the range of 10 μg/ml for immunephelometric assays (compared with 1 μg/ml for in-gel methods). It appears to matter little whether laser or tungsten light is used, and the angle at which scattered light is read is not of major importance. Recent studies (41,42) have confirmed earlier reports of agreement between immunonephelometric and in-gel methods and stress the excellent reproducibility of results obtained with immunonephelometric techniques, particularly when automated. Use of nephelometry in the determination of rheumatoid factor (43) is interesting and should prove successful. The closely related turbidimetric approach has been used in the rapid screening of serum samples for cryoglobulins (44). The method had been applied earlier in the detection of monoclonal cryoglobulins but it works as well for mixed cryoprecipitates, including those containing immune complexes. Results are obtained in one hour.

Radioimmunoassay

For greater sensitivity in the detection of proteins occurring in concentrations in the ng/ml or lower range, radioimmunoassay or, when applicable, other radioligand assays, have been the techniques of choice. In recent years, methods using non-radioactive markers such as enzymes or fluorescent tags have been introduced. These are relevant to serum protein measurements in only a few instances where proteins occur at very low levels in normal adult serum but under certain pathological circumstances are found in substantial concentrations, such as α-fetoprotein or C-reactive protein (45).

C2 Assay

Finally, a method that required only readily available reagents for the measurement of C2, based upon assay of its biological function, was reported (46). This is of some clinical interest because C2 deficiency is the most common complement deficiency state and because 14-40% of persons homozygous for C2 deficiency have systemic lupus erythematosus.

MISCELLANEOUS

This section will deal briefly with a number of articles which we felt were important, but which did not logically fit into any of the previous three selected categories. As will be seen, we have chosen a very few papers from among several different topics, some because they seemed to be of significant general interest and some because they involve subjects of potential growing importance (for example, the various papers dealing with fibronectin).

Proteins in Amyloidosis

A clue to the pathogenesis of amyloidosis was obtained from studies of the degradation of serum amyloid A (SAA) protein by surface-associated serine proteases of peripheral blood monocytes (47). Monocytes from 8 of 20 normal subjects degraded SAA protein completely with no detectable intermediates. Monocytes from an additional eight normal individuals produced a transient intermediate AA protein with the characteristics of tissue AA protein but could degrade the tissue AA protein. Monocytes from the remaining four subjects degraded the SAA protein to a persistent AA protein like that of the tissue protein. Monocytes from these four individuals failed to degrade tissue AA protein. The monocytes of all ten patients with amyloidosis behaved like this last small normal group. This suggests that the defect in amyloidosis may be related to abnormal catabolism of SAA protein by monocytes.

In familial amyloidotic polyneuropathy as opposed to other forms of amyloidosis, extraction of amyloid fibrils from kidney, thyroid, and peripheral nerves with 6M guanidine hydrochloride yielded a unique 14,000 dalton protein (48). Antisera made against the amyloid fibrils of this disease reacted with this protein and thyroxin-binding prealbumin in serum. Conversely, antiserum to thyroxin-binding prealbumin reacted with the fibrils. These amyloid fibrils appeared not to contain amyloid L protein (immunoglobulin light chain) or amyloid A protein.

Hemopexin

The measurement of serum hemopexin has by and large been disappointing as an index of hemolysis *in vivo*. Since hemopexin binds heme and heme is an integral part of myoglobin as well as hemoglobin, it seemed logical to investigate serum hemopexin concentrations in fulminant muscle disease and in other neuromuscular disease. In fulminant rhabdomyolysis, there is myoglobinuria, normal haptoglobin concentrations and some fall in serum hemopexin (49). Paradoxically, it has previously been noted that serum hemopexin levels are *increased* in patients and carriers of Duchenne muscular dystrophy, and after repeated injections of heme. Other neuromuscular diseases, such as dermatomyositis/polymyositis and myasthenia gravis, but not amyotrophic lateral sclerosis were characterized by similar increases (50).

Fibronectin

Fibronectin (cold-insoluble globulin, α_2-surface binding glycoprotein, large external transformation sensitive protein or LETS) not only has numerous synonyms but has also been the subject of numerous publications in the past year. There is evidence that it is synthesized by cultured human endothelial cells (51), by T lymphocytes but not B lymphocytes (52), and perhaps other cells as well. Fibronectin adheres to a wide variety of compounds and surfaces, and appears to be required for cell adhesion and spreading in culture (53). Under certain conditions, it is capable of opsonizing particles to which it is attached, and it apparently binds to Staphylococcus aureus (54). Fibronectin has been highly purified from serum and shown to be a 220,000 dalton molecule after reduction (55). There is intriguing recent evidence that fibronectin is in fact the collagen receptor on platelet membranes (56). In studies of human-mouse hybrids, the gene for human fibronectin segregated with that for human glutathione reductase, known to be on chromosome 8 (57). It is intriguing that rearrangements and translocations of chromosome 8 are associated with the development of certain tumors. Since fibronectin is thought to be involved somehow in tumorigenicity and cellular transformation, this new evidence provides further support for such involvement.

REFERENCES

1. Mittman C: *Am Rev Respir Dis* 118:649, 1978.

2. Sveger T: *Pediatrics* 62:22, 1978.

3. O'Brien ML, Buist NR, Murphey WH: *J Pediat* 92:1006, 1978.

4. Shocat D, et al: *J Biol Chem* 253:5630, 1978.

5. Owen MC, Lorier M, Carrell RW: *FEBS Letts* 88:234, 1978.

6. Jeppsson J-O, Laurell C-B, Fagerhol M: *Eur J Biochem* 83:143, 1978.

7. Yoshida A, Wessels M: *Biochem Genet* 16:641, 1978.

8. Kueppers F, Christopherson MJ: *Am J Hum Genet* 30:359, 1978.

9. Fretz JC, Gan CJ: *Arch Biochem Biophys* 188:226, 1978.

10. Clemmensen I: *Thromb Haemost* 39:616, 1978.

11. Kauffmann RH, et al: *Am J Med* 65:607, 1978.

12. Koie K, et al: *Lancet* 2:1334, 1978.

13. Podack ER, Kolb WP, Müller-Eberhard HJ: *J Immunol* 120:1841, 1978.

14. Podack ER, Müller-Eberhard HJ: *J Immunol* 121:1025, 1978.

15. Bhakdi S, et al: *J Immunol* 121:2526, 1978.

16. Bhakdi S, Tranum-Jensen J: *Proc Nat Acad Sci USA* 75:5655, 1978.

17. Ziccardi RJ, Cooper NR: *J Immunol* 121:2148, 1978.

18. Laurell A-B, et al: *Acta Pathol Microbiol Scand* 86:299, 1978.

19. Pitts JS, et al: *J Lab Clin Med* 92:501, 1978.

20. Scharfstein J, et al: *J Exp Med* 148:207, 1978.

21. Fujita T, Gigli I, Nussenzweig V: *J Exp Med* 148:1044, 1978.

22. Fernandez HN, Hugli TE: *J Biol Chem* 253:6955, 1978.

23. Fernandez HN, et al: *J Immunol* 120:109, 1978.

24. Levinsky RJ, Harvey BA, Paleja S: *J Immunol Methods* 24:251, 1978.

25. Yamamura M, Valdimarsson H: *Immunology* 34:689, 1978.

26. Davis AE, III, et al: *Clin Immunol Immunopathol* 11;98, 1978.

27. Daha MR, Austen KF, Fearon DT: *J Immunol* 120:1389, 1978.

28. Brennan SO, Carrell RW: *Nature* 274:908, 1978.

29. Gigli I: *Nature* 272:836, 1978.

30. O'Neill GJ, Yang SY, Dupont B: *Proc Nat Acad Sci USA* 75:5165, 1978.

31. O'Neill GJ, et al: *Nature* 273:668, 1978.

32: Hobart MJ, Joysey V, Lachmann PJ: *J Immunogenet* 5:157, 1978.

33. Lachmann PJ, Hobart MJ, Woo P: *Clin Exp Immunol* 33:193, 1978.

34. Lee TJ, et al: *J Infect Dis* 138:359, 1978.

35. Constans J, et al: *Hum Genet* 41:53, 1978.

36. Thymann M: *Hum Genet* 43:225, 1978.

37. Kühnl P, Spielmann W: *Hum Genet* 43:91, 1978.

38. Daiger SP, et al: *Am J Hum Genet* 30:202, 1978.

39. Kristensen BO, Petersen GB: *Circulation* 58:622, 1978.

40. Prause JU, et al: *Anal Biochem* 85:564, 1978.

41. Daigneault R, Lemieux D: *Clin Biochem* 11:28, 1978.

42. Bruver RM, Salkie ML: *Clin Biochem* 11:112, 1978.

43. Virella G, et al: *J Immunol Methods* 22:247, 1978.

44. Kalovidouris AE, Johnson RL: *Ann Rheum Dis* 37:444, 1978.

45. Siboo R, Kulisek E: *J Immunol Methods* 23:59, 1978.

46. Thompson RA: *J Immunol Methods* 24:223, 1978.

47. Lavie G, Zucker-Franklin D, Franklin EC: *J Exp Med* 148:1020, 1978.

48. Costa PP, Figueira AS, Bravo FR: *Proc Nat Acad Sci USA* 75:4499, 1978.

49. Adornato BT, et al: *Arch Neurol* 35:547, 1978.

50. Adornato BT, Engel WK, Foidart-Desalle M: *Arch Neurol* 35:577, 1978.

51. Jaffe EA, Mosher DF: *J Exp Med* 147:1779, 1978.

52. Hauptman SP, Kansu E: *Nature* 276:393, 1978.

53. Grinnell F, Hays DG: *Exp Cell Res* 115:221, 1978.

54. Kuusela P: *Nature* 276:718, 1978.

55. Blumenstock FA, et al: *J Biol Chem* 253:4287, 1978.

56. Bensusan HB, et al: *Proc Nat Acad Sci USA* 75:5864, 1978.

57. Owerbach D, Doyle D, Shows TB: *Proc Nat Acad Sci USA* 75:5640, 1978.

Index